PRACTICAL DATA ANALYSIS

CASE STUDIES IN BUSINESS STATISTICS

INSTRUCTOR'S EDITION

PRACTICAL DATA ANALYSIS

CASE STUDIES IN BUSINESS STATISTICS
INSTRUCTOR'S EDITION
VOLUMES I, II, AND III

SECOND EDITION

Peter G. Bryant
Marlene A. Smith
College of Business and Administration
and
Graduate School of Business Administration
University of Colorado at Denver

Irwin
McGraw-Hill

Boston Burr Ridge, IL Dubuque, IA Madison, WI New York San Francisco St. Louis
Bangkok Bogotá Caracas Lisbon London Madrid
Mexico City Milan New Delhi Seoul Singapore Sydney Taipei Toronto

Irwin/McGraw-Hill

A Division of The ***McGraw-Hill*** *Companies*

PRACTICAL DATA ANALYSIS: CASE STUDIES IN BUSINESS STATISTICS INSTRUCTOR'S EDITION

This book is printed on acid-free paper.

1 2 3 4 5 6 7 8 9 0 DOC/DOC 9 3 2 1 0 9 8

ISBN 0-07-304010-X

Vice president/Editor-in-chief: *Michael W. Junior*
Publisher: *Jeffrey J. Shelstad*
Senior sponsoring editor: *Scott Isenberg*
Developmental editor: *Wanda J. Zeman*
Marketing manager: *Zina Craft*
Senior project manager: *Susan Trentacosti*
Senior production supervisor: *Heather D. Burbridge*
Designer: *Jennifer McQueen Hollingsworth*
Cover designer: *Steven Vena/SrV Unlimited Design*
Supplement coordinator: *Mark Sienicki*
Compositor: *Carlisle Communications, Ltd.*
Typeface: *10/12 Times Roman*
Printer: *R. R. Donnelley & Sons Company*

Library of Congress has cataloged the Student Edition as

Library of Congress Cataloging-in-Publication Data

Bryant, Peter G.
Practical data analysis : case studies in business statistics /
Peter G. Bryant and Marlene A. Smith.—2nd ed.
p. cm.
Includes bibliographical references (p.).
ISBN 0-256-23871-5 (vol. 1). —ISBN 0-256-23872-3 (vol. 2.) —
ISBN 0-07-365488-4 (vol. 3)
1. Commercial statistics—Case studies. 2. Commercial statistics-
–Data processing. 3. Statistics—Data processing. I. Smith,
Marlene A. II. Title.
HF1017.B79 1999
650'.01'5195—dc21 98-43159

http://www.mhhe.com

To L.S.B. and L.G.B.

In memory of Reuben and Charlotte Aschenbach

And he who is versed in the science of numbers can tell of the regions of weight and measure, but he cannot conduct you thither.

Kahlil Gibran
The Prophet

Was man nicht nützt, ist eine schwere Last.

Johann Wolfgang Goethe
Faust I, 684

Some of the blocks to dealing comfortably with numbers . . . are due to quite natural psychological responses to uncertainty, to coincidence, or to how a problem is framed. Others can be attributed to . . . romantic misconceptions about the nature and importance of mathematics.

John Allen Paulos
Innumeracy

PREFACE

CASES, DISCUSSIONS, AND STATISTICS CLASSES

Practical Data Analysis: Case Studies in Business Statistics (PDA) provides real data analysis problems for students to analyze and discuss. They must:

- Understand the problem and the context.
- Choose a method of analysis.
- Discover what the data have to say.
- Communicate that message to you and/or their classmates.

We organize our business statistics courses around three components: lectures, cases, and discussion.

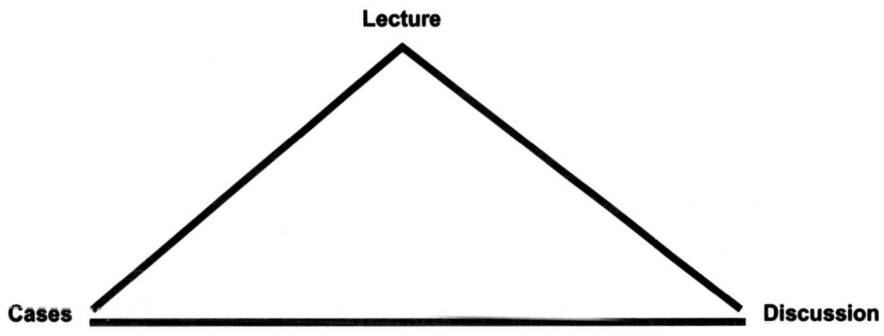

Lecture or its equivalent serves to present new ideas and techniques, reinforce or extend material from a textbook, and establish tone and expectations. Cases provide a chance for students to practice the techniques from the text or lecture, of course, but they provide more. They provide a basis for discussion with classmates and with the instructor. *PDA's* cases are specifically intended for courses incorporating such discussion elements.

Combining discussion with the lecture permits students to articulate different approaches, uncover alternative sets of assumptions (and the corresponding effects on the analysis), and discover mistakes and misunderstandings and correct them while they're "live" issues. They learn the underlying *intents* of the methods of statistics as well as the formulas. They learn that to understand statistics, you must actually analyze data. They also learn that statistics is a rewarding activity, not just a course to be endured. The instructor learns which messages and techniques have sunk in and which haven't. Finally, teaching is more fun using discussion. Lectures have a way of being the same from semester to semester, but no two group discussions are ever the same. In short, neither statistics nor statistics instruction should be a spectator sport.

In the rest of this *Instructor's Edition,* we explain our ideas in greater detail, and give some specific guidelines on using *PDA* cases in class.

ABOUT THIS *INSTRUCTOR'S EDITION*

This *Instructor's Edition* of *PDA,* offers the following features:

1 An *Instructor's Guide* section, which contains:
   - **a** An *Introduction for the Instructor,* with our general approach to using cases in statistics classes and advice for instructors just beginning a case-based approach.
   - **b** *Directories* of the cases provided, by application area, level, and technique.
   - **c** *Instructor's Notes* for each case, containing specific summaries of the case, questions for assignment, suggestions for leading classroom discussion, and typical results and approaches used to work the case.
   - **d** *Sample exams* for case-based courses.
   - **e** *A sample case write-up with a critique* and an improved write-up, designed to help your students develop good analysis and writing skills.
   - **f** A brief description of ways to convert the data files for *PDA* to various statistics programs.

2 Cases for students, arranged in three volumes of 25 cases each. For most courses, you will specify that the students use one of these volumes (or you may order individual cases from the publisher). The complete texts of all volumes follow the instructor's guide section. Each volume includes introductory material for the student, examples of good written reports and write-ups, and brief instructions for using the data files provided on the disk that comes with the book.
   - **a** Many of the cases come from past student projects. They reflect topics of genuine concern and interest to today's business students.
   - **b** Most cases are set in a business context, but a few reflect nonbusiness and general interest applications; they can be assigned early in the course to motivate those who haven't studied much business yet.
   - **c** Each volume contains one or more cases *without* data. They focus attention on aspects of data analysis such as survey design or implementation of quality measurements in an organizational setting.

3 A data disk, with data sets for the cases in ASCII, MINITAB, EXCEL, JMP, and LOTUS 1-2-3 worksheet formats.

HOW TO USE THE *INSTRUCTOR'S EDITION*

1 Read the *Introduction for the Instructor.*
   - You'll learn our intent for these cases, ways in which we've used them, and ways in which *you* might use them.
   - You can then assess how these cases fit into what you're doing or what you might try in the future.

2 Skim the *Directories of Cases.* The tables there give an idea of the kinds of features the cases offer.

3 Select a few appealing cases, and read them. All the cases are in this *Instructor's Edition.*

4 Read the *Instructor's Notes* for those cases.
   - You'll find specific suggestions for each case, questions to assign, approaches, ways to lead the classroom discussion, and typical results.
   - You'll then have a good sense for what a course organized around cases might feel like.

5 Look at the Appendixes for sample exams.

6 Review the *Introduction for the Student,* found in each volume.

Now plunge in. Experiment with cases in *your* class, and develop the way that works for you. Please let us know how it goes.

CHANGES IN THE SECOND EDITION

Volume III (Cases 51–75) is new in this edition. Cases 64 *(The Advertising Experiment),* 69 *(Alicia's Union Data),* and 72 *(Blood Pressure)* were formerly in the *Instructor's Edition,* held "in reserve" for use in exams. The remaining cases in Volume III are new.

We have also made a few changes to Volumes I and II. In Volume I, we have revised the data sets in Cases 3 *(Spot Prices),* 6 *(Stocks A and B),* 9 *(Healthy Lifestyles),* 23 *(Money Supply and Interest Rates),* 24 *(SAT Scores),* 26 *(Dave's Truck Washing Service),* 34 *(Leading Indicators),* and 46 *(Natural Gas Prices),* and updated the texts of the cases accordingly. Case 12 (*Strep Infections* in the first edition) has been replaced by Kerrville Rainfall. In Volume II, Case 29 (*Medical Laboratories* in the first edition) has been replaced by *Books,* Case 36 *(Defense Contractors)* has been eliminated, Case 31 *(Violence on TV)* has been split into two cases, Case 38 *(AIDS Rates)* has been completely changed, and in several other cases we have randomized the order of the data observations, at the request of several previous users. We have also expanded the text of Case 44 (*Performance Evaluations*) to include the evaluation forms and internal memos used to generate the data for that case.

Peter G. Bryant

Marlene A. Smith

ACKNOWLEDGMENTS

Many colleagues have reviewed individual cases and the manuscript of *Practical Data Analysis.* We thank specifically:

Sung K. Ahn, *Washington State University*
Randy Anderson, *California State University at Fresno*
Bridget G. Heidemann, *Seattle University*
Raj Jagannathan, *University of Iowa*
Jerzy J. Letkowski, *Western New England College*
Walter J. Mayer, *University of Mississippi*
Dale McFarlane, *Oregon State University*
J. B. Orris, *Butler University*
Paul Paschke, *Oregon State University*
John R. Pickett, *Georgia Southern University*
Kipling M. Pirkle, *Washington and Lee University*
Al Schainblatt, *San Francisco State University*
Milo Schield, *Augsburg College*
William L. Seaver, *University of Tennessee*
Herbert Spirer, *University of Connecticut*
Jeffrey W. Steagall, *University of North Florida*
Stanley A. Taylor, *California State University at Sacramento*
Betty M. Thorne, *Stetson University*
Thomas K. Tiemann, *Elon College*
Wayne Winston, *Indiana University at Bloomington*

We acknowledge also the founders, organizers, and participants in the annual conference entitled *Making Statistics More Effective in Schools of Business* for their hard work and interest in changing the way in which we teach statistics in our business classrooms.

For specific permissions we thank the American Statistical Association for permission to reproduce items from *The American Statistician;* the *Rocky Mountain News* for copyright permission; and Stanley D. Elofson for assisting us in obtaining copyright permission for the 1992 Arapahoe County property assessment study. Further acknowledgments are listed with individual cases.

Several friends, colleagues, and relatives helped in various ways. Edward J. Conry advised us on the preparation of our prospectus. Lucinda Bryant, Louise Bryant, and Lynwood Bryant read the book and commented on content, style, and presentation. Kathryn E. Chmelir and Adriano Pimenta helped us with research, data input and analysis, and other related tasks. The staff at Irwin/McGraw-Hill, including Scott Isenberg and Wanda Zeman, have also been helpful.

Finally, we thank our students and other contributors to the cases in the book. Where appropriate, we acknowledge them individually in the source notes of their respective cases. All of our students contributed indirectly to this project: they told us what did and didn't work, complained about bad cases (and rightly so!), and had fun with the good ones.

Peter G. Bryant

Marlene A. Smith

CONTENTS

NOTE ON PAGE NUMBERS

Page numbers for the *Instructor's Edition* are listed in Arabic numbers (1, 2, . . .). Page numbers for the volumes of cases are given in the form I–1 (Volume I, page 1) or II–1 (Volume II, page 1). This ensures that the page numbering for the cases is the same for both student and instructor.

Practical Data Analysis: Case Studies in Business Statistics, Volume II

Practical Data Analysis: Case Studies in Business Statistics, Volume III

PRACTICAL DATA ANALYSIS

CASE STUDIES IN BUSINESS STATISTICS

INSTRUCTOR'S EDITION

INTRODUCTION FOR THE INSTRUCTOR

WHY USE CASES?

Statistics is the study of data and what we can do with it, but we sometimes lose sight of that fact. It's easy to spend most of a course on the formal methods and the mathematics, giving short shrift to their applications, but it takes the excitement out of things if we do. The statistical theories of this century have brought discipline to the process of analyzing data. For the most part, they've been spectacularly successful, but without the application and without the context in which the material is to be used, students lose motivation. That's where cases can help.

Cases provide real situations with real data. Students learn that

- Real statistics problems don't come as antiseptic little modules at the end of the chapter.
- The interesting parts of the problem are
  - deciding what the problem really is,
  - deciding what method might help shed some light on it, and
  - learning to interpret and communicate the data's message.

The details of the calculations are important; they must be well chosen and correctly carried out. Yet all that can take place in a context, a context that shows how the formulas may be applied and why the results might matter.

Students who analyze cases select for themselves the methods they think will be helpful. Because they discover the answers for themselves, they remember them and take a new level of ownership in the learning process. They defend their approaches in class discussion and present them orally and in writing. All this is training for the workplace. Our experience indicates that students also remember the theory better than when you just lecture, even though you may lecture very well.

In the following sections, we address some of the methods of using cases in your class and some of the issues that arise when you use them, but in a sense, those are just details. The main point is philosophical and simple: you should use cases or their equivalents in statistical classes because students learn better by *doing* statistics than by listening to it, and analyzing data is what statistics is all about.

USING CASES

The Case

In our scheme of things, the case is a brief description of a situation, usually accompanied by some data. The data presumably shed some light on the situation; we ask the student to figure out what light the data shed. Usually the student must also present his or her conclusions in written and/or oral form, and the class as a whole may discuss them.

Cases may come from almost anywhere. Most of the cases in *Practical Data Analysis: Cases Studies in Business Statistics (PDA)* come from business situations our students have dealt with personally. Others come from our various consulting experiences or from other sources. Some are "general interest" cases not requiring specific business knowledge. Almost all share one characteristic: a real person really wants to know what the data might say, though names and places have often been disguised. The few cases have some other particular purpose, described in the *Instructor's Note* for the case.

PDA cases may differ from what you think of as a "case:"

- *PDA* cases are generally not as complex as classical business cases. A business case might have 15 pages of description. Ours generally have one to three.
- The solution to a classical business case may involve some subtle tone or details in the material. A solution to a *PDA* case may involve a pattern in the data or details revealed by the *data* rather than by the words.
- Cases in, say, law or medicine are usually vehicles leading a student to a single correct answer. Business cases often have competing solutions. *PDA* cases are closer to business cases than to cases in law or medicine. There are usually competing, legitimate ways of analyzing a *PDA* data set. We designed them to be discussed, not just treated as homework exercises. Students will have to make choices, wrestle with some ambiguities and incomplete information, and choose from among competing techniques whose assumptions may be only approximately satisfied. They will arrive at a variety of approaches of varying qualities, not answers to be characterized as "right" or "wrong." Instructors will be surprised at how much progress students can make using techniques completely different from those the instructor had in mind, and at how much the students may have misunderstood the techniques the instructor intended the students to apply.
- Just as a solution to a business case is insufficient unless it is communicated in some way, so *PDA* cases are designed to be written up, not treated as algebra exercises.

The cases basically provide the students with a chance to look at a situation in which the data may be helpful and to wrestle with such things as defining the problem, choosing and executing an analysis method, and interpreting and communicating the results as would be appropriate in a business setting. The data play the central role; the context provides some motivation for looking at the data.

The pedagogical context for *PDA* cases is different from that in most business courses in that most students take their required statistics course *before* they've studied much in the way of business. Because of this, we find it wise to include a mix of business and nonbusiness cases. This helps capture the attention of those who need to see how it applies in business and of those who (for the moment) need to relate it to other areas because they haven't studied business yet. Most *PDA* cases require little specific functional area knowledge. The context may be oriented to marketing, finance, or personnel—it makes it easier to think about the analysis—but we make few assumptions about course backgrounds in those fields.

The Assignment Questions

When you assign a case to your students, you may wish to assign specific questions along with it. These *assignment questions* serve three purposes.

1 *They guide the students into the case.* The questions suggest aspects they should consider and methods of attack or things to look out for. They get the student going.

2 *They help you to control the analysis.* For example, many of the cases may be attacked in different ways. If you want the students to emphasize a particular approach, you may use the assignment questions to guide them in that direction.

3 *They allow you to adjust the difficulty of the case.* Many of the cases are deliberately a bit vague. They may have missing assumptions or terms not fully defined. The business problem itself may not be well defined.

- If your students are mature, they may be able to define the problem for themselves, select an analysis method and carry it out. If they can, we recommend that you let them do it; they'll get more out of it.
- With a less mature class, though, you may need to give guidance, help them define the problem or suggest possible approaches. Assignment questions are your vehicle for doing this. In the *Instructor's Notes,* we suggest assignment questions for you to consider.

Discussing the Case

Students arrive in class with a lot invested in their work on the case you assigned. The discussion serves several purposes.

1 It's an opportunity for students to articulate and present their conclusions as they might in a business setting. They're usually eager to do so.

2 The different approaches that emerge often highlight both the individual interpretations of the problem and the various methods used.

3 Students can see how the assumptions they made about the problem translate into different methods of attack.

- They discover new ways of thinking about the problem.
- The dialogue with others helps to clarify and delimit the results as well as the strengths and weaknesses of the various approaches. At the end of the discussion, many aspects of the problem and of the methods will have been illuminated, and students will have discovered most of it by themselves.

We give suggested ways of leading the discussion in the *Instructor's Notes* for each case. As a general rule, though, probably the biggest mistake instructors make in leading these discussions is to adopt too high a profile.

1 Let the students own the discussion and limit your role to guiding it.

2 Keep a list of questions in reserve, so that when an important point hasn't been brought out, you can elicit it without giving the game away.

3 Be stingy about giving out "right" answers, too, either during or after the discussion. Students like to know what really happened, but once they become convinced that you will give them the right answer, they start devoting less energy to their own analyses.

Lectures and Cases

We find that dividing class time about equally between discussion and lecture works well.

1 For classes meeting once per week, we begin with a discussion of the case, take a break, and then use lecture and discussion to address new materials or methods. Students are fresher at the beginning and come ready to discuss the case. If you lecture first, they tend not to listen, and they are frustrated to have to wait for the case discussion.

2 For classes meeting two or three times a week, you can devote one or two class meetings to the cases and the other(s) to more classical approaches.

We find the most effective learning comes when students work in small groups on the cases. Most of the education happens in those groups, as they try to decide what their answer should be. We require that each group provide a one-page summary of the case and come prepared to discuss the case in class. Other approaches can be effective, too, but that's the one we've become comfortable with. You will want to let your own approach evolve. Be sure to assign the case far enough in advance, so that students have enough time to try various approaches, discuss it with group members, and so forth. We find that one case a week is about right. Unless you have a very disciplined, mature class, you will probably want to devote some time to presentation of new material and discussion of questions about the specific techniques involved.

USING CASES: OTHER ISSUES

Sequence and Timing

1 Don't be afraid to assign a case *before* you've covered the "official" solution technique. The students may solve most of it using other methods. They may run multiple simple regressions, for example, and discover the *need* for multiple regression, even if they don't know how to do it yet. Of course, you shouldn't penalize them in grading for not using a technique they don't know about yet. Once they've wrestled with the problem, though, they're ready to understand the technique when it's presented—they will know what it's for.

2 You may assign a case several times, as you develop new techniques. The *Housing Prices*—cases (numbers 27 and 28 in Volume II) illustrate this possibility well. You may also choose one or more cases as the basis of your own presentations during lecture. Each new technique can be illustrated on the basis of one of the cases with which they're already familiar.

3 Start early in the course. For many students, the level of involvement required in the case approach is new. If they're used to sitting in class and taking notes, you will need to indicate that something else is expected. The longer you wait, the harder this gets. We've found it useful to break students into groups and analyze data in class for the first or second class meeting.

Term Papers and Exams

For term papers or projects you can assign undergraduates one or two of the cases, but it's usually better (and almost always better for graduate students) to assign a project of gathering and analyzing some data of interest to them. This works particularly well when they work in groups.

If you've just spent a large part of the semester analyzing cases, you probably don't want to turn around and give the students a standard sort of exam. To base an exam on cases, several methods work.

1 Use one or more cases as the basis of a take-home examination. This allows time for full analyses.

2 Invent a scenario such as: "The board of directors would like you to present your summary of these data to them next Tuesday. Please prepare a summary and be prepared to respond to their questions." They can turn in their summary at the exam and answer written questions (simulating the questions from the board). This scenario seems real to many students, and they respond well.

3 Provide a case and some sample printouts for an in-class exam, and ask the students to provide a written report. If the students have their own laptop computers, of course, they can do the analysis without your printouts.

You can combine these approaches, too. Appendix A gives two sample exams for you to consider.

Large Classes

Large classes are a fact of life for many of us. The lot of the policeman who collects and grades all those cases is not a happy one.

1 If your class is organized in lecture-recitation format, use the recitation sections for discussing the cases. This assumes the section leader has some training in leading discussions.

2 Break the students up into groups of 4 or 5. Have each group compare results with a neighboring group, and have them provide a combined solution. Finally, call on two or three combined groups to report their findings. This is in the active learning tradition and, if nothing else, makes a good break from straight lecture.

DIRECTORIES OF CASES

DIRECTORIES BY *PDA* VOLUME

On the following pages, we give directory tables for the cases in *PDA* Volumes I, II, and III. The first three tables present the cases in Volumes I, II, and III, respectively. Each of the first group of columns is labeled by a kind of statistical technique often used in working the case: Descriptive Statistics, Simple Linear Regression, and so forth. Subsequent columns are labeled by application areas: Processes and Operations, Health and Medicine, Marketing, and so forth. Each row in the table corresponds to a case. The case name is listed in the first column. Subsequent columns contain a bullet (•) if the case involves ideas, concepts, or applications from the application area at the top of the column.

In providing these tables, we do not mean to suggest that you will necessarily feel that the techniques listed (or the application areas) are necessarily the way *you* would label the case. Most of the cases can profitably be attacked from different points of view or using different techniques. We merely suggest that if you are looking for cases involving ideas in a particular area, or for which particular techniques might play a role, these are ones to consider, at least as a start.

DIRECTORY BY LEVEL

The next directory lists the cases separately by level of complexity: beginning, intermediate, and advanced. In classifying the cases into levels, we have used the following general guidelines.

- *Beginning cases* are suitable for undergraduates. A technique for solving the problem is usually apparent and it solves the problem pretty well. The problem is reasonably well-defined and the results are straightforward to interpret. Students who devote appropriate time to the problem will likely find an appropriate solution.

- *Intermediate cases* require more general experience or maturity than beginning cases. There may be multiple approaches to the problem, or the approaches may not be obvious. Assumptions underlying the techniques may be satisfied only approximately. Careful reading, interpretation, or resolution of ambiguity may be needed. Techniques may depend on those readings or interpretations. Data may have missing values or require transformations. More care in describing and presenting results will be needed than for beginning cases.
- *Advanced cases* are more suited to graduate students. The case may have a lot of ambiguity. The hardest part of the analysis may be going from the description of the business situation to a statistical analysis. The techniques may give only approximate answers, and interpreting the results in the business context may be tricky. Outliers or missing data may affect the analyses. Transformations may be essential. Sometimes knowledge of a certain amount of jargon from the field involved may be needed to do a good job.

Some cases fit in more than one category. Beginning students may be able to attack part of a problem, while more advanced students may be able to do more. A case may be easy if you give a lot of guidance and difficult if you don't. We've put such cases in both categories.

DIRECTORY OF CASES WITH SPECIAL FEATURES

Finally, we've listed cases with special features:

- Cases that do not require the use of a computer.
- Cases with missing data.
- Cases without data (usually cases about survey design or implementation issues).
- Cases that require little business knowledge or few statistical prerequisites, and hence can profitably be assigned early in a course.
- Cases that refer to the Internet or World Wide Web, either as the source of the data provided or as a place for the students to seek data.
- Cases that are part of multiple-case sequences (where one case is assigned and discussed, and then a case related to the first may follow).

DIRECTORY OF VOLUME I

Case	Case Name	Analysis Technique(s)											Application Area(s)													
		Statistical Process Control	Confidence Intervals, Tests	Time Series	ANOVA Techniques	Multiple Linear Regression	Simple Linear Regression	Cross Tabulations	Nonparametric Methods	Survey Design	Plots	Descriptive Statistics	Quality Management	Real Estate	Other/Nonbusiness	Public Affairs	Forecasting	Information Systems	Mgt, Personnel, Legal, Policy	Accounting	Marketing	Finance	Economics	Health and Medicine	Processes and Operations	Ethical Issues
1.	Waste Paper	•			•	•							•												•	
2.	Dome Loading				•	•					•		•												•	
3.	Spot Prices			•		•											•					•	•			
4.	Joanne's Forecasting Problem					•						•						•								
5.	T-Shirt Designs		•					•	•	•	•										•					
6.	Stocks A and B										•	•										•				
7.	Tom's Used Mustangs		•			•															•		•			
8.	American History Illustrated							•			•	•			•						•					•
9.	Healthy Lifestyles										•	•			•	•								•		
10.	Lake County Lunches										•	•				•										
11.	Machine Production Records		•								•	•	•												•	
12.	Kerrville Rainfall			•			•				•	•			•		•									
13.	Workload Ratings		•			•					•	•			•				•							
14.	Roaring Fork Transit			•		•					•	•				•										
15.	Sturgel Division									•			•					•	•							
16.	Photocopying Abuses			•							•	•							•						•	•
17.	AgriComp		•					•	•		•	•							•						•	
18.	Tips					•						•							•							
19.	Rubbergate		•					•	•			•			•	•										•
20.	HiTech Engineering		•	•	•																•					
21.	401(k)					•											•					•				
22.	Late Charts	•									•	•	•											•	•	
23.	Money Supply and Interest Rates			•			•					•											•			
24.	SAT Scores					•					•	•			•											
25.	Emergency Admissions	•	•	•		•					•	•					•							•		

DIRECTORY OF VOLUME II

		Analysis Technique(s)											Application Area(s)													
Case	Case Name	Statistical Process Control	Confidence Intervals, Tests	Time Series	ANOVA Techniques	Multiple Linear Regression	Simple Linear Regression	Cross Tabulations	Nonparametric Methods	Survey Design	Plots	Descriptive Statistics	Quality Management	Real Estate	Other/Nonbusiness	Public Affairs	Forecasting	Information Systems	Mgt, Personnel, Legal, Policy	Accounting	Marketing	Finance	Economics	Health and Medicine	Processes and Operations	Ethical Issues
26.	Dave's Truck Washing Service			•		•											•		•							
27.	Housing Prices—I		•								•	•		•									•			
28.	Housing Prices—II					•					•	•		•									•			
29.	Books			•		•														•						
30.	ELISA Plate Uniformity				•	•					•	•	•											•		
31.	Violence on TV—I									•						•										•
32.	Overdue Bills						•				•	•			•						•					
33.	The Brazing Operation	•				•					•	•	•												•	
34.	Leading Indicators			•		•					•												•			
35.	Interrupting Tina	•									•	•	•		•										•	
36.	Violence on TV—II							•	•	•						•										
37.	Commercial Property					•								•		•										
38.	AIDS Rates							•			•	•				•								•		
39.	Software Service Calls			•		•					•		•												•	
40.	Titanic Shipping Company										•	•	•					•								
41.	Campaign Statistics							•			•	•				•										•
42.	Hospital Charges		•			•						•												•		
43.	Coal Testing		•		•	•						•	•		•										•	
44.	Performance Evaluations		•			•													•							
45.	Amchem's Tax Credit			•		•					•	•								•					•	
46.	Natural Gas Prices			•		•											•						•			
47.	Westar Beverage Sales			•		•					•	•					•									
48.	Price Fixing Cases	•	•	•							•	•				•							•			•
49.	Property Crimes					•									•								•			
50.	Rockville Physicians' Group	•									•	•	•											•		

DIRECTORY OF VOLUME III

		Analysis Technique(s)											Application Area(s)													
Case	Case Name	Statistical Process Control	Confidence Intervals, Tests	Time Series	ANOVA Techniques	Multiple Linear Regression	Simple Linear Regression	Cross Tabulations	Nonparametric Methods	Survey Design	Plots	Descriptive Statistics	Quality Management	Real Estate	Other/Nonbusiness	Public Affairs	Forecasting	Information Systems	Mgt, Personnel, Legal, Policy	Accounting	Marketing	Finance	Economics	Health and Medicine	Processes and Operations	Ethical Issues
51.	Abandoned Calls—I			•	•		•				•							•							•	
52.	Abandoned Calls—II			•	•				•		•							•							•	
53.	Everybody Does It									•			•					•	•							
54.	Marriage PREP											•			•											
55.	Predicting Snowfall			•			•				•	•			•		•									
56.	Cigarette Rankings							•			•	•									•					
57.	Utility Bills	•		•		•					•						•		•	•					•	
58.	Job Hopping at Chips, Inc.		•					•	•			•							•							
59.	Medication Errors	•						•			•	•	•											•		
60.	Clinical Treatment of Depression		•		•	•									•	•								•		
61.	Deer-Train Collisions										•	•			•										•	
62.	Public Debt in Germany							•			•	•				•							•			
63.	Jobs in Denver			•		•					•					•	•						•			
64.	The Advertising Experiment					•						•									•					
65.	Staffing the Phone Desk			•	•						•	•	•		•				•							
66.	Home Equity Loans		•					•	•			•			•				•	•						
67.	Gas Stations		•		•																		•		•	
68.	Attitudes toward Nuclear Power							•	•							•										
69.	Alicia's Union Data					•						•							•							•
70.	Heart Surgery					•		•			•	•	•						•					•		
71.	The Bay Area Athletic Club		•					•	•			•									•					
72.	Blood Pressure		•								•	•							•					•		
73.	Matrimonial Unrest		•					•				•	•		•										•	
74.	Productivity in the WP Center			•							•	•							•							•
75.	Church Attendance			•			•				•	•			•		•									

DIRECTORY OF CASES BY LEVEL

Case	Beginning Level Cases
5.	T-Shirt Designs
6.	Stocks A and B
7.	Tom's Used Mustangs
8.	American History Illustrated
9.	Healthy Lifestyles
10.	Lake County Lunches
11.	Machine Production Records
15.	Sturgel Division
16.	Photocopying Abuses
17.	AgriComp
22.	Late Charts
23.	Money Supply and Interest Rates
27.	Housing Prices—I
32.	Overdue Bills
38.	AIDS Rates
39.	Software Service Calls
40.	Titanic Shipping Company
41.	Campaign Statistics
42.	Hospital Charges
48.	Price Fixing Cases
52.	Abandoned Calls—II
54.	Marriage PREP
56.	Cigarette Rankings
58.	Job Hopping at Chips, Inc.
59.	Medication Errors
61.	Deer-Train Collisions
62.	Public Debt in Germany
68.	Attitudes toward Nuclear Power
71.	The Bay Area Athletic Club
72.	Blood Pressure
74.	Productivity in the WP Center
75.	Church Attendance

Case	Intermediate Level Cases
1.	Waste Paper
3.	Spot Prices
4.	Joanne's Forecasting Problem
7.	Tom's Used Mustangs
12.	Kerrville Rainfall
13.	Workload Ratings
14.	Roaring Fork Transit
15.	Sturgel Division
18.	Tips
19.	Rubbergate
20.	HiTech Engineering
21.	401(k)
23.	Money Supply and Interest Rates
24.	SAT Scores
25.	Emergency Admissions
26.	Dave's Truck Washing Service
28.	Housing Prices—II

Case	Intermediate Level Cases (cont.)
29.	Books
30.	ELISA Plate Uniformity
31.	Violence on TV—1
33.	The Brazing Operation
34.	Leading Indicators
35.	Interrupting Tina
36.	Violence on TV—II
39.	Software Service Calls
42.	Hospital Charges
43.	Coal Testing
45.	Amchem's Tax Credit
46.	Natural Gas Prices
47.	Westar Beverage Sales
49.	Property Crimes
50.	Rockville Physicians' Group
51.	Abandoned Calls—I
52.	Abandoned Calls—II
53.	Everybody Does It
55.	Predicting Snowfall
56.	Cigarette Rankings
57.	Utility Bills
58.	Job Hopping at Chips, Inc.
60.	Clinical Treatment of Depression
61.	Deer-Train Collisions
63.	Jobs in Denver
63.	The Advertising Experiment
65.	Staffing the Phone Desk
66.	Home Equity Loans
67.	Gas Stations
69.	Alicia's Union Data
70.	Heart Surgery
71.	The Bay Area Athletic Club
73.	Matrimonial Unrest
75.	Church Attendance

Case	Advanced Level Cases
2.	Dome Loading
21.	401(k)
24.	SAT Scores
25.	Emergency Admissions
26.	Dave's Truck Washing Service
28.	Housing Prices—II
31.	Violence on TV—1
34.	Leading Indicators
37.	Commercial Property
44.	Performance Evaluations
46.	Natural Gas Prices
47.	Westar Beverage Sales
60.	Clinical Treatment of Depression
63.	Jobs in Denver
65.	Staffing the Phone Desk
70.	Heart Surgery

DIRECTORY OF CASES WITH SPECIAL FEATURES

Case	Case Name	Volume	Special Feature: OK for Hand Calculation	Missing or Miscoded Data	No Data	May Be Assigned Early	Refers to World Wide Web	OK for Multiple Sessions
9.	Healthy Lifestyles	1		•		•		
15.	Sturgel Division	1			•	•		
19.	Rubbergate	1				•		
22.	Late Charts	1	•			•		
23.	Money Supply and Interest Rates	1		•				
25.	Emergency Admissions	1		•				
27.	Housing Prices—I	2						•
28.	Housing Prices—II	2						•
29.	Books	2		•				
30.	ELISA Plate Uniformity	2		•		•		
31.	Violence on TV—I	2			•	•		•
34.	Leading Indicators	2		•			•	
35.	Interrupting Tina	2	•			•		
36.	Violence on TV—II	2		•				•
38.	AIDS Rates	2	•		•	•	•	
39.	Software Service Calls	2	•					
40.	Titanic Shipping Company	2				•		
41.	Campaign Statistics	2	•			•		
42.	Hospital Charges	2				•		
43.	Westar Beverage Sales	2		•				
48.	Price Fixing Cases	2	•			•		
50.	Rockville Physicians' Group	2			•			•
51.	Abandoned Calls—I	3		•				•
52.	Abandoned Calls—II	3						•
53.	Everybody Does It	3			•	•		
54.	Marriage PREP	3	•			•		
55.	Predicting Snowfall	3		•				
56.	Cigarette Rankings	3	•			•		
57.	Utility Bills	3		•				
59.	Medication Errors	3	•			•		
61.	Deer-Train Collisions	3		•		•		
62.	Public Debt in Germany	3	•	•		•		
63.	Jobs in Denver	3					•	
64.	The Advertising Experiment	3	•					
67.	Gas Stations	3		•				
68.	Attitudes toward Nuclear Power	3	•			•		
70.	Heart Surgery	3		•				•
72.	Blood Pressure	3	•			•		
75.	Church Attendance	3	•			•		

INSTRUCTOR'S NOTES FOR INDIVIDUAL CASES

In the following pages, we give an Instructor's Note for each of the cases in *PDA*. Each note is organized as follows:

- *Abstract* gives a quick summary of the main points from an instructor's point of view: what methods might be used and what kinds of surprises the data might hold.
- *Level* of the case (beginning, intermediate, or advanced) comes next. For definitions of the levels, refer to the first page of the *Directories* section.
- *Techniques* that the case illustrates are given in keyword format.
- *Typical Class Discussion Time* gives a rough idea of how complicated the discussion may turn out to be.
- *Data Set Size* gives the nature (time series or cross-sectional) and number of observations and variables in the data set.
- *General Comments* give more details on how the case may be used and any special features we have found important.
- *Assignment Questions* serve to get the student started in thinking about the case and to control the level of difficulty a bit. You can alter them to provide as much or as little structure as you feel you need. For the most part, we recommend providing no more structure than you honestly feel the students require, so as to be sure most of the work is the students'.
- *Suggestions for Leading the Discussion* may get you started thinking about how *you* want the discussion to go. You will probably develop your own approaches over time, but these may help at first.
- *Typical Results and Approaches* give the typical results, usually with a printout or graph that you can use as the basis of class discussion.

CASE **1**

WASTE PAPER, INSTRUCTOR'S NOTE

Abstract: This is a good case for one-way analysis of variance, though it can also serve as the basis for a model-building session based on multiple regression. It allows dummy variables for several of the variables and a chance to transform the dependent variable to percentage or logarithmic terms. Students will need to use that facility of their statistics program that selects certain observations for analysis.

Level:	Intermediate.
Techniques:	One-way analysis of variance, multiple regression, dummy variables, data transformation, selection of observations.
Typical Class Discussion Time:	30–45 minutes.
Data Set Size:	127 cross-sectional observations, 7 variables.

GENERAL COMMENTS

This is a fairly straightforward application of one-way analysis of variance or multiple regression, but there's enough going on that students will have to stay organized to be successful. They will need to compute the dependent variable, try a variety of scales and plots, and select the four-color observations from the data set to analyze. The dependent variable can be either the ratio of WASTE to GROSS (for analysis of variance) or the logarithm of WASTE (for multiple regression, if GROSS is included as an independent variable).

QUESTIONS FOR ASSIGNMENT

1 If you want to give this amount of guidance, ask: "What is the appropriate measure of waste in this problem?"

2 For a one-way analysis of variance approach:

a Should the waste factors be different for the different sizes of books?

b What waste factors should the company use?

3 For a multiple regression approach:

a If you want to give the students an extra variable, you can say, "Consider all the data, not just the four-color cases." This gives them another set of dummy variables to create to assess the effects of COLORS.

b In addition to number of pages, what factors appear to influence waste?

c Write a model or equation describing how waste depends upon the factors you cited in your answer. How well does your model describe the data?

SUGGESTIONS FOR LEADING THE DISCUSSION

You may want to use different methods here, depending on the kinds of methods you feel your students will use. If you have in mind multiple regression and model building, "Well, what factors seem to be involved?" is the obvious start here, and it works fine. Follow-up questions can identify which dependent variable was used and how the model looks. Of course, any logarithmic models should be restated on the original scale. You can then raise the question of criteria for selecting among these competing models. R^2, standard error, and various aspects of the residuals may be mentioned. You could also bring up criteria like Mallows C_p if you're covering them in your course.

On the other hand, as stated, the case cries out for a one-way analysis of variance, applied specifically to just the four-color observations. Here the assignment questions themselves work pretty well. You can start by saying something like, "Should the company be using different factors for different book sizes?" and follow up with, "What sort of analysis led you to conclude that?" That, in turn, will lead into discussion of what the dependent variable ought to be. To emphasize the one-way ANOVA approach, use something like the ratio of WASTE to GROSS, or some transformation of it, because the number of books in the press runs varies so much. It may be helpful to make a transparency of the ANOVA printout like the one that follows and use it as a basis of discussion. You will probably want to point out that, first of all, only the four-color observations, 101 of them, should be used. (As it happens, the conclusions don't change much if you use all of the observations.) You can then go through the parts of the printout one at a time to identify any significant differences in pairs of means or in the aggregate. By most measures, there are no significant differences; the overall *F* ratio has a *P*-value of .07.

TYPICAL RESULTS AND APPROACHES

Typical regression models of the *whole* data set include:

$$\text{WASTE} = 4286.23 + .07398(\text{NET})\ R^2 = .213,\ s = 2687$$
$$\text{WASTE(as percent of NET)} = 52.09098 - .0073(\text{NET})\ R^2 = .402\ s = 16.76$$
$$ln(\text{WASTE as percent of NET}) = 3.95434 - .00002(\text{NET})\ R^2 = .548\ s = .42764$$

Neither PAGES, PAPER, nor COLORS seems to have much effect, though if they are proposed, you should check to see how they were included (unchanged or as dummy variables).

The results from one-way analysis of variance using WASTERAT = WASTE/GROSS as the dependent variable are in the printout that follows, which uses only the data for COLORS = 4.

One-way ANOVA for WASTERAT by PAGES

```
SOURCE     DF        SS           MS          F        P
-------   ----   ---------    ---------    ------   ------
BETWEEN     3     0.08236      0.02745      2.41    0.0702
WITHIN     97     1.10372      0.01138
TOTAL     100     1.18608

                        CHI-SQ      DF        P
BARTLETT'S TEST OF      ------    ------   ------
   EQUAL VARIANCES        2.18       3     0.5358
COCHRAN'S Q                        0.3227
LARGEST VAR / SMALLEST VAR         2.8299
COMPONENT OF VARIANCE FOR BETWEEN GROUPS  7.575E-04
EFFECTIVE CELL SIZE                            21.2
                            SAMPLE     GROUP
  PAGES           MEAN       SIZE     STD DEV
---------     ----------    ------   ----------
    8            0.2713       13       0.1157
   16            0.1991       53       0.1078
   24            0.2454        8       0.0688
   32            0.2491       27       0.1084
TOTAL            0.2254      101       0.1067

TUKEY (HSD) PAIRWISE COMPARISONS OF MEANS OF WASTERAT
BY PAGES
                          HOMOGENEOUS
  PAGES          MEAN     GROUPS
---------     ----------  ------------
    8           0.2713     I
   32           0.2491     I
   24           0.2454     I
   16           0.1991     I
THERE ARE NO SIGNIFICANT PAIRWISE DIFFERENCES AMONG
THE MEANS.

CRITICAL Q VALUE 3.697   REJECTION LEVEL 0.050
STANDARD ERRORS AND CRITICAL VALUES OF DIFFERENCES
VARY COMPARISONS BECAUSE OF UNEQUAL
SAMPLE SIZES.
```

CASE **2**

DOME LOADING, INSTRUCTOR'S NOTE

Abstract: This case can be handled as a two-way analysis of variance or as multiple regression with dummy variables. The dependent variable must first be derived from the data given. The results show substantial differences in one factor, but not in thc other, and students should be encouraged to interpret this in the context given: what does it suggest would improve things? Each part of this case is straightforward; the complexity comes from having to put all of the parts together.

Level:	Advanced.
Techniques:	Analysis of variance, regression.
Typical Class Discussion Time:	45–60 minutes.
Data Set Size:	45 cross-sectional observations (with some time sequencing), 7 variables.

GENERAL COMMENTS

This case is a good example of analysis of variance techniques for fairly advanced students. They will have to think carefully about what the appropriate measurements are (i.e., what the dependent variables are), choose an analysis approach (either two-way ANOVA or multiple regression with dummy variables works fine), and interpret the results in the industrial context from which the data came. One factor seems to have a definite effect, while the other doesn't, although the results are slightly different for the two dependent variables.

QUESTIONS FOR ASSIGNMENT

The question in the case is good enough for advanced students. To provide more structure, you could ask:

1 What are the appropriate variables to consider here?

2 Do you find sheet-to-sheet or dome-to-dome differences on the variables you identified? Summarize any differences you find.

3 Based on your results, what areas of the dome-loading process appear to be candidates for improvement?

SUGGESTIONS FOR LEADING THE DISCUSSION

We like the general approach and begin with something like, "Well, what suggestions do you have for Circle Electronics?" We follow up with questions that probe the reasons for the recommendations, like, "What aspect of the data leads you to think that? What did you compute? Plot?"

If the students seem to be having difficulty, you can provide a bit more focus by using the kind of questions listed above. Ask first what the appropriate measures are and point out that these would be the dependent variables in most analysis approaches. Then begin the discussion of appropriate analysis techniques. The data are a time series only in the sense that the sheets were consecutive (the time sequence of the domes isn't clear from the case), and it's a very short time series, so time series techniques aren't particularly helpful here. The inputs are categorical, so either analysis of variance or some form of multiple regression with dummy variables will likely be suggested as approaches. Either will work, as the experiment is balanced. Finally, it seems critical here to force the students who propose answers to interpret them in the industrial setting. "What aspects of the process seem to be worth looking into?" or, "You found a significant effect due to differences among sheets. What does that suggest about the process? Anything we should be looking at?" can keep the students concerned with the connection between the analysis and the problem. The analysis is not complicated here, but has enough steps to it that students may begin to forget why they were looking at the data in the first place.

TYPICAL RESULTS AND APPROACHES

The most successful approaches to this come from first transforming the measurements given into the corresponding errors: XERROR = XMEAS − XSPEC and YERROR = YMEAS − YSPEC. This makes measurements from one dome comparable to those from the others, at least to some extent. Each of these errors can be analyzed separately and this probably leads to the most interpretable results, as the *x*- and *y*-directions are probably controlled separately in the robot (although the case doesn't say this specifically). Some students may calculate a radial error equal to the Euclidean distance from (XMEAS,YMEAS) to (XSPEC,YSPEC). This corresponds more closely to the stated specification, at least by some interpretations, but doesn't really alter the conclusions much and doesn't correspond as closely to the way the equipment probably works.

Further analysis shows substantial sheet-to-sheet differences for both XERROR and YERROR, and a small effect of one dome on YERROR. A plot like the one that follows can be displayed to illustrate what's going on. Simple plots of XMEAS (or YMEAS) versus SHEET will show this, too, and you can use them to illustrate how ANOVA or regression analysis corresponds to what is visible in those plots.

The main conclusion seems to be that sheet-to-sheet differences are bigger than other differences. That suggests that there might be difficulty (or variation) in the way the sheets were inserted into the dome loader or perhaps in the way the sheets were trimmed afterward. The variations from dome to dome seem much more random; looking at the sheets would seem an appropriate first step. The results show ANOVA ideas in a process setting. They also illustrate the need for experimental design ideas in such settings.

By the way, in plunging into the details, it would be easy to forget a more general conclusion: the process isn't coming close to the stated specifications, even after removing the effects of sheet-to-sheet differences. Considering that the specification "is generally attainable by hand tools," it does seem as though Circle Electronics might do well to be looking for some quality improvement programs and techniques. Improving the management of the individual sheets could be a good first step, but it's not a "solution," apparently.

Plot of XERROR versus Sheet Number

```
XERROR: Error in X placement ( = XMEAS-XSPEC)
 0.022+
      |           *
      |
      |
      |           *
 0.010+           *
      |
      |           *                       *           *
      |           *                       *
      |                                   *           *
-0.003+           *                       *
      |                       *           *           *           *
      |                                                           *
      |           *           *                       *           *
      |                       *                       *           *
-0.015+-----------+-----------*-----------+-----------*-----------*
      0           1           2           3           4           5
                        SHEET: Mylar sheet number
```

CASE **3**

SPOT PRICES, INSTRUCTOR'S NOTE

Abstract: This case is a nice choice for modeling time series and forecasting. There is seasonality in both spot prices (January can be modeled with a dummy variable), and also some cyclical behavior.

Level:	Intermediate.
Techniques:	Time series, regression.
Typical Class Discussion Time:	30 minutes.
Data Set Size:	46 time series observations, 4 variables.

GENERAL COMMENTS

The November 1997 Louisiana spot price is given in the case to be 1.89. The actual value of the Oklahoma spot price in the same month is 1.71.

Although the Oklahoma spot price might just as well be described by an ARIMA model, this case specifically attempts to link the preliminary announcement of the Louisiana spot with a forecast of the Oklahoma spot. To achieve this, we consider models that all attempt to add the Louisiana spot price as a variable in a multiple regression equation. Nonetheless, those of you looking for univariate time series might use either the Oklahoma or Louisiana spot price series as a modeling exercise.

We propose several forecasting models below. However, a simple calculation of ratios ends up forecasting better than some of the multiple regression models. Because the average of the ratio of the Oklahoma to Louisiana spot prices is 0.866 over the time series, one way to forecast the November 1997 Oklahoma spot is to take 86.6% of the Louisiana spot of 1.89, which gives a forecast for the Oklahoma spot of 1.64 and a forecast error of −0.07.

QUESTIONS FOR ASSIGNMENT

1 What is the relationship between the Oklahoma and Louisiana spot prices? How do they each vary over time? How do they vary together over time?

2 Are there any important monthly or seasonal variations in the spot prices?

3 Is there any autocorrelation in the data? If so, how might this pattern be modeled?

4 Is there any cyclical pattern to the spot prices? If so, how is this different from seasonal variations in the data?

SUGGESTIONS FOR LEADING THE DISCUSSION

One way to lead this discussion is to ask: "What is your forecast of the November 1997 Oklahoma spot price, and how did you derive it?" Once you've discussed alternative models and forecasts, bring into the discussion the true Oklahoma spot price of 1.71, and ask the class how they might use the actual value to validate their models. The discussion can then turn to RMSEP and forecasting error.

Another way to begin the discussion might be to present the class with the time series plot of the Oklahoma spot prices (below), and ask: "What information is being conveyed in this graph?" Presumably, any information they see in the graph should ultimately appear (or have been considered) in their forecasting model.

Time Sequence Plot of OSPOT

```
OSPOT
     -                        1
     -
     -
2.000+
     - 2                                 B
     -1           1                     A
     -           B           B
     -  3
1.600+          A              2          1        0
     -         0
     -        9    2   6    A          0
     -   4            5 78
     -    5678      34    90    34567 9           9
1.200+                               8     23456
     -
     -                                          78
     -
     +-----------+-----------+-----------+-----------+
     0          12          24          36          48
                   Observation Number
```

TYPICAL RESULTS AND APPROACHES

Several potential forecasting models follow, along with the November 1997 forecast error (*t* ratios are parenthesized). JANDUM is a dummy variable taking on the value of 1 for January (0 otherwise), LAG4LSPOT is the four-month lag of LSPOT, and TIME_SQ is the square of TIME.

Forecasting Models for Spot Prices

Model	adj R^2	Durbin Watson	Forecast Error
OSPOT = 0.11 + 0.80 LSPOT (27.4)	94.3%	0.93	-0.08
OSPOT = 0.16 + 0.77 LSPOT + 0.08 JANDUM (23.7) (2.2)	94.8%	0.89	-0.09
OSPOT = 0.34 + 0.72 LSPOT - 0.06 LAG4LSPOT + 0.09 JANDUM (26.6) (-2.4) (2.8)	96.4%	1.21	-0.08
OSPOT = 0.11 + 0.85 LSPOT - 0.01 TIME + 0.0003 TIME_SQ (36.0) (-5.6) (6.1)	96.9%	1.72	0.04
OSPOT = 0.15 + 0.82 LSPOT - 0.01 TIME + 0.0003 TIME_SQ + 0.05 JANDUM (30.7) (-5.4) (5.9) (1.7)	97.1%	1.59	0.03

CASE **4**

JOANNE'S FORECASTING PROBLEM, INSTRUCTOR'S NOTE

Abstract: This is a classic problem in a couple of ways. First, it's a standard situation and problem in information systems. Second, multiple regression solves the problem perfectly, or nearly so.

Level:	Intermediate.
Techniques:	Multiple regression.
Typical Class Discussion Time:	30–45 minutes.
Data Set Size:	12 cross-sectional observations, 5 variables.

GENERAL COMMENTS

Joanne needs to find the relationship between what she knows (INVOICE, RESUPPLY, . . .) and what the computer center manager *would like* to know (CPUMIN). So it's tailor-made for multiple regression, and in fact, multiple regression seems to work just fine.

QUESTIONS FOR ASSIGNMENT

1 How do you think Joanne should proceed? Can she use the data she has available?
2 What should Joanne forecast, in CPU minutes?

SUGGESTIONS FOR LEADING THE DISCUSSION

For most classes, the questions above can guide the discussion. You could actually start with the second question—ask for a proposed forecast, then follow up with a request to explain the basis of the forecast. That in turn will lead to discussion of the underlying models.

You could also ask specifically for a regression of CPU minutes on the other variables and for an assessment of which variables have significant effects. This approach turns it into more of an exercise or example of multiple regression than a problem to be solved.

At some point in the discussion, you should bring out the issue of month-to-month fluctuation in CPU usage, and ask how Joanne should incorporate that in her forecasts. Some will suggest that Joanne should simply use the average, others that she should request a lot, since she wants to be sure the capacity is there. Still others will point out that it might depend on who has to pay for any forecast capacity that is unused. That's a common issue in IS departments—whether to operate as a cost center, profit center, staff agency, and so forth, and the choice has important implications for how people actually behave in their use of the department's facilities. Regression is one of the ways in which the users and the IS department can communicate their needs to each other.

You can also use regression printouts as a basis for discussion. Two are given below. The first uses all the variables, while the second omits RESUPPLY and CUSTOMER. For practical purposes the second fits just as well as or better than the first, so you can use the comparison between the two to point out various aspects of model building. R^2 goes down, but trivially, when the two variables are removed, but the standard error of estimate gets better. The remaining coefficients don't change dramatically. All these effects suggest it may be safe to omit the two variables.

TYPICAL RESULTS AND APPROACHES

A quick review of the data shows substantial month-to-month variation in the amount of CPU time Joanne's department uses. That means Joanne probably can't get away with forecasting "3% more than last year" or something equally simple. When workload varies a lot over time but the capacity to handle the workload is spread uniformly over the same time, a forecast that takes no account of the peaks and valleys probably isn't much help.

Most approaches to multiple regression here show ORDERS and INVOICE as the two important variables:

$$\text{CPUMIN} = -53.91 + .264(\text{ORDERS}) + .115(\text{INVOICE})$$

with $R^2 = .995$ and standard error of estimate $= 71.72$ is the typical model. Using Joanne's expected monthly averages as inputs to this equation yields a forecast of $-53.91 + .264(8000) + .115(6500)$, or 2806 CPU minutes per month, up from an average of about 2500 last year. If it were possible for Joanne to develop individual monthly forecasts for ORDERS and INVOICE, she could use the equation to develop corresponding individual monthly forecasts for CPU minutes. If that isn't possible, she should still probably make some adjustment to reflect anticipated month-to-month variations in demand. Last year's peak-to-average ratio was 4016 to 2502, or 1.6, so $2806(1.6) = 4490$ CPU minutes per month might be a reasonable peak-period forecast. The forecasts depend to some extent on how many digits are kept in the calculations.

Incidentally, R^2 is large in this case. That is not atypical in transaction-oriented systems like the one described here. The transactions identified are the only things that use CPU time and each transaction of a given type uses more or less the same amount, so the assumptions underlying a regression model are satisfied well, and an R^2 like this one is not uncommon. If the system were a more general-purpose system, like a time-sharing or inquiry system, lower values for R^2 would be more the rule.

Regression of CPUMIN on ORDERS, INVOICE, RESUPPLY, and CUSTOMER

```
   Dependent Variable:        CPUMIN---CPU MINUTES USED DURING MONTH
   Independent Variable(s):   Constant
                              ORDERS---NUMBER OF ORDERS PROCESSED
                              INVOICE---NUMBER OF INVOICES PROCESSED
                              RESUPPLY---NUMBER OF RESUPPLIES
                              CUSTOMER---NUMBER OF CUSTOMER CHANGES
REGRESSION COEFFICIENTS
IND. VBL.     COEFFICIENT     STD. ERROR     T
Constant         23.56506      129.45932   0.2
ORDERS            0.26725        0.01224  21.8
INVOICE           0.11446        0.01202   9.5
RESUPPLY         -0.11624        0.15117  -0.8
CUSTOMER         -0.79807        1.74582  -0.5

ANALYSIS OF VARIANCE:
SOURCE         DF                 SS                  MS
Regression      4      8394902.61345       2098725.65336
Error           7        42661.63655          6094.51951
Total          11         8437564.25        767051.29545

R-squared: 0.995 Std. error of estimate:           78.06740
```

Regression of CPUMIN on ORDERS, INVOICE

```
   Dependent Variable:        CPUMIN---CPU MINUTES USED DURING MONTH
   Independent Variable(s):   Constant
                              ORDERS---NUMBER OF ORDERS PROCESSED
                              INVOICE---NUMBER OF INVOICES PROCESSED
REGRESSION COEFFICIENTS
IND. VBL.     COEFFICIENT     STD. ERROR     T
Constant        -53.91362       66.83296  -0.8
ORDERS            0.26353        0.01005  26.2
INVOICE           0.11500        0.01033  11.1

ANALYSIS OF VARIANCE:
SOURCE         DF                 SS                  MS
Regression      2      8391269.07870       4195634.53935
Error           9        46295.17130          5143.90792
Total          11         8437564.25        767051.29545

R-squared: 0.995 Std. error of estimate:           71.72104
```

CASE **5**

T-SHIRT DESIGNS, INSTRUCTOR'S NOTE

Abstract: This case offers the students a chance to apply some nonparametric techniques and to criticize the design of a survey instrument. There really isn't much of any relationship or difference to be found in the data. In the context of the case, that's useful information.

Level:	Beginning.
Techniques:	Nonparametric methods, cross-tabulations.
Typical Class Discussion Time:	30 minutes.
Data Set Size:	64 cross-sectional observations, 6 variables.

GENERAL COMMENTS

You can steer this one in a lot of different directions. There's much to criticize in the survey itself. You can use it to bring up the need for nonparametric techniques, since the PREFER score is clearly on an ordinal scale. Or you can use it without a lot of structure to see how well students can define their own problem.

By the way, some truth in advertising might be in order here. This data comes from a survey that we distribute on the first night of class, except for one change. The survey is designed to measure students' anxiety towards statistics. Question 3 is actually:

3 Circle the number that most accurately describes your attitude toward this course:

1 = I am strongly dreading this course
2 = I am somewhat dreading this course
3 = My feelings are neutral
4 = I am somewhat looking forward to this course
5 = I am eagerly looking forward to this course

We use the anxiety survey for a number of different purposes. We've discovered it's a great idea to bring up student anxiety toward the first statistics class (there's lots of it out there!), and this is one way to initiate that discussion. We also tabulate the data and use it throughout the course. Question 6 is used to discuss spurious correlation. You might want to give this a try in your own class. The data set is real and comes from a survey of 64 graduate business students at the University of Colorado at Denver.

QUESTIONS FOR ASSIGNMENT

1 Of the factors measured in the survey, which appear to be related to preference level?
2 What are the appropriate methods for assessing the importance of those factors to preference level?

3 How would you improve the survey instrument itself?
4 Comment on the administration of the survey.

SUGGESTIONS FOR LEADING THE DISCUSSION

You can always open the discussion by saying: "The statistician who analyzed the data gave me this graph, but I don't understand what it means. Can you interpret it for me?" Hopefully, someone will say that not many people are indifferent between the two designs yet not many feel strongly either. Most respondents have a slight preference for one over the other.

Histogram of PREFER for Julio's T-Shirts

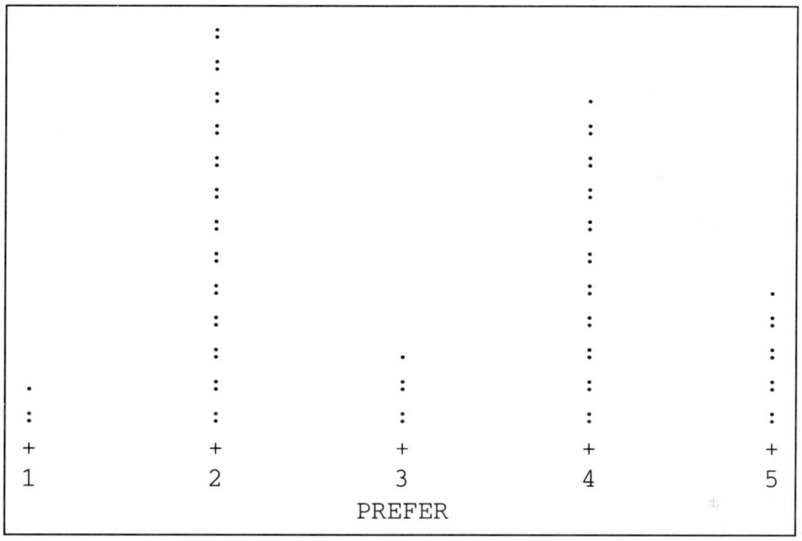

The assignment questions above can also serve to begin the discussion. You can emphasize the ones that fit best with your course. Students with any kind of experience in survey design or market research may have quite a few things they don't like about the survey. Once the discussion gets into this area, it will probably stay there, so you may want to be fairly directive at first, at least until you're satisfied that any technical material you want covered has been covered.

Before you leave the discussion, make sure to wrap it up by returning to the original scenario. After all, we've learned something from the survey: that Julio might as well order an even number of both shirts.

TYPICAL RESULTS AND APPROACHES

About the only factor that shows any hope of a relationship with PREFER is GENDER, and those differences are not significant according to t tests or a variety of nonparametric analogues. Chi-square tests show no pattern, nor does Mann-Whitney, and the like. Still, if someone offers a t test, it's an opportunity for some other student to wonder out loud if that was really the right one, and the need for nonparametric techniques will emerge from the discussion.

The survey has lots of problems: scale, anchors, phrasing of questions. For example, do we care about the handedness, and why is it there? If we want it there, are we after the respondents' *perceptions* of their handedness ("I consider myself right-handed") or do we want something *specific* ("I write with my right hand")? Also, there is some confounding here. We don't know whether a preference is based on the blend of the fabric, the color, or the design.

CASE **6**

STOCKS A AND B, INSTRUCTOR'S NOTE

Abstract: A very easy case for those who follow instructions (compute means and standard deviations), but a surprising number don't follow instructions. Can also serve to introduce the idea of covariance, to be seen again in finance classes.

Level:	Beginning.
Techniques:	Descriptive statistics.
Typical Class Discussion Time:	30 minutes.
Data Set Size:	50 time-sequenced observations, 2 variables.

GENERAL COMMENTS

This case introduces the idea of expected return and risk, as measured by the average and standard deviation. It doesn't follow up on those concepts particularly, though you could make some remarks about markets and random walks if you thought it appropriate. The main idea is that risk is to be measured by standard deviation. The students are left to discover that the variance of a sum is not the sum of the variances.

QUESTION FOR ASSIGNMENT

What do you recommend, all stock A, all stock B, or a 50/50 mixture of both? Why do you say so?

SUGGESTIONS FOR LEADING THE DISCUSSION

Ask the students for their recommendations and why they say so. In each case, you should probe a bit to see if their answers are in fact based on maximizing expected return and/or minimizing standard deviation, as the case specifies.

When the discussion is complete, you may want to point out that these ideas will be further exploited in forthcoming finance classes, but the basis of the formal models will be the same: the trade-off between risk and return ("You heard it here first!"). If you ever *found* a situation like this one, it would be a glorious opportunity to make a lot of money, but that isn't really the point here.

TYPICAL RESULTS AND APPROACHES

As the printout below shows, the means and standard deviations for stocks A and B are nearly identical (mean 0.18, standard deviation .10). (In an earlier version of this case, the values for stocks A and B were the same numbers, just rearranged.) Many students will stop at this point and say it doesn't make any difference which one you invest in. Others will make their recommendations based on the differences in the last two decimal places of the mean, which can lead to some good discussion about whether that would stand up in practice. One step more sophisticated are those who then proceed to invoke every other criterion under the sun except the one specified in the case—standard deviation. Analyses invoking recent trends, head-and-shoulders patterns, discounting by the time value of money, recent up and down patterns, and so on, may be offered.

For those who simply accept the terms of the problem, it's easy. You compute the returns that would have been received for the 50/50 mix (as .5*A + .5*B) and discover that the mean and standard deviation of that sequence are 0.18 and .08, so that the mixture gives the same return with less risk. End of problem. A data-analytic version of "don't put all your eggs in one basket." In our experience, students who didn't see the point (that the standard deviations will be different) can be remarkably defensive about it, so you may want to be careful how you make them see the error of their ways. For all that it's an artificial example, it seems to be quite effective in reminding folks that their intuitions aren't always right, and sometimes it's a good idea to just compute.

Descriptive Statistics for Stocks A, B, and a 50/50 Mix of Stocks A and B

	N	MEAN	MEDIAN	TRMEAN	STDEV	SEMEAN
A	50	0.1800	0.1940	0.1813	0.1000	0.0141
B	50	0.1814	0.1890	0.1780	0.1005	0.0142
MIX	50	0.1807	0.1840	0.1780	0.0787	0.0111

	MIN	MAX	Q1	Q3
A	-0.0350	0.4010	0.1150	0.2527
B	0.0170	0.4280	0.1030	0.2460
MIX	0.0565	0.3670	0.1160	0.2404

CASE **7**

TOM'S USED MUSTANGS, INSTRUCTOR'S NOTE

Abstract: This is a multiple regression model–building exercise with many categorical variables. The existence of the binary information permits beginning classes to perform hypothesis tests and construct confidence intervals for the difference between two population means.

Level:	Beginning, intermediate.
Techniques:	Regression, tests of hypotheses.
Typical Class Discussion Time:	30–45 minutes.
Data Set Size:	35 cross-sectional observations, 10 variables.

GENERAL COMMENTS

This data set can have multiple purposes. The obvious methodological candidate is a multiple regression model–building exercise. The coefficients of the multiple regression model have intuitively appealing interpretations.

The residuals in the multiple regression equation represent estimates of whether a particular car might be under- or overvalued.

Two-sample hypothesis tests can be done for PRICE based on other binary variables in the data set, such as GT, OWNER, and AIR. ANOVA will not be practical, given the limited number of observations in some categories. Some of your students will probably attempt these analyses anyway.

QUESTIONS FOR ASSIGNMENT

For beginning classes, you can perform various two-sample means tests. ANOVA is not practical for COLOR because of the limited number of observations in some of the categories. We offer it as a question below for you to permit a discussion of appropriate sample sizes.

1 Test the hypothesis that mean price does not depend on whether the car is a convertible. Interpret your results in nontechnical terms for Tom.

2 Perform the same hypothesis tests for transmission type, presence or absence of an air conditioner, GT model or not, and private versus dealer ownership.

3 Perform the test that average selling price does not vary with color.

4 How does the number of cylinders influence selling price?

The following questions might give a bit more direction for an intermediate class or exercise:

5 Find a reasonable multiple regression model of PRICE on any of the other variables available in the data set.

6 Plot the residuals from your regression against the fitted values and interpret this plot in nontechnical terms for Tom.

SUGGESTIONS FOR LEADING THE DISCUSSION

We like to begin this case with the second of the two tasks posed in the case write-up: ask the class to explain to Tom in nontechnical terms how to price his cars. Do not permit your students to refer to a model, or an R^2 value, or any such thing. Later, the discussion can turn to the model selection process, the meaning of the residuals, and so forth.

This case is a very nice illustration of the value of residual analysis. If you have time, bring into the discussion the plot of the residuals against the predicted values, and discuss what it means to have a car whose residual is significantly above or below the regression line. If you have not yet discussed standardized residuals in class, this would be a good time to do so, since your students might get stuck on the issue of significantly above or below the regression line.

TYPICAL RESULTS AND APPROACHES

The typical asking price is $8194 with a typical fluctuation of $3783. The cars in this sample aren't really that old: the typical car is about four years old and has about 48,300 miles on it. The majority of the cars have eight cylinder engines. There is only one six-cylinder car, two cars without air conditioners, and only one silver Mustang in the sample.

The two-population hypothesis tests of differences in means for PRICE produce statistically significant differences for CONVERT and GT at the 5% significance level (regardless of whether you use a pooled variance or not). Average prices differ according to ownership at the 10% level only.

An interesting problem with multicollinearity arises in this data set. It is intuitively appealing that AGE and MILES will be related. When each variable appears alone and when both variables appear in a two-predictor multiple regression model, they are individually significant. When other variables, such as GT and CONVERT, are added along with AGE and MILES, MILES becomes insignificant. A fairly simple regression with AGE, CONVERT, and GT produces a reasonable model (*t* ratios are indicated in parentheses).

```
PRICE = 11,516 - 1333 AGE + 3162 CONVERT + 2880 CT         R-squared = 93.2%
                 (-16.7)     (8.1)          (7.9)                  s = 1034
```

For a discussion of which cars are over- or underpriced, here is the plot of the residuals against the fitted values from this regression:

Plot of Residuals versus Fitted Values for Tom's Mustangs

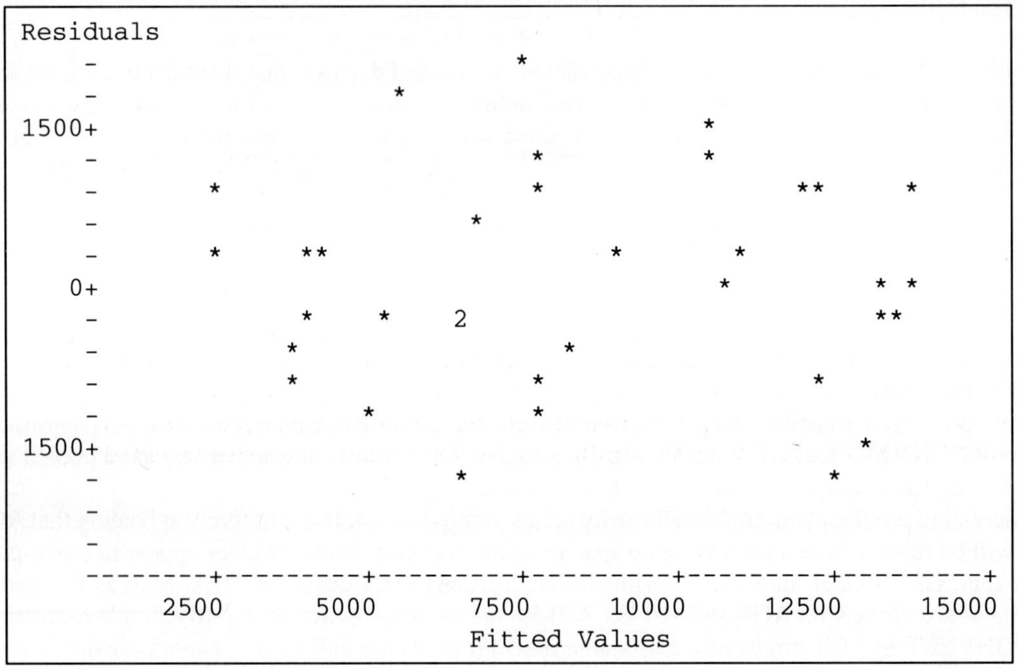

CASE **8**

AMERICAN HISTORY ILLUSTRATED, INSTRUCTOR'S NOTE

Abstract: A classic example of Simpson's paradox. Effective when assigned early in the course. Leads to discussions of issues such as "What's the relevant ratio here?" and a plethora of approaches to summarizing and presenting. Requires no specific business or statistical knowledge, but can repay substantial thought. Points out clearly that data analysis occurs in the context of a problem, not as some isolated exercise. Defining the problem clearly makes the technical part of the analysis easy.

Level:	Beginning.
Techniques:	Computation of ratios, tables, graphs.
Typical Class Discussion Time:	45–60 minutes.
Data Set Size:	4 rows by 5 columns.

GENERAL COMMENTS

This is a good case to assign as the first case in a class. It requires no particular background, and simple arithmetic gives as much insight as does any fancier technique. It will lead to a variety of answers and a variety of summary techniques (tables, graphs, etc.), which can serve as the bases of some good discussions of measurements and ratios.

QUESTION FOR ASSIGNMENT

The case question works fine here: Provide the summary the boss requested and interpret it for him.

SUGGESTIONS FOR LEADING THE DISCUSSION

Playing the role of the boss is effective here: "Well, glad to see you back. What do you have for me? Which department should I be thanking?" Explore whatever answer is offered first to uncover its assumptions, usually revealed as a choice of ratio: renewals as a percent of subscriptions coming due, renewals as a percentage of those renewed, and so forth. "Did everyone use this measure?" is usually a way of eliciting other approaches, which you can then compare and contrast.

When the discussion about various conclusions and approaches dies down, you might shift to the question of how best to summarize and communicate the results. This usually leads to a discussion as students compare the relative merits of tables and percentages versus various kinds of graphs and charts. Almost no single quantity or chart summarizes all of what's going on in this case (that's one of its best features), so this discussion usually provides ample opportunity to show how different techniques illustrate different aspects of the data. A couple of possible graphs follow this note. You could use them as the basis of discussion if the students don't provide appropriate approaches of their own.

TYPICAL RESULTS AND APPROACHES

This is a classic example of Simpson's paradox: within the individual categories, the renewal rate actually decreased, while in the aggregate, it increased. Our sturdy staff assistant will presumably be mildly embarrassed to have to report back to the boss that no department is contributing to the increase, and yet the increase is there. The challenge is to figure out how to explain this in any coherent way.

This happens because the mix of subscriptions coming due changes from an emphasis on categories with generally low renewal rates (mainly subscription agents) to categories with generally high renewal rates (notably, previously renewed and direct mail). The change in mix overwhelms the deterioration of the individual category renewal rates.

All this is hard to describe coherently. Students may dodge the issue by using some definition of renewal rate other than the one given in the case, complaining that more data is needed for a business decision here (true enough), or by focusing on renewals as a percentage of something else. They often try to convey the two phenomena (change in mix and change in within-category renewal rates) on the same graph, usually with confusing results. For students who have studied managerial accounting, you can bring out the analogy with some of the situations described by variance analysis—the problem there is to break deviations from a plan (or budget) into their components, or *variances.* Here it's changes from one month to the next.

All this confusion can be productive. You can point out that different ratios correspond to different aspects of the problem. If we care about the external market, we probably want to see how the mix of subscriptions coming due changed. If we wish to evaluate the performance of the managers of the various categories, the within-category renewal rates will help us more. The case is deliberately vague on the ultimate purpose of the analysis, in the hopes of pointing out how important it is to define the problem clearly. Students may be frustrated that the question wasn't stated more clearly. On the other hand, it's stated about as clearly as would commonly occur in practice. Clarifying the question is part of the statistical analysis. It's important not to penalize anyone for his or her assumptions here (within reason), or you can stamp out participation in a hurry. The real strength of this case is the variety of approaches and answers it generates. Encouragement early on leads to better investigations in more complicated cases later and stimulates more active participation in general.

Renewal Rates for January and February

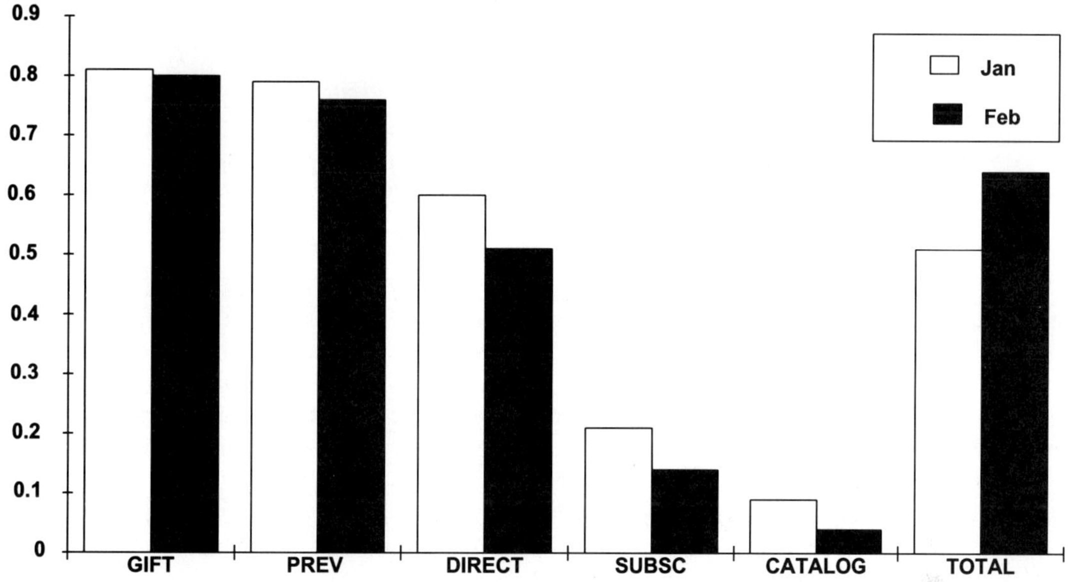

Fraction of Subscriptions Expiring

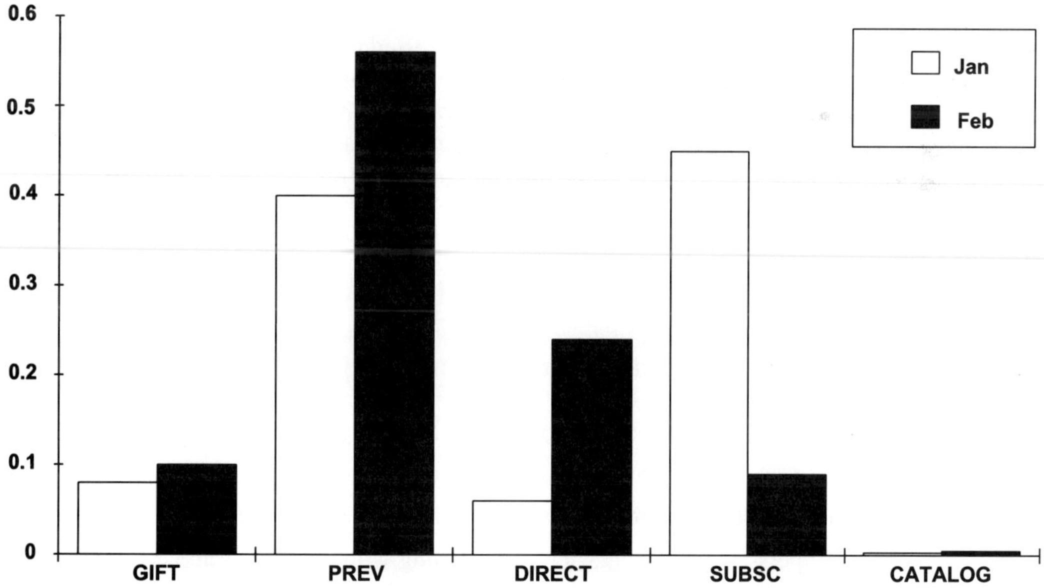

CASE **9**

HEALTHY LIFESTYLES, INSTRUCTOR'S NOTE

Abstract: We assign this case at the beginning of the course. It gives students the opportunity to practice their writing and analytical skills before you've covered much mathematics.

Level:	Beginning.
Techniques:	Descriptive statistics.
Typical Class Discussion Time:	15 minutes.
Data Set Size:	50 cross-sectional observations, 7 variables.

GENERAL COMMENTS

This case is a good one for beginners. Your students will be interested in finding out where their state falls relative to others. There are lots of unexpected results, too. North and South Dakotans, for instance, don't seem to like to wear seat belts. South Dakota also has the highest DWI rate. (But, of course, we don't know whether this is more drinking and driving or a greater willingness to admit to this behavior—maybe people from South Dakota are simply more honest than those in other states!) The western states report more physical activity, led by Colorado, then Washington and New Mexico.

QUESTIONS FOR ASSIGNMENT

1 How does the state of (fill in your state) rate relative to others in terms of these six lifestyle measures?

2 For each of the six lifestyle variables, calculate the mean and standard deviation. What do these numbers tell us about this data set?

3 What, if anything, do these data say about differences in longevity between states?

SUGGESTIONS FOR LEADING THE DISCUSSION

We've found that it helps to personalize the data set by beginning with question 1 above: where do *we* rate? There will probably be at least one surprise in the numbers, no matter where you live.

The suggested question about longevity can be a nice classroom discussion about reading too much into the data. During the classroom discussion, it is likely that at least one student will make the leap from healthy lifestyles to longevity.

Finally, the question of arriving at some overall index will probably create a spirited debate. There is no right or wrong answer, but this could be a marvelous lead-in to weighted averages (or, in this case, weighted proportions).

TYPICAL RESULTS

The means and standard deviations are:

	MEAN	STDEV
BINGE	13.6	3.9
DWI	2.4	1.1
BMI	28.1	3.0
SMOKER	22.7	2.8
SEATBELT	34.3	10.9
EXERCISE	29.1	7.8

Most of the dotplots are fairly symmetric except EXERCISE, which is skewed rightward. Several interesting correlations arise. EXERCISE and SMOKER are positively correlated. (Note that EXERCISE is the percentage of people responding "no" to the exercise question.) The strongest correlation in the data set is between DWI and BINGE (0.74). The correlation matrix is:

	BINGE	DWI	BMI	SMOKER	SEATBELT
DWI	0.738				
BMI	-0.114	-0.086			
SMOKER	-0.139	-0.088	0.582		
SEATBELT	0.234	0.344	0.113	0.089	
EXERCISE	-0.577	-0.440	0.533	0.570	0.090

Before leaving the classroom discussion, be sure that you mention the limitations of looking at correlations among aggregate quantities.

CASE **10**

LAKE COUNTY LUNCHES, INSTRUCTOR'S NOTE

Abstract: A good elementary case, requiring nothing more than computing a few ratios and plotting them. Could be assigned with other cases as a way of getting students used to the idea that they can learn from data.

Level:	Beginning.
Techniques:	Descriptive statistics, data transformation.
Typical Class Discussion Time:	30 minutes.
Data Set Size:	11 time-sequenced observations, 7 variables.

GENERAL COMMENTS

This is an easy case. Compute a ratio or two and plot it (them) over time, and the conclusion is more or less obvious. It's a good one to assign for those students whom you suspect may be afraid of statistics. It shows that simple calculations can be revealing and that ordinary mortals can do them.

QUESTIONS FOR ASSIGNMENT

Probably nothing more than, "What can you advise the program director? Should he be concerned?" is needed here.

SUGGESTIONS FOR LEADING THE DISCUSSION

The same question listed above works fine here, too. You should follow up with requests to describe what plots, computations, and the like, were used, but nothing too sneaky is called for here.

TYPICAL RESULTS AND APPROACHES

The main result is that while lunches have been going down, lunches per enrolled student have been climbing pretty steadily. The data themselves don't suggest any reason for the director to be concerned. Some discussion may occur about what "meeting the public need" means. The case provides no basis for answering this, really, but the discussion may get people thinking about the questions of how to measure aspects of fuzzily defined ideas like "public need."

Students sometimes fit a regression equation or try to see if various ratios are related to unemployment, and so forth. Nothing very striking seems to come up. The time period involved was one of some economic turbulence for Lake County. A major mining employer had big layoffs at a nearby molybdenum mine at Climax, Colorado, and many people left the area in search of employment. The data seem roughly consistent with all this: people left the area, those who stayed depended more on the school lunch program than before, and so forth.

Plot of Lunches Per Student versus Year

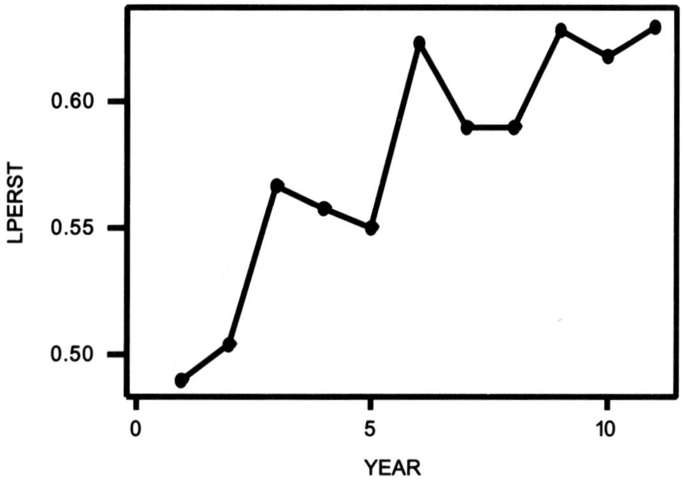

CASE **11**

MACHINE PRODUCTION RECORDS, INSTRUCTOR'S NOTE

Abstract: A good introduction to basic data gathering and analysis. Simple histograms and review of the data give a good overview. Converting some of the variables to rates per minute and treating things separately by machine type show that the three machine types are behaving differently and below their rated performances. It also becomes apparent that some of the data is suspicious, and there are some clues in the case as to where some of these problems might come from.

Level:	Beginning.
Techniques:	Descriptive statistics.
Typical Class Discussion Time:	30 minutes.
Data Set Size:	116 cross-sectional observations, 5 variables.

GENERAL COMMENTS

There's some potential for Helen in all of this, but she'll need to think some more about how her data are being gathered. Students can use histograms, means, and standard deviations to learn a lot here. As long as they review each type of press separately (which should be an obvious thing for most), they will find out more or less what's going on. If they think to convert the production records to production per operating minute, they'll find the comparisons even better. This is a good case for students to use to get familiar with whatever computer program they're using: they have to load the data, select observations to consider, produce a few simple summaries, obtain and interpret their output, and so on, but the actual analyses aren't so complicated as to interfere with their learning about the program.

QUESTIONS FOR ASSIGNMENT

At a general level, you can simply ask, "How should Helen proceed?" or, "Do you think Helen is onto anything useful?" To be more specific, you can try something like:

1 Prepare appropriate summaries of the day's production records (even by machine type).

2 Compare the machines' production rates with their official ratings. Identify any machines that appear to be performing poorly.

3 Determine whether any other patterns are revealed in these data that Helen should be made aware of.

SUGGESTIONS FOR LEADING THE DISCUSSION

Either of the sequences listed under "*Questions for Assignment*" will work for leading the discussion here. This case would normally come fairly early in a course, and most of what's there to be found will be found by most students, so don't do too much for them. As students present alternative summaries, it's good to follow up with questions like "What does that suggest to you?" to get the idea across that the summaries (means, standard deviations, histograms, etc.) are not themselves the goals of the analysis: you need to ask, "what do they say about the situation?" Undergraduates (particularly) may be conditioned to approach statistics as a mathematical exercise, and this is a good chance to get them used to the cycle of "Extract that part of the problem that can be examined with the data," "Examine the data," "Interpret it in the light of the original situation," "Redefine the question and start over again," or whatever your favorite version of this is (Plan-Do-Check-Act, etc.). It may help to display the three histograms that follow as a basis of discussion. Discuss them one at a time and develop conclusions about each machine type.

TYPICAL RESULTS AND APPROACHES

Most students will realize that the three presses should be considered separately. Histograms of the production show a lot of variation, and so do those for HRSWRK. Converting production to a per-minute-of-production-time basis allows more ready comparison with the nominal rates.

For press type 1, the average production per minute is 639 parts per minute, well below the nominal rate of 700, and the histogram reveals a number of machines at the lower end of things, which should presumably be investigated. Several machines have large amounts of downtime, which Helen should probably take a look at, too.

For press type 2, the same kinds of patterns appear. A substantial number of machines have more than four hours of downtime, and the production rates are fairly well centered around their mean of 120.7 parts per minute, compared to a nominal rate of 200.

For press type 3, things are a bit different. A substantial number of these machines had total hours (up plus down) of 3.75, or half a shift. This makes one wonder (or ought to) about the data collection. The total machine hours were manually recorded, according to the case. Did personnel change halfway through the shift? Were some operators not instructed properly about recording the data? This should probably be understood better before any real conclusions are drawn about press type 3. The production rates histogram is dominated by an outlier of over 400 parts per minute, while the others are all around their nominal value of 155. The outlier is observation 108, which lists 16,118 parts produced in 0.58 hour. This seems worth investigating (a data error?). Omitting this observation gives a nice histogram of parts per minute centered around 150.6.

In short, Helen has some work to do to institutionalize her data-gathering process, but it does seem as though there would be some benefit to it. None of the machines seems to be operating very near rated capacity, though type 3 is closest. Whether this is influenced by inaccurate reporting of the up and down times for the machines remains unknown.

The topics covered before this case is assigned and the students' general level of understanding may cause some students to see this as a problem for a control chart; students might not realize that the data aren't a time series, and aren't from the same machine or process, really. This can be a "productive" mistake, a "teachable moment."

Histogram of Parts per Minute for MTYPE = 1

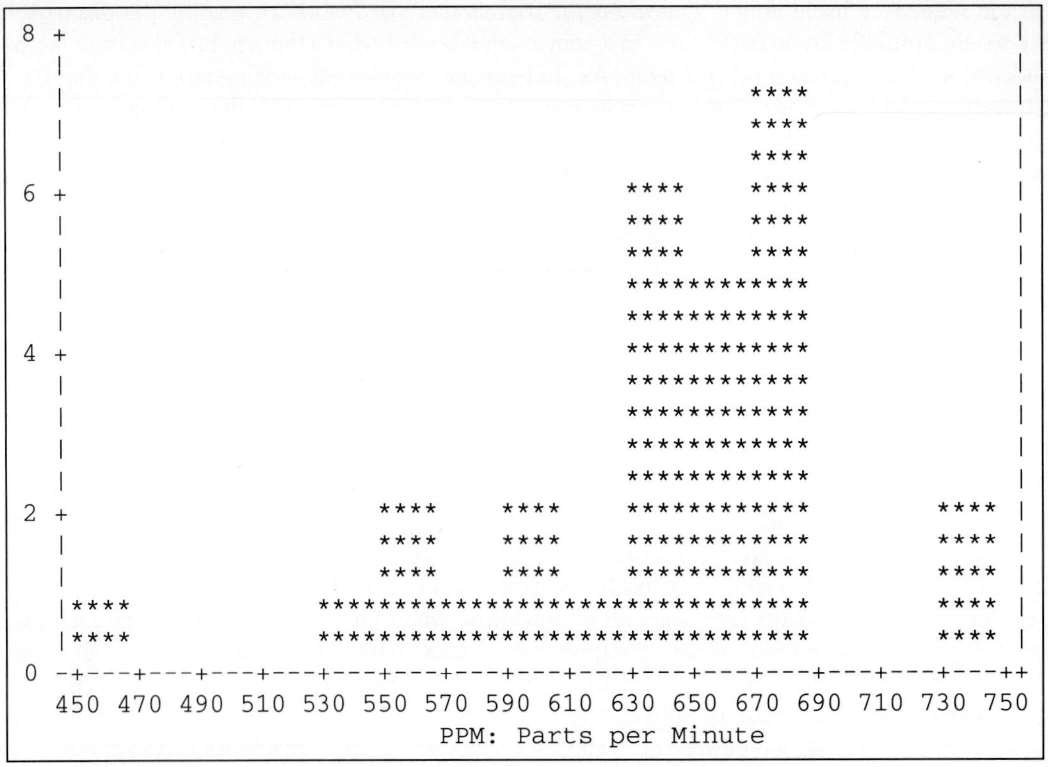

Histogram of Parts per Minute for MTYPE = 2

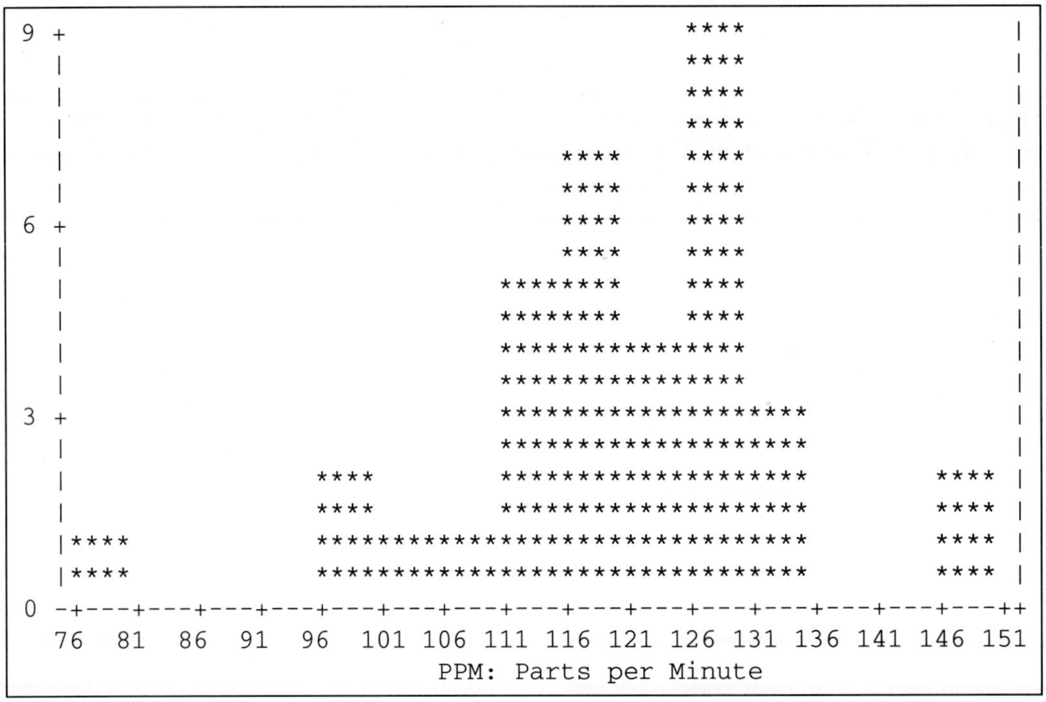

Histogram of Parts per Minute for MTYPE = 3

```
25 +      ***
   |      ***
   |      ***
   |      ***
20 +      ***
   |      ***
   |      ***
   |      ***
15 +      ***
   |      ***
   |      ***   ***
   |      ***   ***
10 +      ***   ***
   |      ***   ***
   |      ***   ***
   |   ************
 5 +   ************
   |   ************
   |   ************
   |*********************                                 ***
 0 -+--+--+--+--+--+--+--+--+--+--+--+--+--+--+--+--+--+--+--++
    90    130   170   210   250   290   330   370   410   450
                        PPM: Parts per Minute
```

CASE **12**

KERRVILLE RAINFALL, INSTRUCTOR'S NOTE

Abstract: The series exhibits a classical random walk pattern. Students can fit a variety of models, naive, autoregressive, random walk, or they can try exponential smoothing. The random walk seems to fit about as well as any of them, at least in the short term.

Level:	Intermediate.
Techniques:	Time series, random walk, regression.
Typical Class Discussion Time:	30 minutes.
Data Set Size:	57 time series observations, 2 variables.

GENERAL COMMENTS

This is a very straightforward case. It illustrates well that for all of the structure that's apparent in a time series plot, the random walk seems to fit about as well as anything, and it doesn't have to be a financial series for that to happen. The naive model ("Next year will be the same as this year") does nearly as well in the short term, and might well do just as well in the long term.

QUESTIONS FOR ASSIGNMENT

1 Derive a model that summarizes the behavior of the TOTAL series.
2 Use your model to derive predictions of rainfall for 1996 if you can.

SUGGESTIONS FOR LEADING THE DISCUSSION

Probably just asking for the various models and listing them on the blackboard works as well here as anything. The corresponding predictions for 1996 can be listed, too, along with a variety of measures of fit. Then the class can take a vote about which model they wish to rely on. You can compare the predictions of their chosen model with the actual values given below.

TYPICAL RESULTS AND APPROACHES

The models that seem to crop up include:

1 The naive model: next year will be the same as this year (forecast = 28.26).
2 Predict by the average (forecast = 31.60, standard deviation 8.87).

	N	MEAN	MEDIAN	TRMEAN	STDEV	SEMEAN
TOTAL	65	31.60	30.90	31.46	8.87	1.10
	MIN	MAX	Q1	Q3		
TOTAL	14.04	55.09	24.68	38.81		

3 Single exponential smoothing, $\alpha = .028$ (forecast = 32.08).

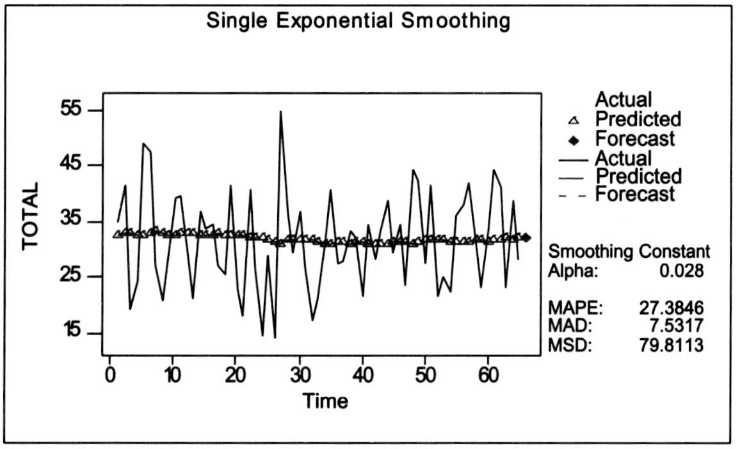

4 Double exponential smoothing, smoothing constants .758 and .046 (forecast = 36.09).

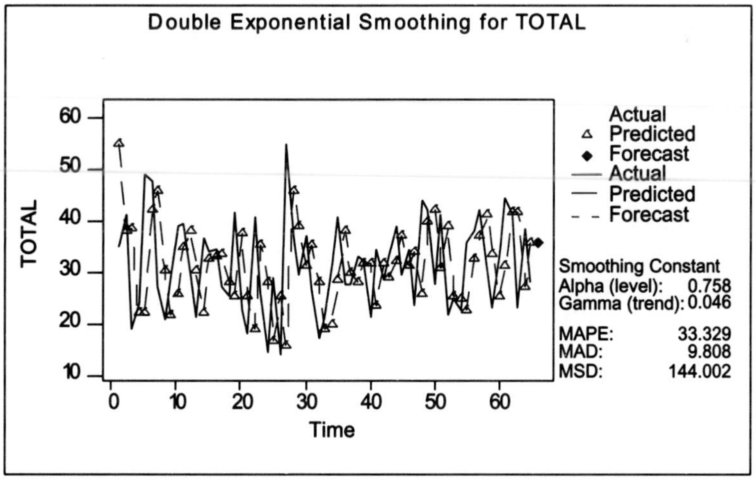

5 Autoregressive: The autocorrelation function suggests that perhaps the first one or two lags might be helpful.

```
ACF of TOTAL

              -1.0 -0.8 -0.6 -0.4 -0.2  0.0  0.2  0.4  0.6  0.8
1.0
               +----+----+----+----+----+----+----+----+----+----+
  1  -0.016                             X
  2  -0.278                      XXXXXXXX
  3  -0.107                          XXXX
  4  -0.026                            XX
  5   0.180                             XXXXX
  6  -0.057                            XX
  7  -0.058                            XX
  8   0.152                             XXXXX
  9   0.012                             X
 10  -0.097                           XXX
 11   0.000                             X
 12   0.005                             X
 13   0.165                             XXXXX
 14   0.007                             X
 15  -0.191                        XXXXXX
 16   0.113                             XXXX
 17   0.057                             XX
 18  -0.086                           XXX
```

On the other hand, a regression using those lags yields an R^2 of only .08 (prediction 29.43).

```
The regression equation is
TOTAL = 41.1 - 0.026 TOTAL1 - 0.282 TOTAL2

63 cases used 2 cases contain missing values

Predictor        Coef       Stdev    t-ratio        p
Constant       41.076       5.624       7.30    0.000
TOTAL1        -0.0256      0.1227      -0.21    0.836
TOTAL2        -0.2817      0.1232      -2.29    0.026

s = 8.688        R-sq = 8.1%       R-sq(adj) = 5.0%


     Fit  Stdev.Fit        95% C.I.               95% P.I.
   29.43       1.47   (  26.50,   32.37)   (   11.81,    47.06)
```

6 Random walk (forecast 28.15). The mean difference is -.11, with the rather large standard deviation of 12.73.

```
                N         N*      MEAN    MEDIAN    TRMEAN     STDEV
SEMEAN
DIFF           64          1     -0.11     -1.46     -0.56     12.73
1.59

              MIN        MAX        Q1        Q3
DIFF       -22.39      41.05    -10.48      8.70
```

7 Moving average (three-year, forecast 30.0267, standard deviation 8.70177).

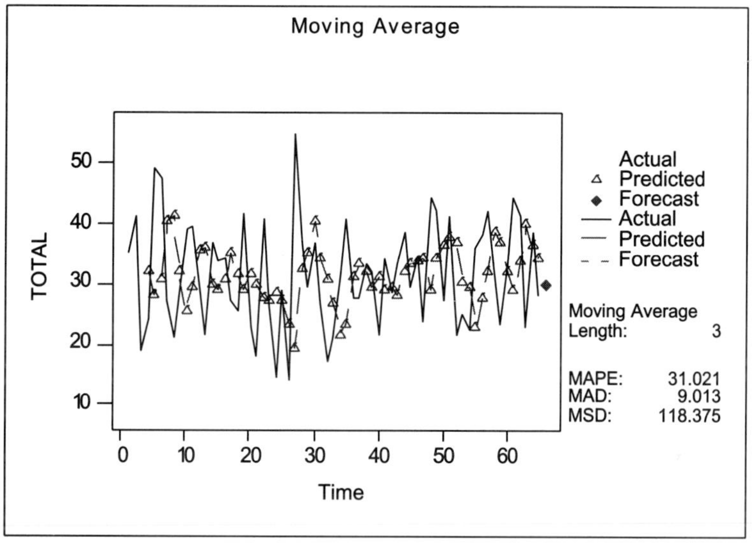

The distributions involved also suggest that looking at logarithmic transformations might be helpful, too. Looking at all these various forecasts, it seems that forecasts around 30 inches would have a lot of support, but that just about any forecasts will have a lot of noise associated with them. The actual rainfalls for Kerrville in 1996 and 1997 were 26.23 and 37.70 inches, respectively.

CASE **13**

WORKLOAD RATINGS, INSTRUCTOR'S NOTE

Abstract: A case where's there's no relationship to be found. With care, simple plots will provide a rough answer. Regression and analysis of residuals will give a more complete answer. It's tailor-made for partial correlation, but can be handled without specifically knowing that technique. Also offers an outlier (which has little effect on the conclusion). Without formal tools, this can be hard (a hard beginning problem). With a good command of formal tools, it's pretty straightforward (an easy advanced problem). Can be used to illustrate the problems of computing means from data that are only ordinal.

Level:	Intermediate.
Techniques:	Regression, residual analysis, correlation, partial correlation.
Typical Class Discussion Time:	30–45 minutes.
Data Set Size:	161 cross-sectional observations, 3 variables.

GENERAL COMMENTS

This case requires nothing more than simple linear regression, as long as the students understand and can interpret residuals. If they know correlation and/or partial correlation, they can also apply those techniques. You can also use it to *introduce* residual analysis or the need for partial correlation.

QUESTIONS FOR ASSIGNMENT

Advanced and/or mature students probably need no more specific questions than those given in the case. To provide more structure, a possible sequence of questions is:

1 Compute the regressions of INSTRTG on CRSRTG and of WORKLOAD on CRSRTG. Compute the residuals from these regressions and describe their meaning in words.

2 Determine whether the residuals from these regressions are apparently related to each other. What methods might assess this relationship? How would knowledge of this relationship help you answer the question in the case, if at all?

3 Comment on the basic data provided.

SUGGESTIONS FOR LEADING THE DISCUSSION

Discussion of this case can start off general ("Well, are the students penalizing the instructors unfairly?") and then proceed to the specific: "Why do you say so? What features of the data support your view? What did you compute? Interpret it in words."

If you assigned the specific questions above, you may wish to organize the discussion around them. The printouts and plots that follow can be a focus for that discussion, if you feel the students need that level of direction. At some point, turn the discussion to the basic data involved. Aside from whatever biases are inherent in the data and the weaknesses of the whole process, students should note that the data consists of means computed from data that are really only ordinal. This is, of course, common. In this case, it's standard operating procedure for the course evaluations. You may wish to have a discussion about what limitations this places on the results. Presumably, *something* can still be learned, but just how *much* is hard to assess. To begin with, the data don't appear to give much support to the claim that workload affects instructor rating; they appear to support it even less after this limitation is recognized.

TYPICAL RESULTS AND APPROACHES

Scatterplots show basically no relationship between INSTRTG and WORKLOAD (or, for that matter, between CRSRTG and WORKLOAD), though INSTRTG and CRSRTG are highly correlated. The correlation between INSTRTG and WORKLOAD is −.03. Most students quickly conclude that the data don't support the claim of any relationship.

The matter of "adjusting for the effect of CRSRTG" is a little more difficult for most students. Three approaches are typical:

1 Those who have seen partial correlation will probably compute the partial correlation of INSTRTG and WORKLOAD, removing the effects of CRSRTG (it comes out −.11), though some may have difficulty deciding which are the variables to be correlated and which effects are to be removed.

2 Others may try to use multiple regression in some way, perhaps modeling INSTRTG as functions of CRSRTG and WORKLOAD and showing that WORKLOAD doesn't really belong in the equation.

3 Finally, those with no knowledge of regression may still be able to break the data set up into subsets based on similar values of CRSRTG, and produce separate scatterplots of INSTRTG and WORKLOAD within each subset, or separate regressions.

All of these approaches are valid, and it helps to compare them. In particular, it helps to see that partial correlation corresponds to something simpler, such as separate scatterplots. Students often get confused about which variable plays which role (e.g., whether to stratify the data based on INSTRTG or on CRSTRTG).

You may also assign this case *after* students are used to assessing relationships, either by plots or by regression, but *before* they've thought about the analysis of residuals. When they've wrestled with the issue of how to "adjust" for something, they're usually more ready to understand that the residuals are the "part" of the dependent variable not explained by the independent variable.

Students often point out outliers: observation 64 (course rating 3.67, instructor rating 0.67!), and observation 145 (course rating 0.89, instructor rating 1.11). Those must have been terrible semesters for all concerned. Observation 64 is an outlier on the residual plots, while observation 145 is not. Running the calculations without observation 64 yields little change (the correlation is even smaller, the regressions are more or less unchanged, the plots are similar), but it offers a good opportunity to bring up the issues of dealing with outliers (Compensate for them with dummy variables? Omit them altogether? etc.)

Regression of Workload on Course Rating

```
    Dependent Variable:          WORKLOAD---WORKLOAD RATING
    Independent Variable(s):     Constant
                                 CRSRTG---COURSE RATING (0-4)

REGRESSION COEFFIEIENTS
IND. VBL.     COEFFICIENT      STD. ERROR      T
Constant          2.27576         0.16815   13.5
CRSRTG            0.02594         0.05799    0.4

R-squared: 0.001 Std. error of estimate:              0.43087
Residuals saved as variable R$WORKLO
```

Regression of Instructor Rating on Course Rating

```
    Dependent Variable:          INSTRTG---INSTRUCTOR RATING
    Independent Variable(s):     Constant
                                 CRSRTG---COURSE RATING

REGRESSION COEFFIEIENTS
IND. VBL.     COEFFICIENT      STD. ERROR      T
Constant          0.18730         0.13140    1.4
CRSRTG            0.96390         0.04531   21.3

R-squared: 0.740 Std. error of estimate: 0.33670
Residuals saved as variable R$INSTRT
```

Partial Correlation of Workload and Instructor Rating

```
PAIRWISE CORRELATIONS:
Effects of the following variables have been removed:
      CRSRTG
Vbl: WORKLOAD, Correlations with:
     INSTRTG  -0.114
```

Plot of Residual Workload versus Residual Instructor Rating

```
R$WORKLO:Residual WORKLOAD
 1.4+
    |                                      * *
    |                                  *        *
    |                                     *    *
    |                                          ***
 0.6*                                      *****
    |                                    ***  ***
    |                                      * *****  *
    |                                  * * *** *
    |                                   *  ******** *
-0.2+                                  *************
    |                                  * * ******** **
    |                            *      *** ****** *
    |                             *         * * *
    |                                          *
-1.0+---------+---------+---------+---------+--*------+
-3.06     -2.33     -1.59     -0.86     -0.12      0.61
                  R$INSTRT:Residual INSTRTG
```

CASE **14**

ROARING FORK TRANSIT, INSTRUCTOR'S NOTE

Abstract: The main message in this case can be found from some simple plots, particularly labeled plots like the MINITAB LPLOT, but it also offers a chance to try time series techniques to characterize seasonal effects and some basic regression models. There's another puzzling trend, too.

Level:	Intermediate.
Techniques:	Descriptive statistics, time series, regression.
Typical Class Discussion Time:	45 minutes.
Data Set Size:	48 time series observations, 3 variables.

GENERAL COMMENTS

This requires nothing more than simple plots. In that sense it's a beginning case, but to figure out exactly *which* plots would reveal the message here is not completely obvious, so we've called it intermediate. The time series and regression techniques are elementary.

QUESTIONS FOR ASSIGNMENT

1 What trends and features of the data do you think RFTA management should be aware of?

2 What's the best way to summarize these trends and patterns for presentation to the RFTA board?

3. If you wish to emphasize the time series/regression approach to analysis, you could add: Prepare a model of INBOUND and OUTBOUND traffic and describe its major features in words.

SUGGESTIONS FOR LEADING THE DISCUSSION

The questions given above will work for discussion, too. The various patterns found can be recorded on the blackboard, as can alternative methods of summarizing them. When enough alternatives have accumulated, the discussion can turn to evaluating those alternatives and their strengths and weaknesses.

TYPICAL RESULTS AND APPROACHES

Ridership has definite seasonal effects (peak ridership in summer and in winter) and a steady long-term increase. You get more or less the same message whether you consider INBOUND, OUTBOUND, or their sum. You can portray all of this by a MINITAB LPLOT. Here's one approach, plotting INBOUND versus month of the year (1–12), and labeling points for the four years by 1, 2, 3, and 4:

Roaring Fork Downvalley Routes Ridership, 1987–1990

```
Plot Symbols correspond to values of YEAR
INBOUND: Ridership inbound to Aspen
30000+
     |                                                4
     |
     |
     |
21000+
     4                                                *
     |        4
     |   4                      4   4             4
     |
12000*   3    *                 3   3    4   4
     |   2        4        4    2   2             3
     |             4   3
     1   1    1   3        2             3   3    2   1
     |            2    3        1   1    2
 3000+---------+--1----*-+-1-------+-----1---*----1----+
   0.9       3.1       5.3       7.6       9.8      12.0
                            MONTH
```

If you connect the dots (connect all of the 4s, all of the 3s, and so on), you see that each successive year is higher than the one before and that the seasonal pattern is roughly the same in each year. Standard time series models for seasonal effects and long-term trend show this, too, though whether they really reveal more than the plots is questionable. Most will feel the plot (or its equivalent) makes the results clearer for the board, anyway.

The ratio of OUTBOUND to INBOUND has decreased steadily over time. At the beginning, OUTBOUND riders were fewer than INBOUND, whereas at the end of the interval more riders went into Aspen than returned. (Where did they go? This is almost exclusively daily commuting, not tourist transportation. Students may have some speculations.) The case gives no real clue as to what might be going on here, but the trend is striking. More detailed records (by individual trip, time of day, etc.) might be revealing here.

CASE **15**

STURGEL DIVISION, INSTRUCTOR'S NOTE

Abstract: A change-of-pace case, containing no data. The organizational situation is described and the student is asked to consider the design and administration of a survey.

Level:	Beginning, intermediate.
Techniques:	Survey design.
Typical Class Discussion Time:	30–45 minutes.
Data Set Size:	No data provided.

GENERAL COMMENTS

In this case, students get to consider how to design and administer a survey. Since no data are provided, they are forced to focus on the basic processes of obtaining the data. Many different results and emphases are possible in this case and the discussions can be revealing, as the students' differing assumptions come to light.

QUESTIONS FOR ASSIGNMENT

You can pick and choose from the questions below, depending on the emphasis you wish to give:

1. Should Martha do a survey? Why or why not?
2. If Martha should do a survey, should it be one or many surveys?
3. How often should Martha run the survey?
4. How should the survey be administered?
5. Design the survey form for Martha's survey.

SUGGESTIONS FOR LEADING THE DISCUSSION

You can pretty well use the questions you selected for assignment. Follow up to find out the basis for the students' answers; this will stimulate worthwhile discussion.

TYPICAL RESULTS AND APPROACHES

Students will use a variety of approaches, most of which can be well defended. They should at least consider:

1. Is a common survey appropriate for the varied services the IS department provides?
2. How often should it be administered? Monthly? Yearly?

3 How should it be administered: mail, on the computer itself, by phone?
4 What should a form look like? What questions should be asked?
5 What kinds of techniques will be used to *analyze* the data once it's obtained?

Just about all of the alternatives can be defended, depending on the students' assumptions about the nature of the service and the firm. You may need to remind them that a *simple* survey is more likely to be answered than a more complicated one. Martha would probably be well advised to rely more on repeated use of the survey (focusing on changes from period to period) than on any single administration.

You may wish to point out that students may see this kind of problem again in the realm of market research, where survey design and administration play such a large role. Martha is doing market research, basically; *internal* market research, to be sure, but market research just the same.

CASE **16**

PHOTOCOPYING ABUSES, INSTRUCTOR'S NOTE

Abstract: Sequence plots reveal information about the difference in usage rates over time that descriptive statistics do not. This case sends the lasting message that graphics are crucial tools to effective data analysis.

Level:	Beginning.
Techniques:	Time series plots, descriptive statistics.
Typical Class Discussion Time:	15 minutes.
Data Set Size:	50 time series observations, 3 variables.

GENERAL COMMENTS

This case is a lesson in the importance of undertaking initial graphical analysis. Students who examine only the means and standard deviations of the 3rd- and 10th-floor usage discover that the summary statistics are about the same; these students conclude that the implicit monitoring technique has had no influence on usage. It is only upon examination of the sequence plots that a difference emerges. The 10th-floor usage declines fairly consistently over the 50-day test period, whereas the 3rd floor remains relatively flat over time.

This case uses artificial data, and is a variation on the theme of Anscombe, who manufactures four different (Y, X) pairs of observations for simple regression analysis. In his data set, the regression coefficients and summary statistics are identical in all four cases, and it is only the residual plots (or scatterplots of Y versus X) in which differences emerge. We suggest looking at that data for your classroom discussion of simple regression. The citation is F. J. Anscombe, "Graphs in Statistical Analysis," *The American Statistician* 27 (1973) pp. 17–21.

Students need not get on a computer at all to analyze this problem. Just looking down the list of numbers for the variables TENTH and THIRD is revealing.

One of our students reported the following incident that arose at her place of business. Apparently her firm attempted a similar experiment of distributing cards and monitoring daily usage. Management apparently failed to properly motivate the experiment, though—employees were so angry at the implicit suggestion of misuse of the photocopying machines that they organized a revolt and *increased* their usage. This could lead to a discussion of experimenting with human subjects in the workplace.

QUESTIONS FOR ASSIGNMENT

1 What is the best statistical technique for determining whether the card accounting system will effectively lower photocopying usage?

2 How might we measure the rate of change in photocopying for both groups over the 50-day experiment?

3 What are the pros and cons of the design and implementation of this particular experiment?

SUGGESTIONS FOR LEADING THE DISCUSSION

We've always begun the discussion by asking for an informal survey of the students: "How many of you discovered that the card accounting system produced lower photocopying?" Then, "How many did not?" Ask someone from the NO group to explain their methodology. The other students who discovered the "answer" in the time series relationship will then immediately jump into the discussion with no prompting from you.

One interesting aspect of this data is that you don't know whether the 10th-floor employees are eliminating extraneous copying (personal income tax forms or letters to grandma and grandpa) or whether they are simply intimidated by the system and are eliminating necessary business copying. If employees are skipping important use of the machine (not copying legal briefs, let's say), then this experiment could be a disaster. You might attempt a discussion of how to redesign the experiment to determine which of the two types of usage (unnecessary versus necessary) is being eliminated by the 10th-floor employees.

We have also had great fun relating the story in the "General Comments" section above about the employee revolt. Ask students how managers might assist the company statistician in motivating employee participation. This might include employee contests, for example.

If this case happens to appear at a juncture in your course in which you are about to enter a discussion of simple regression, the question about measuring the rate of change in usage is obviously motivated by determining the slope of the line through the sequence plot: a natural lead-in to regression.

TYPICAL RESULTS AND APPROACHES

There are only two outcomes that we've had in class—the YES group and the NO group. The YES group will point to the sequence plots, whereas the NO group stopped their analysis at the descriptive statistics.

The descriptive statistics for the 3rd and 10th floors are:

	N	MEAN	MEDIAN	TRMEAN	STDEV	SEMEAN
TENTH	50	248.1	250.0	247.6	143.3	20.3
THIRD	50	247.9	250.0	247.6	143.6	20.3

	MIN	MAX	Q1	Q3
TENTH	20.0	500.0	127.5	362.5
THIRD	20.0	500.0	127.5	362.5

However, the sequence plots are as given in what follows.

Time Sequence Plot of TENTH

```
TENTH
525+
   - 1  4
   -   3 512
   -  2     345
   -           1  4
350+            23 5 23
   -                1  45
   -                     123
   -                        451234
   -                              5
175+                               1 3
   -                                2  51234
   -                                  4     51 3
   -                                          2 4
   -                                             512345
  0+
    +----+----+----+----+----+----+----+----+----+----+
    0    5   10   15   20   25   30   35   40   45   50
                             Time
```

Time Sequence Plot of THIRD

```
THIRD
525+
   -                                2                4
   - 1           3              3             2
   -        3   2 4                     1
   -                              5    5
350+   3  1             5      2                      5
   -         4                1                  5
   -           1                 4   3
   -     5           2 4  23                5
   -  2
175+                1              1
   -               5     1            4  2        1 3
   -    4                   4             3        2
   -       2                               4
   -          5       3      5               1 34
  0+
    +----+----+----+----+----+----+----+----+----+----+
    0    5   10   15   20   25   30   35   40   45   50
                             Time
```

CASE **17**

AGRICOMP, INSTRUCTOR'S NOTE

Abstract: A case for simple cross-tabulations and/or chi-squared. It can also be summarized by simple histograms or dotplots.

Level:	Beginning.
Techniques:	Cross-tabulations, histograms, chi-squared.
Typical Class Discussion Time:	30 minutes.
Data Set Size:	292 cross-sectional observations, 2 variables.

GENERAL COMMENTS

This case offers a chance for students to deal with the data before it's cross-tabulated. Nothing very fancy is required, but the students will need to recognize that a cross-tabulation is in order (or at the very least that some separate tabulations are needed). The dealer preferences are different for those who have used the existing appeals process than for those who haven't. The more they've used the process, the less they perceive a need for change. If the data are tabulated in the aggregate, this trend is not apparent.

QUESTIONS FOR ASSIGNMENT

1 What do the dealers think about the warranty appeals process?
2 Should it be replaced?

SUGGESTIONS FOR LEADING THE DISCUSSION

You can start with a question like, "How do the dealers feel?" and follow it up with, "Do *all* of the dealers feel that way?" This will normally bring out a cross-tabulation or something like the series of MINITAB dotplots given below. Then you should turn the discussion to how best to summarize or display the conclusions. Either some sort of cross-tabulation (using appropriate percentages instead of counts) or plots like those below will work. The main point is to be sure students don't simply declare a "significant" lack of independence and let it go at that. They should have to say something about what kind of *dependence* they find, not just assert the absence of *independence.*

TYPICAL RESULTS AND APPROACHES

The more the dealers have used the existing appeals process, the less they agree with the statement that it should be replaced, so it appears that Jody was hearing from some malcontents. The cross-tabulation of responses by number of uses is given on the next page. The corresponding chi-squared is 82.16 on 12 degrees of freedom, so something is clearly going on. Students may offer a variety of summaries of just *what* is going on, and you should prompt them for such summaries if all they offer is chi-squared. The general trend is illustrated by such diagrams as the dotplots given on the page following the cross-tabulation.

Cross-Tabulation of REP by USE

	USE				
REP	0	1	2	3	
1	12	12	4	6	34
2	6	27	18	12	63
3	8	18	16	18	60
4	4	9	8	52	73
5	1	12	12	37	62
	31	78	58	125	292

Dotplot of REP by USE

```
USE 0
            :
            :                       :
            :           :           :
            :           :           :           :
            :           :           :           :           .
            1           2           3           4           5
USE 1
                        .
                        :
                        :
                        :
                        :
                        :           :
                        :           :
                        :           :
            :           :           :                       :
            :           :           :           .           :
            :           :           :           :           :
            :           :           :           :           :
            :           :           :           :           :
            :           :           :           :           :
            1           2           3           4           5
USE 2
                        :
                        :           :
                        :           :
                        :           :                       :
                        :           :                       :
                        :           :           :           :
                        :           :           :           :
            :           :           :           :           :
            :           :           :           :           :
            1           2           3           4           5
USE 3 (Each dot represents 3 points)
                                                :
                                                :
                                                :           .
                                                :           :
                                                :           :
                                                :           :
                                    :           :           :
                        :           :           :           :
            :           :           :           :           :
            1           2           3           4           5
                                    REP
```

CASE **18**

TIPS, INSTRUCTOR'S NOTE

Abstract: A case for multiple regression, including dummy variables and possibly transforming the dependent variable to a percentage or logarithmic scale. No particular complications result. One factor appears to influence things; the others don't.

Level:	Intermediate.
Techniques:	Multiple regression, data transformations.
Typical Class Discussion Time:	30–45 minutes.
Data Set Size:	244 cross-sectional observations, 8 variables.

GENERAL COMMENTS

You can approach this as a multiple regression, with TIP as the dependent variable, and the result isn't too bad. For students who can handle a bit more, it may make sense to consider the tip as a percentage of the total bill (whose distribution is a bit skewed) or the logarithm of this ratio (whose histogram is nicely bell shaped).

QUESTIONS FOR ASSIGNMENT

1 What factors appear to influence the amount of the tip?
2 Why do you say so?

SUGGESTIONS FOR LEADING THE DISCUSSION

Starting with the assignment question, solicit lists of factors that appear to influence the amount of the tip. For each one, ask for the basis of the conclusion, looking for the underlying model that was used, which can be recorded on the blackboard. If the class is slow to respond, you can prompt them by putting up something like the chart of possible models (given below) and asking for additions to it. If the idea of transforming the dependent variable in some way doesn't appear, you could raise the issue indirectly by asking, "What's your general rule for tipping in restaurants?" Sooner or later, someone will state it in terms of percentages, and you can then see which models have been stated in percentage terms and which in absolute terms. This in turn can lead to a discussion of how to choose model form. In this case, the models don't fit dramatically differently, so the eventual choice must be influenced to some extent by our prior feeling and preferences for model form.

TYPICAL RESULTS AND APPROACHES

The regression models usually include TOTBILL, which has a strong effect, and sometimes SIZE, which has a smaller, questionable effect. The *t* statistic for SIZE comes out about 2 or a bit bigger, so on these degrees of freedom, it would be considered significant at conventional significance levels. On the other hand, omitting it from the model has only a very marginal effect on the fit and is at least as interpretable. TOTBILL and SIZE are, of course interrelated (correlation 0.6), which provides an opportunity to mention how multicollinearity can interfere with the interpretation of the individual *t* statistics on a regression printout.

Alternative Models for TIPS

TIP = .66894+.09271(TOTBILL)+.19260(SIZE),	R^2=.468, s = 1.01351
TIP = .92027+.10502(TOTBILL)	R^2=.457, s = 1.02205
ln(100*TIP/TOTBILL)=2.95013-0.01865(TOTBILL)+.05304(SIZE)	R^2=.164, s = .32163
ln(100*TIP/TOTBILL) = 3.01933-.01526(TOTBILL)	R^2=.150, s = .32351

The last model translates back to

$$\text{TIP as percent of TOTBILL} = 20.477 * (.9849)^{\text{TOTBILL}} * (\text{factor between } .7236 \text{ and } 1.382)$$

and it's worth pushing those students who propose a logarithmic model to express it back in these original terms. This offers two benefits: (1) it forces those who propose logarithmic transformations in a knee-jerk way to see the (tacit) implications for the original model form, and (2) it offers practice for beginning students or those whose algebra and logarithms may be rusty. It's easy to let the logarithm be just another transformation on the computer screen, or just another calculator key, and not understand what it denotes.

CASE **19**

RUBBERGATE, INSTRUCTOR'S NOTE

Abstract: Many different types of one- and two-population hypothesis tests can be performed, as well as tests of association.

Level:	Intermediate.
Techniques:	Confidence intervals, tests of hypotheses, nonparametric methods.
Typical Class Discussion Time:	45 minutes.
Data Set Size:	146 cross-sectional observations, 8 variables.

GENERAL COMMENTS

In the problem described in the case we seek to discover whether congresspeople might be more (or less) inclined to bounce checks depending on the size of their constituency, percentage of votes won in the last election, years of service, or age. We might suspect that the actual determinants are more subtle and complex, and probably unmeasurable—some psychological test of honesty or responsibility or some such thing might be needed here! In any case, many different types of statistical methodologies can be demonstrated with these data—tests of difference between means, both parametric and nonparametric, and chi-square tests of association. This is an extremely rich data set amenable to all sorts of statistical methodologies.

One additional lesson to be learned here is that sometimes data don't really tell us much of anything. Students learn that statistically insignificant results are nonetheless informative.

QUESTIONS FOR ASSIGNMENT

1 Determine the nature of the distribution of the variable BADCHKS. How or why might this distribution influence your study of check-bouncing behavior?

2 Perform the test that the mean number of bad checks written is the same for both political parties. Should you use a test that presumes equal variances or not? Should you use a parametric or nonparametric test? How should you decide these issues?

3 List other interesting differences that arise by party affiliation of the congressperson.

4 Perform appropriate tests of association and independence between party affiliation and check-bouncing behavior.

SUGGESTIONS FOR LEADING THE DISCUSSION

Because this data set is so rich with possible applications, the way in which you might guide a classroom discussion is enormously variable. Regardless of the approach to this case, we try to focus the classroom discussion around appropriate communication skills.

We stipulate at the beginning of the discussion that students (and the instructor, presumably) are not allowed to use any technical terms during the discussion. All statistical terms, ideas, and methodologies must be translated into a common vocabulary. Bring a whistle to class and blow it each time someone uses jargon. This case is a good one on which to perform this experiment, because much of this material is pretty introductory.

The rule that we establish for this game is that you may not use a word in the class discussion if you did not use it on a regular basis before coming to this class. That means that "average" is OK, but "standard deviation" is not. And, of course, "chi-square" is cause for a blow of the whistle! This will greatly slow down the classroom discussion—openly invite other students to assist struggling speakers with difficulties in communication.

TYPICAL RESULTS AND APPROACHES

In the overall sample, the average number of bad checks written is 50, but this distribution is skewed, with a median number of bad checks written of only 3. Of the sample, 62% engaged in check-bouncing, defined as writing one or more bad checks. Democrats comprise 63% of the sample. Our typical congressperson is 53 years old, has clocked in 11 years of service in Congress, received 69% of the counted votes in the last election, and represents an average of 573,000 district constituents.

The highest correlation between BADCHKS and all other metric variables (in the full sample of 146 observations) is with DISTRICT. This correlation is $-.2$.

The tendency to bounce checks relates to young congresspersons, small district populations, larger state populations, and larger percentages of votes in the last election. Only SERVICE has statistically different means by check-bouncing behavior. The *t* ratios for the two-sample *t* test of the equality of means are: AGE, -0.49; DISTRICT, -0.75; SERVICE, -2.18; STATE, 0.82; and VOTES, 1.28. (The *F* test for equality of variances is not rejected for all of these variables at the 10% significance level.)

The answer to the second question above (and question 3 in the case), about differences in the metric variables by political affiliation and the need for parametric versus nonparametric tests, is summarized by the tables below. (The *F* test for the equality of variances by political affiliation is rejected for the variables BADCHKS, DISTRICT, and SERVICE at the 10% level. In the table below, the *t* test for these variables presumes unequal variances.)

Tests for Differences between Parties

	Parametric Two-Sample t-test		Non-Parametric Median Test	
	t-ratio	p-value	chi-square	p-value
AGE	0.76	0.45	2.55	0.11
BADCHKS	-1.77	0.07	2.83	0.09
DISTRICT	2.71	0.01	1.88	0.17
SERVICE	-2.13	0.04	2.94	0.09
STATE	0.89	0.38	3.84	0.05
VOTES	-1.18	0.24	3.39	0.07

Question 4 in the case asks about summarizing the BADCHKS variable only for those individuals who wrote bad checks to begin with. This question can lead to a discussion of censored data. For the 90 observations in which BADCHKS>0, the following results are obtained:

Statistics on Bad Checks and Other Variables

	BADCHKS	CORRELATIONS WITH	BADCHKS
N	90	AGE	-0.0023
MEAN	81.356	DISTRICT	-0.2536
SD	176.63	PARTY	0.1306
MINIMUM	1.0000	SERVICE	0.0243
MEDIAN	8.5000	STATE	-0.0916
MAXIMUM	878.00	VOTES	0.0169
SKEW	2.9504		
KURTOSIS	8.4506		

DISTRICT is most highly related to number of bad checks written. A "kitchen sink" regression of BADCHKS on the six variables listed above generates an R^2 value of 6.9% with only the DISTRICT variable being statistically significant at the 5% level. (By the way, the logistic regression of CHECKS on AGE, DISTRICT, PARTY, SERVICE, STATE, and VOTES using all 146 observations finds AGE and SERVICE to be statistically significant at the 5% significance level.)

The answer to the question of whether check-bouncing behavior is statistically independent of political affiliation is unclear. Pearson's chi-square is 3.48 (*P*-value = 0.06), and Yates's corrected chi-square is 2.85 (*P*-value = 0.09). The cross-tabulation of these two variables is given below.

Cross tabulation of CHECKS by PARTY

CHECKS	PARTY 0	PARTY 1	
0	26	30	56
row %	46.4	53.6	38.4
1	28	62	90
	31.1	68.9	61.6
	54	92	146

CASE **20**

HITECH ENGINEERING, INSTRUCTOR'S NOTE

Abstract: A straightforward ANOVA exercise that isn't all that straightforward in the end.

Level:	Intermediate.
Techniques:	Analysis of variance.
Typical Class Discussion Time:	30 minutes.
Data Set Size:	85 time-sequenced observations, 8 variables.

GENERAL COMMENTS

The assignment given in the case is to find the most effective advertising type. The answer depends on how students interpret the word *effective.* If the goal is to find the media with the highest average number of responses, then postcards are the way to go. If the goal is to minimize costs per response generated, then news releases, postcards, and literature all appear to be viable.

Results are suspect for several reasons. Variances are unequal between groups and some of the data might be wrong. These issues can generate lively classroom discussions.

A very important point about this case, which only a few of your students will pick up, is that the data set doesn't really answer the interesting question. The bottom line to management is whether spending translates into *sales,* not just requests for more information. This issue cannot be addressed with the given data set.

The data set is also amenable to ANCOVA methods.

QUESTIONS FOR ASSIGNMENT

1 What is the most effective type of advertising for HEI to undertake? What method should you use to answer this question? Are there any shortcomings of the technique you've chosen to analyze the data?

2 What are your recommendations to management about levels and types of advertising that should be pursued in the future?

3 What recommendations do you have for improving the data collection?

SUGGESTIONS FOR LEADING THE DISCUSSION

A nice lead-in to the discussion of this case is to ask: "What do you think management had in mind when they used the word *effective,* and how did your interpretation factor into your analysis?" Let the students discuss the issue of responses versus cost per response.

At some point, lead the discussion toward viability of ANOVA with unequal variances. Ask at some point whether there are any problems with the data. (Postcards have three zero values, which is unusual given the apparent effectiveness of postcards. One might suspect that these were actually missing observations replaced with zeroes in the coding of the data set.) Ask how these problems did, or should have, influenced the chosen statistical methodology.

It is crucial that you (or someone) mention that these data do not answer the real question that should be asked—how many dollars in sales, not just leads, are being derived from HiTech's advertising expenditures? The leads data only relates to potential customers, not dollars realized. It would probably be wise for the company statistician to recommend abandoning the collection of leads data and designing a survey of actual, not potential, customers.

TYPICAL RESULTS AND APPROACHES

ANOVA rejects equal means for both RESPONSE and $/LEAD (a generated variable defined to be SPEND/RESPONSE). Average RESPONSE is greatest for postcards, and average $/LEAD is lowest for postcards, news releases, and literature, as illustrated by the printouts that follow.

Analysis of Variance Printout for RESPONSE

```
ANALYSIS OF VARIANCE ON RESPONSE
SOURCE      DF        SS         MS         F        p
MEDIA        4   1885716     471429     35.47    0.000
ERROR       80   1063246      13291
TOTAL       84   2948962
                                     INDIVIDUAL 95 PCT CI'S FOR MEAN
                                     BASED ON POOLED STDEV
LEVEL        N      MEAN      STDEV  -+---------+---------+---------+-----
 MAGS       17      74.4       34.7     (---*---)
 POST       17     445.9      239.3                               (---*--)
 EDIT       17      46.0       28.7  (---*---)
 NEWS       17      94.6       43.7       (--*---)
  LIT       17      88.8       72.5      (---*---)
                                     -+---------+---------+---------+-----
POOLED STDEV =     115.3              0       150       300       450
```

Analysis of Variance Printout for $/LEAD

```
ANALYSIS OF VARIANCE ON $/LEAD
SOURCE      DF        SS         MS         F        p
MEDIA        4     73254      18313      7.22    0.000
ERROR       80    203043       2538
TOTAL       84    276297
                                     INDIVIDUAL 95 PCT CI'S FOR MEAN
                                     BASED ON POOLED STDEV
LEVEL        N      MEAN      STDEV  -------+---------+---------+---------
 MAGS       17     61.89      62.10                    (------*------)
 POST       17     12.58      11.03      (------*------)
 EDIT       17     77.11      82.28                          (------*------)
 NEWS       17      2.69       1.59   (------*-------)
  LIT       17     18.68      44.04       (------*------)
                                     -------+---------+---------+---------
POOLED STDEV =     50.38                    0        35        70
```

CASE **21**

401(K), INSTRUCTOR'S NOTE

Abstract: The task assigned in this case is to find a way to improve upon the forecasts provided by an underwriter. Almost any regression model will do the job. Students need to think about the role of residuals in regression to solve the case.

Level:	Intermediate, advanced.
Techniques:	Regression.
Typical Class Discussion Time:	30 minutes.
Data Set Size:	92 cross-sectional observations, 10 variables.

GENERAL COMMENTS

The underwriter is doing a pretty poor job. The total amount of error made (which was calculated as the absolute value of the estimated amount minus the eventual contribution) was $6.9 million for these 92 different cases. The average error was about $75,000, which is pretty bad given that the average contribution level of these 92 accounts was about $87,000. The underwriter appears to be severely overestimating the actual amounts, as shown below.

Dotplot of (ESTIMATE-CONTRIB)

```
                  .
                  :
                  :  :    .
               : .:.::: : :
               :.:::::: ::: :     .
         . . ..::::::::::::::.....:  :    .             .     .
      +---------+---------+---------+---------+---------+------
-100000         0    100000    200000    300000    400000
                          ESTIMATE - CONTRIB
```

It is possible to derive a number of different regression models that lower the average error rate to under $35,000 while also eliminating the skewness of the dotplot shown above. An error of $35,000 per account may not be acceptable to the company, but it is a significant improvement over the underwriter's guess.

QUESTIONS FOR ASSIGNMENT

1 Can you determine from the data how the underwriter is forming his/her estimates?

2 In attempting to beat the underwriter's estimate, what statistic or summary measure should be used to gauge whether your methodology is better (or worse) than the underwriter's estimate?

3 After constructing a regression model of CONTRIB on any of the relevant variables in the data set, tell how well your model predicts actual contributions.

4 What is the best possible regression model for beating the underwriter's estimate? Is this the best possible regression model based on the conventional statistical criteria? Should the company train its representatives with the kinds of skills and expertise that Susan has acquired?

SUGGESTIONS FOR LEADING THE DISCUSSION

When leading the discussion of this case, it is probably important to establish at the very beginning the question of what statistic will be used to decide which methodology produces the best estimate. Presuming that the company weighs overestimates as equally bad as underestimates (meaning that only the absolute value of the error is important), your students will likely use the sum or average of the absolute value of the regression residuals. Other creative ideas may be developed by your students.

There is an interesting dichotomy here between the quality of the statistical results and the value of the results to the company. One model developed below meets the usual criteria for statistical adequacy with an R^2 value of 48%, statistically significant predictors, and relatively random residual patterns. Yet, the company is still faced with an average absolute error of $40,000 on a typical $87,000 account. You might use this outcome to talk about the potential conflict between statistically significant results and economically meaningful results.

TYPICAL RESULTS AND APPROACHES

The underwriter is apparently relying heavily on ELIGIBLE to do the estimation. The correlation between ESTIMATE and ELIGIBLE is 80.6%.

Here are several possible regression equations that might be used as alternatives to the underwriter. In what follows, *t* ratios are parenthesized and the term TOTAL ERROR is the sum of the absolute value of the residuals (which is to be compared to the error of $6,896,742 of the underwriter), and AVERAGE ABS ERROR is the average of the absolute value of the residuals (to be compared to $74,964 of the underwriter).

Possible Prediction Models for CONTRIB

```
CONTRIB = 41331 + 484 ELIGIBLE
                  (7.2)

                      R-sq = 36.6%
                         s = $58,440
               TOTAL ERROR = $3,949,202
        AVERAGE ABS ERROR = $42,926


CONTRIB = -16,468 + 537 ELIGIBLE + 1.46 SALARY + 51440 SUSAN
                    (8.5)          (3.1)         (2.8)

                      R-sq = 47.9%
                         s = $53,610
               TOTAL ERROR = $3,671,577
        AVERAGE ABS ERROR = $39,908


CONTRIB = 15,080 + 0.46 ESTIMATE
                   (10.9)

                      R-sq = 56.9%
                         s = $48,224
               TOTAL ERROR = $3,218,535
        AVERAGE ABS ERROR = $34,984
```

The last model listed is based on having the underwriter prepare his or her estimate, taking 46% of that value, and adding about $15,000. This could be viewed as a corrected estimate. Of course, this wouldn't make much *business* sense, because you would still have to pay the underwriter's salary or fees.

CASE **22**

LATE CHARTS, INSTRUCTOR'S NOTE

Abstract: A simple quality control case, calling for a *p*-chart (or *Np*-chart). The process seems to be under control, although the *result* is probably not satisfactory.

Level:	Beginning.
Techniques:	Control charts, plots.
Typical Class Discussion Time:	20 minutes.
Data Set Size:	20 time series observations, 2 variables.

GENERAL COMMENTS

This is a pretty straightforward case. The percentage of charts late fluctuates around 11%. While there's a bit of a surge in late charts in the last three or four weeks, standard control chart techniques don't indicate anything out of control. This case gives an opportunity to discuss the difference between specification limits and control limits, for 11% late charts hardly seems satisfactory from a *customer* viewpoint.

QUESTIONS FOR ASSIGNMENT

1 Is the records delivery system at the medical center operating under control?
2 Is the records delivery system producing satisfactory results?

SUGGESTIONS FOR LEADING THE DISCUSSION

The assignment questions work pretty well here. After asking if the system is under control, you may wish to ask, "Why do you say so?" or something similar, to prompt a discussion of the appropriate techniques. Then you can follow up with the second question: "Is the result satisfactory?" Most students will feel that an 11% defect rate is not satisfactory. At this point, you can mention that statistical control simply means the process is doing the best it can, and any improvements will need to come from fundamental changes to the process. This kind of defect rate is (unfortunately) not uncommon in the administrative aspects of some medical settings.

TYPICAL RESULTS AND APPROACHES

Most students will produce some sort of control chart like the one below. According to most normal criteria, the process is operating under control.

***P*-Chart of NLATE**

```
Fraction Nonconforming
        +---------------------------------------------------------+
   0.21 +                                                         |
        |---------------------------------------------------------| 0.2039
        |                                                         |
        |                                                         |
        |                                                         |
   0.16 +                                                     +   |
        |                                                         |
        |+                                                        |
        |        +             +                                  |
        |      +             +          +                +        |
   0.11 +--------------+------------------------+-----------------| 0.1100
        |   +             +       +          +     +       +      |
        |           +                      +                      |
        |                            +                +           |
        |                                                         |
   0.06 +                                                         |
        |                                                         |
        |                                                         |
        |                                                         |
        |---------------------------------------------------------| 0.0161
   0.01 +                                                         |
        ++-------------+-------------+-------------+-------------++
         1             6            11            16            21
                                Case Number
```

CASE **23**

MONEY SUPPLY AND INTEREST RATES, INSTRUCTOR'S NOTE

Abstract: A series of simple regression models, or correlations, can be used to answer the questions posed in the case. Students are asked to go to the Web to update the data set.

Level:	Beginning, Intermediate.
Technique:	Simple regression, correlation.
Typical Class Discussion Time:	30 minutes.
Data Set Size:	214 monthly observations, 7 variables.

GENERAL COMMENTS

This is a straightforward case that requires students to look at correlations to understand relationships between various macroeconomic variables. As part of the exercise, they are encouraged to go to the Web to find the latest data. Going to the Web will be the most frustrating part of the case. Because data are continually revised, students will find the numbers on the diskette to be different from the ones on the Federal Reserve Board's Web site. Some will take the initiative to replace data from us with the latest updates. Others will not.

The case familiarizes the students with publicly available macroeconomic data sources and forces them to deal with changing data from federal government sources.

QUESTIONS FOR ASSIGNMENT

1 Run a regression of PRIME on M1. Repeat the exercise for all pairs of money supply and interest rate variables in the data set.

2 Compare the regression models from question 1 based on R^2. How do the R^2 values apply to the problem posed in the case?

3 Compare the regression models from question 1 based on the size of the estimated regression slope coefficients. Relate your findings to the initial problem posed in the case.

4 Compare the regression models from question 1 based on the statistical significance of the estimated slope coefficients. Apply your findings to the case problem.

5 Decide whether you should be looking at R^2, the size of the slope coefficients, the significance of the slope coefficients, or all of these to determine whether a relationship is "good."

SUGGESTIONS FOR LEADING THE DISCUSSION

Begin the discussion with the general question, "What has Professor Hicks discovered from the data set?" At some point, the matter of revisions, and how they were incorporated into the data set, will arise. Expect the discussion to focus on the issues posed by the "Questions for Assignment" above. Because students have various ways to evaluate the relationships between money supply and interest rate, some will undoubtedly reach different conclusions.

TYPICAL RESULTS AND APPROACHES

The regression equations are quite similar, regardless of the definition of money supply or interest rate that is used. All slopes are statistically significant and negative (*t*-ratios are parenthesized in the exhibits that follow), and R-squared values run from 58% to 85%. Goodness of fit is lowest for the Prime interest rate equations, and highest for the mortgage loan (Home) equations.

Regression Equations for Prime

```
Prime = 18.1 - 0.01M1          R-sq = 57.6%
                 (-17)          RMSE = 2.2

Prime = 20.9 - 0.004M2         R-sq = 62.8%
                 (-19)          RMSE = 2.1

Prime = 20.7 - 0.003M3         R-sq = 60.1%
                 (-18)          RMSE = 2.2
```

Regression Equations for Discount

```
Discount = 14.6 - 0.01M1        R-sq = 76.5%
                   (-26)        RMSE = 1.4

Discount = 17.1 - 0.004M2       R-sq = 79.2%
                   (-28)        RMSE = 1.3

Discount = 16.9 - 0.003M3       R-sq = 75.6%
                   (-26)        RMSE = 1.4
```

Regression Equations for Home

```
Home = 18.9 - 0.01M1           R-sq = 82.6%
               (-32)           RMSE = 1.2

Home = 21.4 - 0.004M2          R-sq = 85.1%
               (-35)           RMSE = 1.1

Home = 21.3 - 0.003M3          R-sq = 82.4%
               (-31)           RMSE = 1.2
```

Regression Equations for T-Bill

T-Bill = 14.7 − 0.01M1	R-sq = 65.9%
(−20)	RMSE = 1.8
T-Bill = 17.0 − 0.004M2	R-sq = 68.2%
(−21)	RMSE = 1.7
T-Bill = 16.8 − 0.003M3	R-sq = 64.6%
(−20)	RMSE = 1.8

Of course, these equations will differ depending on whether or not the students update the data set as requested. Be prepared for some students to interchange the roles of interest rates and money supply, which is perfectly reasonable since money and interest rates are endogenous in the larger scheme of things. The data lend themselves to logarithmic transformations, allowing elasticities to be estimated directly.

CASE **24**

SAT SCORES, INSTRUCTOR'S NOTE

Abstract: At first glance, this appears to be a straightforward model-building exercise. The results are counterintuitive, though, unless the student has given some thought to what factors, or transformations of the factors, are appropriate. Because the data are state level, this case also allows you to discuss problems of aggregation and the ecological fallacy.

Level:	Intermediate, advanced.
Technique:	Regression.
Typical Class Discussion Time:	30 minutes.
Data Set Size:	51 cross-sectional observations, 8 variables.

GENERAL COMMENTS

This is a case in which the statistical results defy expectations. SAT scores and teacher salaries are inversely related! It's a fun case for those of us who like to insist that good statisticians bring expertise, intuition, and knowledge to an analysis.

In terms of the overall question about resources and educational success, it is only after correcting for the size of the states that an intuitively appealing answer arises. There are some potential regional differences; SAT scores are generally higher in the North Central Region and parts of the Midwest. Additionally, the data could give us a completely different answer if all dollar values were adjusted for cost of living.

QUESTIONS FOR ASSIGNMENT

1 What factors appear to influence SAT scores of high school students?
2 Are there any regional differences in SAT scores? If so, how might this influence your analysis?
3 Which states are most out of line with the others in terms of their SAT scores?

SUGGESTIONS FOR LEADING THE DISCUSSION

When we do this case, we find it's best not to lead the discussion at all. Open the discussion with something noncommittal like, "Well, what determines SAT scores?" and let students wrestle with the measurement issues. If your students use only correlation on this case (no regression), play the devil's advocate and ask them if the results mean that if we cut teachers' salaries, SAT scores will increase.

Don't leave this case until you've discussed aggregation and measurement issues.

TYPICAL RESULTS AND APPROACHES

The average SAT score is 963, with a typical fluctuation of 75.6 points. Iowa, North Dakota, and Minnesota lead the pack. Teachers were paid an average of $36,000 during the 1995–96 school year, although there is a great deal of variation in this figure—the typical fluctuation in the average teacher salary is $6400.

SAT scores correlate negatively with EXPEND, TAKERS, TEACHERS, SALARY, and ENROLL. Since the variable RATIO is pupils taught per teacher, we would probably expect a negative correlation with SAT scores, but it is positive.

Students who throw all of the original data into a multiple regression model obtain the following results (*t* ratios are parenthesized):

```
SAT = 1038 + 5.33 EXPEND - 2.79 TAKERS - 0.0313 ENROLL
              (1.2)         (-12)          (-1.1)

           - 0.105 TEACHERS - 0.99 RATIO + 1.33 SALARY
             (-0.2)           (-0.3)        (1.1)


     s = 34.40        R-sq = 81.8%      R-sq(adj) = 79.3%
```

With only TAKERS significant, the more astute students will attempt to understand a bit more about the variables and how they relate to SAT scores. Larger states will have larger numbers of expenditures, enrollments, and more teachers. Adjusting for the size of the state by calculating:

$/STUDNT = EXPEND/ENROLL
SAL/STNT = SALARY/ENROLL

gives a different outcome:

```
SAT = 1032 - 2.83 TAKERS - 1.60 RATIO + 11197 $/STUDNT
             (-13)         (-0.7)         (2.5)

           - 41.2 SAL/STNT
             (-0.9)


     s = 33.98        R-sq = 81.4%      R-sq(adj) = 79.8%
```

Your more confident students may delete TAKERS from the equation, on the grounds that it likely covers for a lurking variable of some sort.

CASE **25**

EMERGENCY ADMISSIONS, INSTRUCTOR'S NOTE

Abstract: This case involves the construction of a forecasting model from a univariate time series. Nine missing observations will command some of the classroom discussion time.

Level:	Intermediate, advanced.
Techniques:	Forecasting, regression.
Typical Class Discussion Time:	30–45 minutes.
Data Set Size:	241 time series observations, 1 variable.

GENERAL COMMENTS

The time series plot of emergency department admissions demonstrates some daily and seasonal variation. Winter months are typically lower than others, and certain days of the week are busier, depending on the month of the year. We have used regression analysis to model the data.

Several alternative methodologies can be demonstrated with the data. Statistical quality control charts might be constructed from the time series. It is possible to test hypotheses about whether the mean levels of emergency admissions differ from month to month and by day of the week. Box-Jenkins modeling can be performed on this time series.

The actual values of emergency department admissions for August 29–31 were 125, 116, and 142, respectively. Since, January 1 was a Wednesday, August 29 was a Saturday (unless it was leap year).

QUESTIONS FOR ASSIGNMENT

1 Determine what you should do about the nine missing values in the data set. Should these be ignored or replaced by some "reasonable" values? What might "reasonable" be in this instance?

Those of you looking for applications of hypothesis tests might consider the following:

2 Test the hypothesis that average weekday emergency department admissions are the same.

3 Test the hypothesis that average monthly levels of admissions are the same.

SUGGESTIONS FOR LEADING THE DISCUSSION

As with most of our forecasting cases, we organize the discussion around whose forecasts are the best. Take the actual values to class with you.

Some of the discussion about this case should involve the handling of those missing values. If you have not spent any class time discussing missing data prior to this case, most of your students will probably not give this issue much consideration or will simply choose to ignore the problem.

If you specifically ask them to consider the problem of missing data, but don't tell them how to deal with it, you will get a wide range of responses. Some will do nothing, which will likely have consequences if you are doing Box-Jenkins analysis—lots of missing data will arise from the lagging required to undertake ARIMA modeling. Some students may decide to replace missing values with averages of one sort or another.

Make sure that you or someone mentions that forecasts from September to December are suspect.

TYPICAL RESULTS AND APPROACHES

In what follows, we have replaced each missing value with the average of the same days of the week available elsewhere in the data set. That is, missing Mondays are replaced with the average of other Mondays in the data set, and so forth. With missing values the average level of admits is 127.7 (standard deviation of 18.1), whereas the average admits after replacing missing values was 127.7 (standard deviation of 17.8), suggesting that our missing value replacement method might be reasonable for this case. You and your students will get different results if you use different approaches, but what follows will give you a general feel for the patterns in the time series. The monthly patterns in the data appear to be most pronounced for January through March, as suggested by the following dummy variable regression model (*t* ratios are parenthesized):

```
ADMITS = 134 -26.5 JANDUM -11.5 FEBDUM -15.4 MARCHDUM -4.5 APRILDUM -2.2 MAYDUM
              (-6.8)       (-2.9)        (-3.9)         (-1.2)        (-0.6)

         + 6.1 JUNEDUM + 0.8 JULYDUM            R-sq = 31.5%
          (1.6)          (0.2)                     s = 14.9
```

A fancier model includes a Sunday dummy variable, a weekly lag of admits, and an adjustment for Saturdays in winter (JANFEBSAT = 1 if that day is a Saturday in January or February, 0 otherwise):

```
ADMITS = 112 - 22.9 JANDUM - 11.4 FEBDUM - 12.5 MARCHDUM + 0.18 LAG7ADMITS
              (-6.3)        (-3.8)         (-4.5)           (2.8)

             - 14.0 SUNDAYDUM + 12.2 JANFEBSAT              R-sq = 43.1%
              (-5.1)            (2.3)                          s = 13.4
```

CASE **26**

DAVE'S TRUCK WASHING SERVICE, INSTRUCTOR'S NOTE

Abstract: This case is primarily designed to be a modeling exercise resulting in short-term forecasts. Along the way to those final forecasts, you can use this case to illustrate multicollinearity, influential observations, and appropriate versus inappropriate independent variables.

Level:	Intermediate, advanced.
Techniques:	Time series, regression.
Typical Class Discussion Time:	45 minutes.
Data Set Size:	46 monthly observations, 8 variables.

GENERAL COMMENTS

Much of the learning, as well as classroom discussion, will likely revolve around the intended use of the variable SALES2. In this data set, SALES is total sales, the sum of SALES1 (not directly available in the data set) and SALES2. Needless to say, SALES2 is the most highly correlated variable in the data set with SALES, and your students will probably put it in the regression equation. This doesn't make much sense, since SALES2 is only a component of SALES, and doesn't serve to explain much to us from a behavioral point of view. This point will probably become obvious to your students when they try to forecast SALES with SALES2.

Even though this is a time series, it will be difficult (without SALES2) to achieve an R-squared statistic in the 90% range. Much of the variability in SALES appears to be related to time, and many of the time-dependent predictors are collinear (PREC, TEMP, etc.).

The case requires students to forecast two months past the end of the data set. The actual values of SALES for these two months are $9557 and $8066. The corresponding values of the potential predictors are:

	SALES2	TEMP	PREC	CAT	SCHOOL	EMPLOY	AUCTION
March	3090	43.0	3.5	0	0	2.5	0
April	1645	54.8	0.5	0	1	2.5	1

Use these values to explore with your students various out-of-sample forecasting summary statistics during the classroom discussion of this case.

QUESTIONS FOR ASSIGNMENT

1 Which variables seem to be the important predictors of monthly sales of TWS in some behavioral sense? Which variables seem to be the important predictors of monthly sales of TWS in a statistical sense? Should your answer to these two questions be the same? Why or why not?

2 How should the variable named SALES2 be used in this analysis? What is its purpose from an accounting perspective? What is its purpose from a statistical perspective?

3 Are there any monthly fluctuations in sales worthy of note? How might monthly fluctuations influence your model's forecasting ability?

SUGGESTIONS FOR LEADING THE DISCUSSION

Focus the classroom discussion of this case on the forecasts. Presumably each student or group will come to class with their forecasts for the two months past the end of the data set. Begin the discussion of how each of those forecasts was derived. On the board or overhead, keep track of the competing forecasting models or techniques and the numerical forecasts that derive from them.

Some of the uncertainty in the students' minds will come from the need to forecast the actual values of the predictor variables (if regression was used). Move the discussion in the direction of how various groups solved this problem. Then, give them the actual values of the predictor variables (above), and ask how they might have changed their forecasts had they had these true values prior to coming to class. Have students recompute their forecasts from the actual values of the predictors, and keep track of these on the board.

Finally, give them the actual values of the response variable SALES (above). Given the actual values, have the students discover for themselves which of the competing models is best. We often use class discussion to develop the notion of and formula for the MSEP during this portion of the class discussion.

Having gone through these steps, you've demonstrated for the students various out-of-sample forecasting methods and why we sometimes withhold data to validate forecasting models. Close the class discussion with a summary of these ideas.

TYPICAL RESULTS AND APPROACHES

Monthly sales tend to be higher in the winter and lower in the summer, but the highest monthly average sales is recorded in August ($7330) and the lowest in December ($4881). The precipitation variable has an extreme value of almost 10 inches in the first September in the data set.

The sequence plot of SALES is:

Time Series Plot of Sales

```
         -
    10000+                                           8
         -
 SALES   -
         -                            5  8
         -
     7500+
         -              3              67 9             B
         -          A          0            A 12
         -       8       4         23       0    3 5  9    2
         -      6           67   A 1  4           4  7    1
     5000+                   89                   6    A
         -        9   1                   B
         -    5    0 B 2         B                    0
         -      7
         -                5
         +-----------+-----------+-----------+-----------+-----------+
          0         12          24          36          48          60
```

One of the models we like requires creating a time trend and two monthly dummy variables. The model:

```
SALES = 4481 + 69.7 TIME + 1594 AUGUST - 1134 DECEMBER - 1986 SCHOOL - 1108 CAT
               (5.7)        (3.3)          (-2.3)            (-3.8)          (-3.4)
```

has an adjusted R-squared of 53.5% and a Durbin-Watson of 2.09 (t-ratios are shown in parentheses). The forecasted value of SALES from this model are \$7755 and \$5838. The RMSEP = \$2026 for the two projected months.

A simple regression of SALES on time gives:

```
SALES = 4745 + 39.6 TIME
               (2.8)
```

with an R-squared value of 15.4% and a DW of 1.96. The forecasted values from this model are \$6608 and \$6648 with RMSEP = \$2314.

CASE **27**

HOUSING PRICES—I, INSTRUCTOR'S NOTE

Abstract: This case is best used for simple graphics and summary statistics. Students quickly discover that univariate analysis is not appropriate or not revealing for this data set. A nice data set for demonstrating the notion of lurking variables.

Level:	Beginning.
Techniques:	Descriptive statistics, hypothesis tests, confidence intervals.
Typical Class Discussion Time:	30 minutes.
Data Set Size:	108 cross-sectional observations, 11 variables.

GENERAL COMMENTS

This case was designed to accommodate a discussion of descriptive statistics and simple graphics. However, it is a perfect lead-in to multivariate analysis, because the simple univariate statistics and even the bivariate relationships will leave many questions in your students' minds, and they should then be receptive to a discussion of multivariate modeling (as in Housing Prices—II).

For instance, in this data set, comparing the means of the selling price grouped by houses with and without fireplaces might lead one to conclude that the existence of a fireplace adds $30,000 to the price of a house, which averages only $98,000!

Those of you looking for applications of hypothesis tests and confidence intervals can amend the questions below to ask whether means are statistically different, rather than just numerically different, from each other.

QUESTIONS FOR ASSIGNMENT

1 How much more can you expect to get for your house, which has a fireplace, over one that does not?

2 How much more can you expect to get for your house, which has four bedrooms, as opposed to one that has three?

3 If your house has no basement, how much lower do you expect your selling price to be?

4 What about these descriptive statistics makes you wary or uncomfortable? In other words, if you were a real estate agent, would you go around spouting the answers that you gave above? Explain.

SUGGESTIONS FOR LEADING THE DISCUSSION

Our classroom antics come out with this case. Act shocked and pleasantly surprised when a student quotes the $30,000 fireplace figure. "You mean, since the typical house is $98,000, a fireplace accounts for a third of the price? Since my house doesn't have a fireplace, should I go home and knock a hole in the wall, add some bricks, and put it on the market for $30,000 above the price I bought it for?" Hold them responsible for a cohesive and logical discussion of alternative explanations of this phenomenon. Don't let them get away from the issue until something has been resolved about the influence of the other variables in the data set.

The three-way graph below might be used to guide the discussion. In the graph, the graph symbols refer to the numerical code for FIRE (asterisks mean more than one observation at that point). Clearly houses without fireplaces tend to be smaller.

Scatterplot of PRICE versus SQ_FT

```
PRICE
200+                                                     1
   -                                                     1
   -
   -
   -                                           1  1
150+                                        1      1
   -                                        1
   -                       1          *    1*  *
   -                                 1  111
   -                            1* 11  ***   1 1
100+                 1   * 11 1  11  * 1  1
   -           1  11 1*** 1 **  111 * 1
   -      1       0 0 *00    *1*     1   1
   -    0 1 *  111 0*0        000
   -   0  1 0     1           1
 50+
   -
     --+---------+---------+---------+---------+---------+
     800      1200      1600      2000      2400      2800
                           SQ_FT
```

You might also take the correlation matrix to class. It can be used to guide the discussion in a number of different ways, including a lead-in to partial correlation if appropriate.

TYPICAL RESULTS AND APPROACHES

In the overall sample of 108 houses, selling prices average $98,000 (median of $92,500) with a standard deviation of $26,400. The minimum and maximum prices are $59,000 and $195,000, respectively. Selling prices are skewed rightward.

The 39 tri-level houses in the sample average $88,000 with typical fluctuation of $24,500. The 25 two-story houses average $86,400 (standard deviation of $19,300), but the 44 ranch houses sell at $113,300 on average (standard deviation of $24,300). The highest selling house in the sample is a tri-level.

Houses with and without basements differ by about $19,000 on average, but only 17 of the 108 houses do not have a basement. Similarly, houses with and without fireplaces differ by about $30,000 on average, but only 12 of the houses don't have fireplaces. The square footage variable more likely explains this difference. If your software has the ability to prepare three-way plots, the plot of square feet, price, and fireplace is very revealing.

PRICE correlates most highly with square feet. PRICE tends to decline with AGE. There appears to be a bit of diminishing returns to bathrooms; the fourth bathroom doesn't add as much to the median price of the home as the third one does.

It will be difficult to get an accurate measure of the relationship between type of heat in the house and price, since there are only four electrically heated houses in the sample of 108. The same applies to GARAGE—most houses have two-car garages. At first glance, the Apple Valley school district appears to be enjoying a premium of about $17,000 on average over the Eastville school district houses.

Correlation Matrix for Housing Variables

	PRICE	SQ_FT	BEDS	BATHS	HEAT	STYLE	GARAGE	BASEMENT
SQ_FT	0.828							
BEDS	0.254	0.239						
BATHS	0.561	0.614	0.393					
HEAT	-0.086	0.071	-0.029	-0.053				
STYLE	0.428	0.315	0.093	0.496	-0.122			
GARAGE	0.548	0.433	0.066	0.275	-0.022	0.186		
BASEMENT	0.267	0.085	0.196	0.186	-0.050	0.342	0.202	
AGE	-0.391	-0.292	0.279	-0.124	0.054	-0.162	-0.120	0.060
FIRE	0.341	0.321	0.010	0.310	0.069	0.254	0.128	-0.072
SCHOOL	0.319	0.161	0.024	0.236	-0.141	0.389	0.148	0.220

	AGE	FIRE
FIRE	-0.281	
SCHOOL	-0.188	0.134

CASE **28**

HOUSING PRICES—II, INSTRUCTOR'S NOTE

Abstract: This is primarily a model-building exercise using multiple regression. The data can also be used to demonstrate certain advanced topics, such as curvilinear functional forms (a logarithmic transformation on the dependent variable is a possibility) and interactive terms.

Level:	Intermediate, advanced.
Techniques:	Regression.
Typical Class Discussion Time:	45 minutes.
Data Set Size:	108 cross-sectional observations, 8 variables.

GENERAL COMMENTS

This case can be made as simple or as complex as you need. It's a data set that nicely demonstrates model building with regression, prediction, and inherent multicollinearity. The regression coefficients are intuitively appealing, highlighting the notion of the role of expected signs and magnitudes of coefficients in regression analysis.

Multicollinearity arises in these data. If your students have already completed Part I of this case, they are concerned about fireplaces accounting for a third of the selling price of a house. Fireplaces become statistically insignificant when included in a regression with square feet. Similarly, BEDS and BATHS lose their significance when included in a regression with square feet. This last example is intuitively appealing; adding more bedrooms and bathrooms necessarily increases square feet.

The data accommodates certain advanced topics. The dependent variable is skewed rightward; we sometimes use a logarithmic transformation of PRICE. Numerous interactive terms can be created; interpreting such a model can be a very challenging exercise for your students. One such model is shown below.

QUESTIONS FOR ASSIGNMENT

1 Use regression analysis to derive a model of selling prices of houses in Eastville. Interpret your final model and its coefficients within the context of this problem.

2 Prepare a spreadsheet or formula to use in predicting selling prices. Explain your formula in nontechnical terms.

3 Tell what additional information you'd like to have to predict the selling prices of these homes more accurately.

SUGGESTIONS FOR LEADING THE DISCUSSION

We like to turn this discussion over to our students. The class meeting before this case is due, we mention that one or more individuals or groups will be asked to give a formal presentation, as if to a Board of Realtors. We ask them to prepare a summary of their findings, as well as a nontechnical discussion of their forecasting formula. Our students have done a nice job; we believe that this is partly because the coefficients of the regression models have appealing, real-life interpretations (e.g., add $49/square foot, or a basement adds $17,000 to the price of a house).

Another great pedagogical tool is looking at the residuals to determine which houses might be underpriced or overpriced. The residual plot below may be used to demonstrate this point. It is from the multiple regression of PRICE on SQ_FT, AGE, GARAGE, BASEMENT, FIRE, and SCHOOL (see "Typical Results and Approaches" section). Houses with standardized residuals exceeding two are listed. (A logarithmic transformation of price alleviates some of the curvilinear pattern to this graph.)

Unusual Residuals for Housing Prices

Obs.	SQ_FT	PRICE	Fit	Stdev.Fit	Residual	St.Resid
12	1162	72.80	49.49	4.56	23.31	2.08R
13	1996	75.21	98.74	3.60	-23.53	-2.04R
21	2167	84.90	110.87	2.58	-25.97	-2.20R
39	2809	195.00	156.81	4.54	38.19	3.40R
104	2282	150.58	124.84	2.20	25.74	2.16R
106	2380	155.00	131.16	2.56	23.84	2.01R
108	2804	192.00	167.00	4.38	25.00	2.21R

Plot of Residuals versus Fitted Values for PRICE

```
Residuals
   -                                                  *
   -
   -
 25+       *                             * *              *
   -      *            *          **         *
   -      * *        2*                       *
   -           * *    *    *  *    **   *2
   -          * ** *   * 2** 2  *  *      *
  0+          *  **  * *3  2 *   **32*  *         *
   -              *  * *   *22** * **22   **
   -                  *  2 *2  **22*           **
   -                   ****         *  * *
   -                       * *  *  *
-25+-------+---------+--------*+---*-----+---------+---------+
          50        75       100        125       150       175
                           Fitted Values
```

TYPICAL RESULTS AND APPROACHES

The most important determinant of price in the data set is square feet. Depending on other determinants in the model, selling prices are estimated to be as high as $49/square foot. BEDS and BATHS are individually important, but not in a regression equation with square feet. The type of heating in the house is unimportant, although a caveat about small sample size of houses with electric heat is appropriate.

A partial explanation of the "fireplace problem" (see Part I of this case) follows. Fireplaces become statistically insignificant when combined in a regression with square feet, as shown below (*t* ratios are parenthesized).

```
PRICE = 12.2 + 0.0491 SQ_FT          R-sq = 68.5%
               (15.18)

PRICE = 72.6 + 28.6 FIRE             R-sq = 11.6%
               (3.74)

PRICE = 8.75 + 0.0475 SQ_FT + 7.04 FIRE     R-sq = 69.1%
               (13.99)        (1.47)
```

Here's one regression using only the metric predictors. BEDS and BATHS are individually significant, but become insignificant when included in a regression with SQ_FT.

```
PRICE = -25.3 +0.0401 SQ_FT +2.06 BEDS +2.90 BATHS +18.7 GARAGE      R-sq = 73.6%
               (9.88)         (0.96)      (0.89)       (4.19)
```

Next is another possible regression utilizing only categorical predictors:

```
PRICE = 51.9 - 9.4 HEAT + 17.5 BASEMENT + 28.0 FIRE +11.4 SCHOOL      R-sq = 25.2%
              (-0.78)    (2.75)           (3.86)      (2.37)
```

One regression that combines some metric and categorical predictors is shown below. Note what happens to fireplaces once again:

```
PRICE = - 8.94 + 0.0385 SQ_FT - 0.851 AGE + 15.3 GARAGE + 11.7 BASEMENT
                 (12.32)        (-3.12)     (3.84)        (3.46)

              + 5.23 FIRE + 6.13 SCHOOL             R-sq = 80.2%
                (1.30)      (2.44)
```

A more complex model utilizing some interactive terms is shown below. In this regression equation, the variable SQ*BASE is square feet times basement, and so forth.

```
PRICE = 89.6 - 0.0273 SQ_FT - 1.36 AGE + 19.0 GARAGE - 41.5 BASEMENT -49.9 FIRE
              (-1.94)        (-3.05)     (5.29)        (-2.50)         (-3.48)

     - 8.84 SCHOOL + 0.0311 SQ*BASE + 0.0392 SQ*FIRE + 1.33 AGE*SCH
     (-1.40)         (3.23)           (3.85)           (2.67)

R-sq = 85.0%
```

CASE **29**

BOOKS, INSTRUCTOR'S NOTE

Abstract: This case is a model building exercise using multiple regression. The trick is that the three potential independent variables yield a model with an R^2 of 100%.

Level:	Intermediate.
Technique:	Regression.
Typical Class Discussion Time:	30 minutes.
Data Set Size:	38 pooled cross-section, time series observations; 6 variables.

GENERAL COMMENTS

The trick to this data set involves the second task requested in the case—to build a model of OpAssets. As it turns out, the multiple regression model is an accounting identity. The OpAssets variable was created by the equation OpAssets = PPE + Assets − Liab. For firm 1 in 1991, the values for Assets and Liab were changed from 306.169 to 306.168, and 322.337 to 332.338, respectively, to permit calculation of the regression diagnostics. Your students will get an answer from the multiple regression, but should be tipped off by *t* ratios for the estimated slope coefficients of several hundred thousand or the zero error sum of squares.

The reason we created this case is that every semester when students are asked to design their own project and collect their own data, one or more students invariably comes up with a deterministic model. For instance, one group proposed studying total electricity consumption in Denver as a function of consumption from residential, commercial, and manufacturing sectors, a tautology. The confusion likely arises from the message that higher R^2 values are better, yet we don't generally talk about what it means to have an R^2 of 100%. This case will allow you to rediscover with your students the difference between stochastic and deterministic models and the fundamental nature of residuals and residual analysis to multiple regression model building.

QUESTIONS FOR ASSIGNMENT

1 Characterize the trends in each of the financial variables in a time series plot. What do these plots reveal?
2 What are the major financial determinants of OpAssets?

SUGGESTIONS FOR LEADING THE DISCUSSION

The interesting part of the class discussion involves the matter of the 100% R^2. Some students will overlook the perfect R^2 value and simply write up the report, bragging about what a fine model they have derived. Others will finally discover the trick to the case. The best way to lead the class discussion in our experience is to plead total ignorance to the issue of the accounting identity. Act as if you are completely surprised by the outcomes being reported by your students. Then let them take control of the discussion.

When the trick is finally uncovered, ask your students about the general nature of the error term in regression analysis. Ask them the purpose of the error term, where and how the error term comes into play in generating the regression statistics, and what it means to have an equation with zero residual values.

TYPICAL RESULTS AND APPROACHES

Operating assets are up for Barnes & Noble, Books-a-Million, and Borders over the time series available for each corporation. Crown and Village are relatively flat. Expect one or more of your students to report that the book industry is in great decline in the later years of the time series. They will have wrongly reached this conclusion by failing to recognize that a straight sequence plot of OpAssets is a repeated time series of the five firms, with the smallest (Village Green) at the end of the list.

On the model-building portion of this case, some students may actually discover the accounting identity inherent to the multiple regression model of OpAssets on PPE, Assets, and Liab, and solve the puzzle by leaving out one or two of the predictors. Here's a printout that you can take to class that shows the essence of the case.

```
OpAssets = 0.000082 + 1.00 PPE + 1.00 Assets - 1.00 Liab

36 cases used 2 cases contain missing values

Predictor        Coef        Stdev      t-ratio        p
Constant   0.00008230   0.00007317         1.12    0.269
PPE           1.00000      0.00000    558060.88    0.000
Assets        1.00000      0.00000    903669.06    0.000
Liab         -1.00000      0.00000   -773517.13    0.000

s = 0.0002906    R-sq = 100.0%     R-sq(adj) = 100.0%

Analysis of Variance

SOURCE        DF          SS          MS          F        p
Regression     3     1168723      389574  4.613E+12    0.000
Error         32           0           0
Total         35     1168723

Unusual Observations
Obs.    PPE   OpAssets        Fit  Stdev.Fit   Residual   St.Resid
  1      84     57.933     57.932      0.000      0.001      5.66R
  2      93     64.747     64.748      0.000     -0.001    -3.18RX
  7     435    647.466    647.466      0.000      0.000     1.07 X

R denotes an obs. with a large st. resid.
X denotes an obs. whose X value gives it large influence.
```

CASE **30**

ELISA PLATE UNIFORMITY, INSTRUCTOR'S NOTE

Abstract: A process is described, with three possible sources of variation to identify. One of them seems worth pursuing. The process is not currently meeting its stated objectives. Simple scatterplots suggest multiple regression models with dummy variables, two- or three-way ANOVA (although it's not quite balanced).

Level:	Intermediate.
Techniques:	ANOVA, multiple regression.
Typical Class Discussion Time:	30–45 minutes.
Data Set Characteristics:	480 cross-sectional observations, 5 variables.

GENERAL COMMENTS

Many student version statistics packages make no provision for three-way ANOVA (or even two-way ANOVA), so you may have to tackle this case by using multiple regression with dummy variables. Actually, you can discover a lot from some scatterplots, like the one that follows, at least enough to suggest some further investigation. That may be all that matters. In the spirit of quality management, dealing with the primary source of variation now is usually more important than understanding the details of the other sources of variation later.

QUESTIONS FOR ASSIGNMENT

If you feel a need to structure this problem any more than the case provides, you could simply ask whether injectors, columns, and/or plates appear to be sources of variation, and ask for suggestions for further action for ELISAs, Inc.

SUGGESTIONS FOR LEADING THE DISCUSSION

"What do you conclude?" or "What do you recommend?" seem to work pretty well here. You should ask for the reasoning behind the conclusions and recommendations, and require that they be stated in coherent ways. You may need to be sure that students interpret the specification correctly, too, and translate from standard deviations on a printout to coefficients of variation.

If you prefer a more structured approach, you can put up for discussion the scatterplot that follows. With that as starting point, you can ask the class to suggest formal models that would describe the patterns in the plot and permit us to assess their significance. You can record the suggested models on the board for further comparison and analysis. This kind of approach leaves less for the student to figure out, but may be a good way to suggest the need for models involving dummy variables and how they are interpreted.

TYPICAL RESULTS AND APPROACHES

The process doesn't come close to specification. The biggest single source of variation is from column to column. After column 1, the readings drop, then rise again (see the scatterplot below), and these effects appear significant by conventional measures. ELISAs, Inc., should try to identify any sort of start-up effect as the bank of injectors begins filling wells on each new plate, or perhaps the "blueness" measuring process is faulty. There is variation attributable to other causes, too, but the column-to-column changes have some pattern to them, so they're probably the ones to pursue first.

Scatterplot of OPTDEN versus COL

```
OPTDEN
2.4 +                                                 |
    |+                           +               +    |
    |                                            3    |
    |+                               +                |
2.1 +                        2       +   +            |
    |5                       +   2   2       +   4    |
    |3   +   +           +   +   +   +   2       3    |
    |2   +               2   +   4   3   4   2        |
1.8 +4       2   +   3   3   4   3   2   2   7   7    |
    |+   +   2   3   4       3   2   5   5   2   5    |
    |4   4   +   2   +   3   2   3   7   5   4   +    |
    |3   2   6   +   +   3   3   5   6   7   5   4    |
1.5 +3   +   2   2   6   5   5   5   3   +   5   2    |
    |6   +   3   3   5   7   5   4   3   3   6   +    |
    |+   +   2   2   6   7   2   4   +   5   +   4    |
    |    9   8   *   6   +   5   2   3   3   4   2    |
1.2 ++   6   8   8   4   +   4   2   2   2   3   3    |
    |    9   4   3   +   5   +   2                    |
    |    +   +   +   3   +   +                        |
    |    3       +       +                            |
0.9 +                                                 |
    ++---------------+---------------+---------------++
     1               5               9              13
                            COL
```

CASE **31**

VIOLENCE ON TV—I, INSTRUCTOR'S NOTE

Abstract: A practical situation, one that can be profitably approached on many levels. Part II of the case (Case 36) provides a follow-on, with one possible survey approach and some data to analyze.

Level:	Intermediate, advanced.
Techniques:	Survey design.
Typical Class Discussion Time:	30–45 minutes.
Data Set Size:	None.

GENERAL COMMENTS

For sophisticated students, you can probe for a lot of aspects of survey design here. It might be helpful, for example, if you were emphasizing market research or political polling. For all that the situation is simple, it embodies many real and difficult practical problems: type of survey, costs, phrasing questions, and so on. You can also use this case to set up *Case 36, "Violence on TV—II."*

QUESTION FOR ASSIGNMENT

The case lists a number of questions to stimulate discussion of data-gathering techniques. Almost any of them can lead to good (although often lengthy) discussion. If you ask too much of any answer, they may turn into questions that are "too easy for the beginners, too hard for the experts," but if you use them to raise awareness rather than with a view to eliciting fixed answers, they will be productive.

SUGGESTIONS FOR LEADING THE DISCUSSION

Pick one of the case questions and see where it leads. You can emphasize the practical aspects of data gathering, the design and precision issues, or the ethical issues, as suits your purpose. The follow-on case also provides a means of comparison: students will often disagree strongly with the approach taken there.

TYPICAL RESULTS AND APPROACHES

No one set of recommendations seems to emerge, but the practical aspects pretty much dictate phone or mail surveys rather than shopping malls or the like. Many students will be unaware that there *are* ethical issues to consider.

CASE **32**

OVERDUE BILLS, INSTRUCTOR'S NOTE

Abstract: The moral of the story is that failure to perform graphical analysis on your data can lead to misleading results. The data encompass two distinct subgroups; failure to recognize and model the two groups results in incorrect policy recommendations.

Level:	Beginning.
Techniques:	Descriptive statistics, regression.
Typical Class Discussion Time:	30–45 minutes.
Data Set Size:	96 cross-sectional observations, 2 variables.

GENERAL COMMENTS

The key to this case is doing background graphical analysis of your data before performing summary statistics. The correlation between DAYS and BILL is -0.04, and the simple regression model of DAYS on BILL gives an estimated slope coefficient of -0.01 with a *t* ratio of -0.4. At first glance, there appears to be no relationship between the size of the bill and the number of days the payment is late.

However, a simple scatterplot of DAYS against BILL gives a completely different story. The case indicates that half of the bills are from residential customers and half are from commercial accounts; these two groups show up clearly on a scatterplot. The relationship between BILL and DAYS is positive for residential accounts and negative for commercial accounts.

This case can be a natural transition for regression models with dummy variables. Once the students see that two distinct lines need to be modeled, you can introduce the notion of dummy variables as intercept shifters and interactive terms as slope shifters. Some students will attempt to split the data into two samples, which can motivate a discussion of the gain in degrees of freedom resulting from using regression on the complete data set.

To address the issue of the money-back guarantee, students can calculate regression lines of DAYS on BILL and use the line to find the bill sizes that meet the 60-day guideline.

QUESTIONS FOR ASSIGNMENT

1 What recommendation do you have about the feasibility of the 60-day money-back guarantee?

2 Does the size of the bill somehow relate to the number of days the payment is late? If so, how?

3 Does your answer to either of the above questions depend upon whether the customer is a residential or commercial customer? If so, how?

Scatterplot of DAYS versus BILL

```
DAYS
105+
   -     *   commercial
   -  **    * *
   -        ** ***
   -                *     * **
 70+                       ***    **   *
   -                      *         * **
   -                                    ***  *      *
   -                                        *  *  *   *  *  *
   -                                    *  ****  *  *         * *
 35+                               **   **
   -                     *****
   -              **          *
   -  *  * ***    *  residential
   -  --*-------**+---------+---------+---------+---------+
       50       100       150       200       250       300
                           BILL
```

SUGGESTIONS FOR LEADING THE DISCUSSION

It is probably best not to ask the last question above, since this leads directly to the answer. Rather, you might ask for a show of hands of who believes that the variables DAYS and BILL are unrelated. Then ask who believes that they are related. Pick one person, or perhaps assemble a group of people, for each category and ask the groups to lead the class discussion for the day.

Everyone in the class may discover the key. If this happens, there is still a good chance that different students took different approaches to solve the problem. You might just ask, "Well, what did you do to account for the differences in the residential and commercial customers?"

TYPICAL RESULTS AND APPROACHES

The mean number of late days for the group as a whole is 50, but it is 31 for the residential customers, and 68.5 for the commercial customers. Your students may discover that the variable BILL is repeated and reordered for the residential and commercial groups; the mean and standard deviation for BILL are \$174 and \$78 for both the commercial and residential accounts. The regression equations for the overall sample and the residential and commercial subsamples are

		N	t-ratio	R^2	s
POOLED:	DAYS = 52 - 0.013 BILL	96	-0.4	0.002	23.7
RESIDENTIAL:	DAYS = 2.2 + 0.17 BILL	48	25.3	0.933	3.5
COMMERCIAL:	DAYS = 102 - 0.19 BILL	48	-31.8	0.957	3.2

CASE **33**

THE BRAZING OPERATION, INSTRUCTOR'S NOTE

Abstract: A case to be taught as a regression problem or as a process control problem or (better still) as both. By most measures, the process is out of control at the beginning. By running a regression relating the output to the inputs, we identify what seems to be causing the variability and quantify how much improvement could be obtained if certain inputs to the process could be better controlled.

Level:	Intermediate.
Techniques:	Statistical process control, run charts, multiple regression.
Typical Class Discussion Time:	45 minutes.
Data Set Size:	53 observations (time-sequenced, but unevenly spaced), 6 variables.

GENERAL COMMENTS

This case is attractive in several ways. You can view it as process control: Is grain count under control? (No, it isn't.) By using regression, you can relate the variation in output (grain count) to some of the other quantities, temperature and cooling rate, specifically. The temperature input, viewed as a process, is not under control, and this in turn sets the grain count process out of control. The student must define the "process": Is it the output, one or more inputs, or their combination?

QUESTIONS FOR ASSIGNMENT

The case asks only for recommendations and this may be too vague for some classes. A possible set of assignment questions might be:

1 Is the process under control? Why do you say so?

2 What factors appear to affect the grain count? If those factors could be controlled, would grain count be more stable? To what extent?

If you wish to provide even more structure:

3 Consider the process producing the grain count results. Is it under control?

4 Which of the variables TIMEAVE, TEMPAVE, COOLRATE, RUN, and/or VACLEVEL appear to be influencing the grain count? To what extent?

5 If these factors could be controlled, how much difference would it make?

SUGGESTIONS FOR LEADING THE DISCUSSION

"Well, is the process under control?" works as lead-in here. You should follow up whatever answer you get to make sure it's somehow justified. Usually, students have multiple views, based on different underlying definitions of the process; it's good to bring these out.

Most will run some sort of regression that links grain count to the other variables. When it's presented, follow up with some question like, "What does that model really tell us, though?" to be sure the implications of the model are clear (these inputs, if better controlled, would lead to reduced variation in grain size) and that it doesn't degenerate into just a mathematics exercise.

It's also a good chance to force students to consider the *residuals,* which measure the part of the process *not* explained by the input variables. It may not yet have dawned on the students that they could analyze *them* as a process, too.

If you want to make this case the basis of a mini-lecture, you could review the plots that follow, and follow through the process of observing changes in GRNCOUNT, then changes in TEMPAVE, plotting one versus the other and noting the relationship between them. The multiple regression that follows is then a natural model.

TYPICAL RESULTS AND APPROACHES

Most statistical process control approaches conclude the process is out of control. GRNCOUNT shifts downward after observation 18. An I-chart with the first 18 observations as its base shows the remaining observations with many out-of-control points, using usual kinds of criteria. From case 19 on, the process is much more nearly in control, so most of the problem seems to be associated with this shift.

TEMPAVE also shifts at observation 19, and this will usually suggest to students that they should examine the relationship between these quantities. The regression equation that results is:

$$\text{GRNCOUNT} = 623.470 - 0.02721(\text{COOLRATE}) - 0.29680(\text{TEMPAVE})$$

with R^2 of .92 and standard error of estimate of .28926 (compared to the original standard deviation of 1.00085). You can also make a marginal case for including VACLEVEL in the model. That case is weak, because it isn't corroborated by any external knowledge.

Students may have difficulty seeing the connection between the ideas of a process and the regression tools for examining relationships. Encourage them to define which process they're talking about: the *input* process, the *brazing* process itself (assuming fixed inputs) or the *combination.* The input processes are described by variables like TEMPAVE (which do not appear under control). The brazing process itself is described by the regression equation, and its variability is described by the residuals from the regression (the residuals, viewed as a process, are nearly under control, with no apparent autocorrelation or other structure).

Of course, the data don't form an equally spaced time series, and some students will (properly) object that control charts, etc., are not really applicable. In any case, they should be interpreted only approximately.

The change in temperature at observation 19 may have been an artifact of the measuring scheme: the engineers discovered a defective thermocouple and replaced it. The change may simply have been, in effect, a change in the calibration of the thermometer. The apparent relationship between TEMPAVE and GRNCOUNT puzzled the engineers, but regression omitting the first 18 points yields essentially the same relationship between GRNCOUNT and the other variables: the phenomenon wasn't *completely* an artifact.

Scatterplot of GRNCOUNT versus RUN

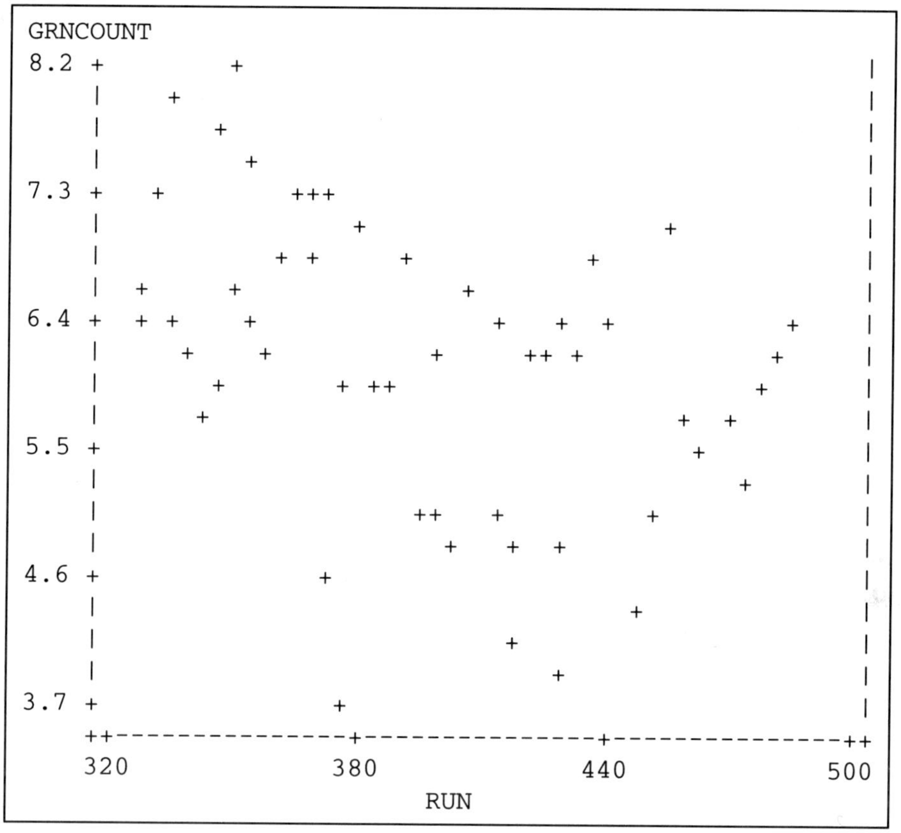

Scatterplot of TEMPAVE versus RUN

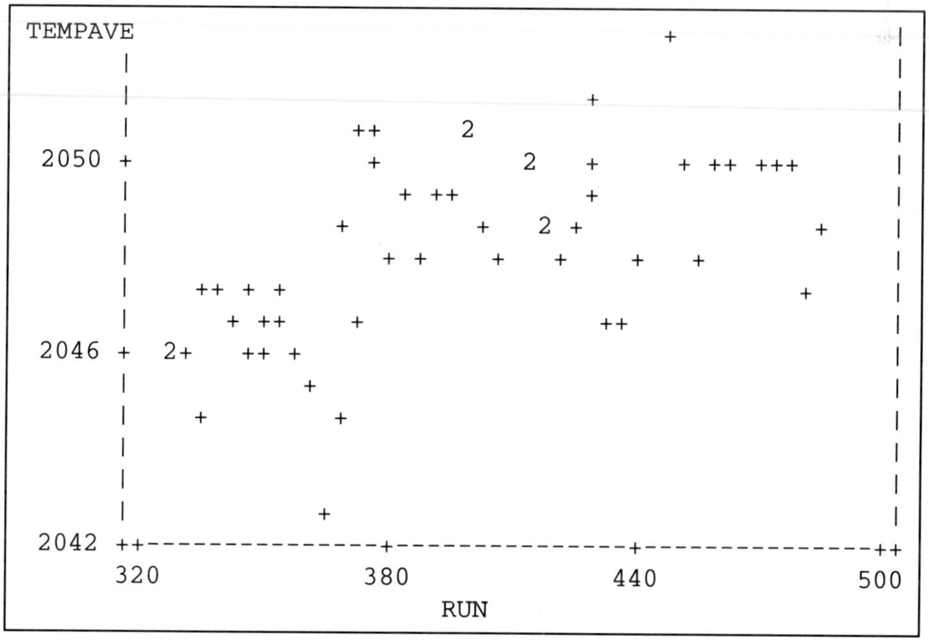

Regression of GRNCOUNT on COOLRATE and TEMPAVE

PREDICTOR VARIABLES	COEFFICIENT	STD ERROR	STUDENT'S T	P
CONSTANT	623.470	40.8736	15.25	0.0000
COOLRATE	-0.02721	0.00158	-17.27	0.0000
TEMPAVE	-0.29680	0.01998	-14.85	0.0000

R-SQUARED	0.9209	RESID. MEAN SQUARE (MSE)	0.08367
ADJUSTED R-SQUARED	0.9177	STANDARD DEVIATION	0.28926

SOURCE	DF	SS	MS	F	P
REGRESSION	2	48.7049	24.3524	291.06	0.0000
RESIDUAL	50	4.18345	0.08367		
TOTAL	52	52.8883			

Scatterplot of GRNCOUNT versus TEMPAVE

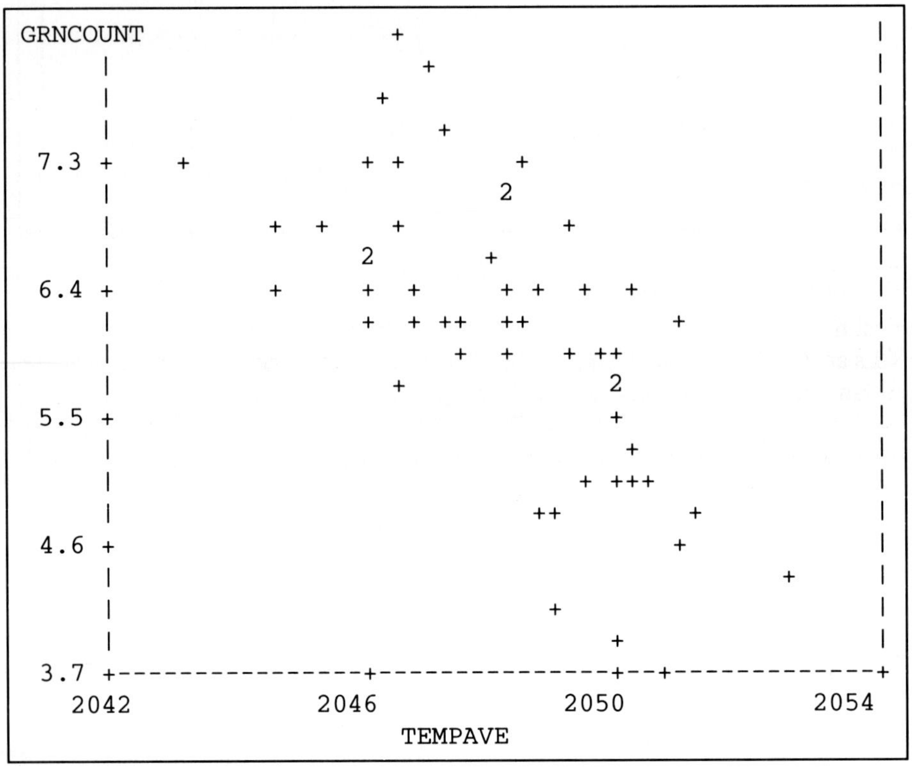

CASE **34**

LEADING INDICATORS, INSTRUCTOR'S NOTE

Abstract: This case will get your students to think about issues that arise in macroeconomic time series, such as seasonal adjustments and lagged variables. Students are asked to get updated data from the Web.

Level:	Intermediate, advanced.
Technique:	Forecasting, regression.
Typical Class Discussion Time:	30–45 minutes.
Data Set Size:	82 monthly observations, 7 variables.

GENERAL COMMENTS

We devised this case because our students seem to enjoy designing course projects that use macroeconomic data and need to know a bit about seasonally adjusted data and lagged relationships to do a reasonable job. This case will get them thinking a bit about some of these issues. It will also introduce them to leading indicators, a practical thing for all business students to know.

Web site URLs are provided for each variable in the data set, and the student is encouraged to update the time series. This will familiarize them with on-line data from government sources such as the Census Bureau and the Federal Reserve Board. If you ask your students to include recent data from the Web into their data sets, their answers will surely differ from the information below, which comes from the data available on the data diskette.

The data set contains information on seasonally adjusted and unadjusted data. Some of your students will devise forecasting models that use seasonally adjusted data as a predictor of its unadjusted analogue without thinking about the utility of such a model.

QUESTIONS FOR ASSIGNMENT

1 Before looking at the data, decide which of the variables in the data set might serve as leading indicators of Const_NA. Why do you think these factors are likely to be important predictors?

2 Which of the variables in the data set are *statistically* related to Const_NA? Const_SA?

3 Does it make any sense to consider lagged values of the predictor variables? What sort of statistical tools might help you uncover appropriate lag structures?

SUGGESTIONS FOR LEADING THE DISCUSSION

This case lends itself to comparing the different models developed by the students. Ask that students put competing models on the board. Some students will have modeled both Const_NA (the unadjusted series) and

Const_SA (the seasonally adjusted series). You could then ask who might be interested in models of the unadjusted variable (presumably business forecasters) and who might be more concerned with understanding the adjusted series (macroeconomists and policy makers, among others). You may also wish to prepare a mini-lecture on how and why macroeconomic data are seasonally adjusted. Students are likely to ask, so be prepared!

TYPICAL RESULTS AND APPROACHES

The unadjusted construction orders series is seasonal, as would be expected in this industry, and growing over time:

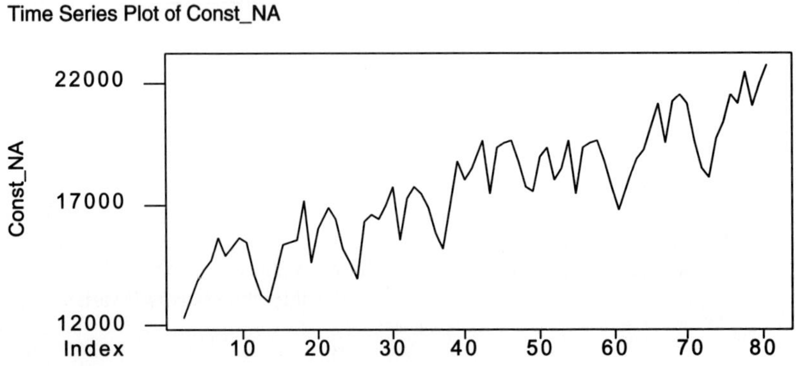

The seasonality adjustment does its job.

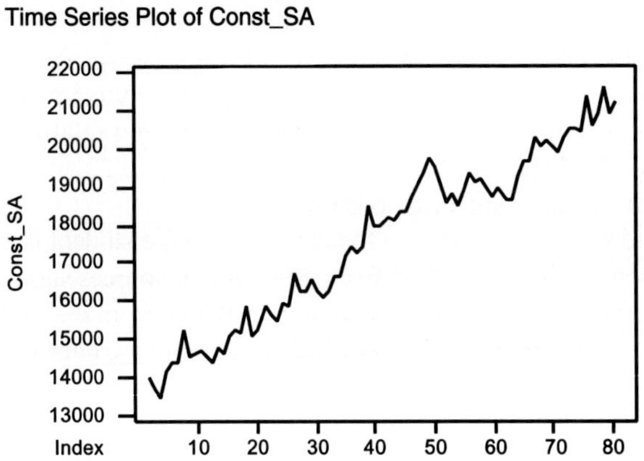

For these data, population is highly correlated with both Const_SA and Const_NA, but really serves as nothing more than a proxy for time. The mortgage rate displays an interesting structural change in its relationship with both Const_NA and Const_SA, as seen in the scatterplot of Cost_SA against MortRate:

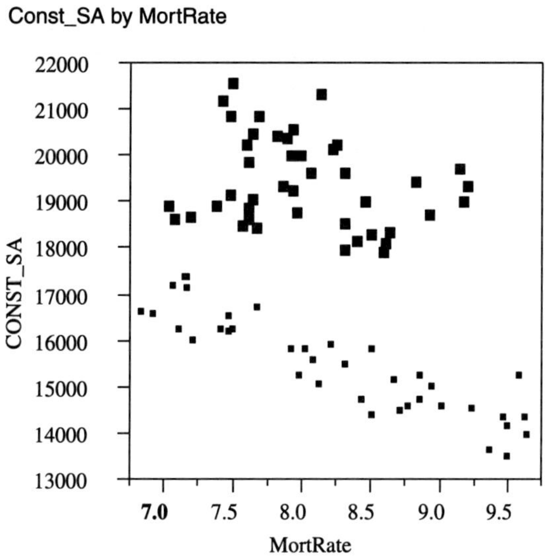

The upper cloud of points represents all observations after February 1994 when mortgage rates began to climb. Apparently mortgage rates could have been a useful leading indicator of construction activities prior to the second quarter of 1994, but appear to have little predictive ability thereafter.

Forward stepwise regression of Const_NA against PermitNA, its 12 lags, and of MortRate and its 12 lags, results in a model containing 2nd-, 5th-, 11th-, and 12th-order lags of PermitNA. Below we use $p<0.004$ as an entry significance level to protect against overfitting:

$$\text{Const\_NA}_t = 4171 + 20.4\ \text{PermitNA}_{t-2} + 35.6\ \text{PermitNA}_{t-5} + 32.4\ \text{PermitNA}_{t-11} + 45.9\ \text{PermitNA}_{t-12}$$

$R^2 = 0.89$, RMSE = 750.4

For the seasonally adjusted data:

$$\text{Const\_SA}_t = 3365 + 5.0\ \text{PermitSA}_{t-4} + 3.5\ \text{PermitSA}_{t-10} + 3.3\ \text{PermitSA}_{t-12}$$

$R^2 = 0.94$, RMSE = 468.6

Mortgage rates enter neither equation with such stringent entry requirements. Loosening the criterion will hint at a relationship between Const_NA and mortgages rates approximately one quarter later.

The autocorrelation function of Const_NA hints at a 12th-order relationship:

ACF of Const_NA

```
          -1.0 -0.8 -0.6 -0.4 -0.2  0.0  0.2  0.4  0.6  0.8  1.0
             +----+----+----+----+----+----+----+----+----+----+
 1   0.841                            XXXXXXXXXXXXXXXXXXXXXX
 2   0.734                            XXXXXXXXXXXXXXXXXXX
 3   0.666                            XXXXXXXXXXXXXXXXXX
 4   0.594                            XXXXXXXXXXXXXXXX
 5   0.560                            XXXXXXXXXXXXXXX
 6   0.563                            XXXXXXXXXXXXXXX
 7   0.508                            XXXXXXXXXXXXXX
 8   0.497                            XXXXXXXXXXXXX
 9   0.519                            XXXXXXXXXXXXXX
10   0.542                            XXXXXXXXXXXXXXX
11   0.564                            XXXXXXXXXXXXXXX
12   0.584                            XXXXXXXXXXXXXXXX
13   0.439                            XXXXXXXXXXXX
14   0.347                            XXXXXXXXXX
15   0.308                            XXXXXXXXX
16   0.253                            XXXXXXX
17   0.227                            XXXXXXX
18   0.249                            XXXXXXX
19   0.191                            XXXXXX
```

One possible autoregressive model for Const_NA is:

$$Const\_NA_t = 1111 + 0.26\ Const\_NA_{t-1} + 0.72\ Const\_NA_{t-12}$$

$R^2 = 0.90$, RMSE = 703.4

CASE **35**

INTERRUPTING TINA, INSTRUCTOR'S NOTE

Abstract: Many of the TQM tools can be applied to these data, once the students figure out what the interesting or useful question is.

Level:	Intermediate.
Techniques:	Control charts, Pareto diagrams.
Typical Class Discussion Time:	45 minutes.
Data Set Size:	No data set.

GENERAL COMMENTS

We have purposely not prepared a data file for this case. The students will be forced to decide what interesting information can or should be pulled from the worksheet. Do we want to study number of interruptions, or length of interruptions, by type? Do we want to study interruptions per day or per work session? Or maybe per week? What exactly should be considered an "interruption"? Ambiguity of this type should lead to lively classroom discussions of the case.

QUESTIONS FOR ASSIGNMENT

1 What do you think should be considered an "interruption"?

2 Prepare a summary of the number of interruptions by type. Should Tina be concerned about the number of interruptions or the length of interruptions?

3 Do you have any comments on the data collection? Any suggestions for improving future data collection exercises?

4 Construct a control chart of these data, and interpret this chart for Tina. What type of chart is appropriate? Having constructed and studied the chart, do you have any recommendations?

5 Construct a Pareto diagram of interruptions. Interpret this chart for Tina. Do you have any recommendations?

SUGGESTIONS FOR LEADING THE DISCUSSION

We suggest you begin the discussion with the first question above. Given the description of Tina's work responsibilities in the case, it could be that none of the incidents are controllable interruptions (with the possible exception of the personal phone call, and maybe helping someone move furniture). Or maybe scheduled appointments (types G and I) cannot really be considered interruptions. A fishbone diagram might help to clarify some of these issues.

TYPICAL RESULTS AND APPROACHES

Walk-in appointments are both the most frequent and biggest time-consuming incidents, but the case clearly states that walk-in students must be served on demand. Does this count as a controllable interruption?

Interruptions by Type

		Number			Minutes
F	walk-in	25	F	walk-in	381
D	phone (unsched)	16	G	appointment	270
G	appointment	14	H	meetings	80
E	clerical	7	E	clerical	78
A	SICS	6	D	phone (unsched)	63
B	coworkers	6	A	SICS	36
C	supervisor	4	M	other	25
I	phone (scheduled)	2	I	phone (scheduled)	22
M	other	2	C	supervisor	19
H	meetings	1	B	coworkers	15
	TOTAL	83			989

The U-chart on incidents per minute for all incidents recorded in the worksheet has points out of statistical control limits. Incidents per minute is tending downward over time, possibly because Tina's coworkers stopped interrupting so much when they noticed her recoding items on the worksheet.

U-Chart of Interruptions per Minute

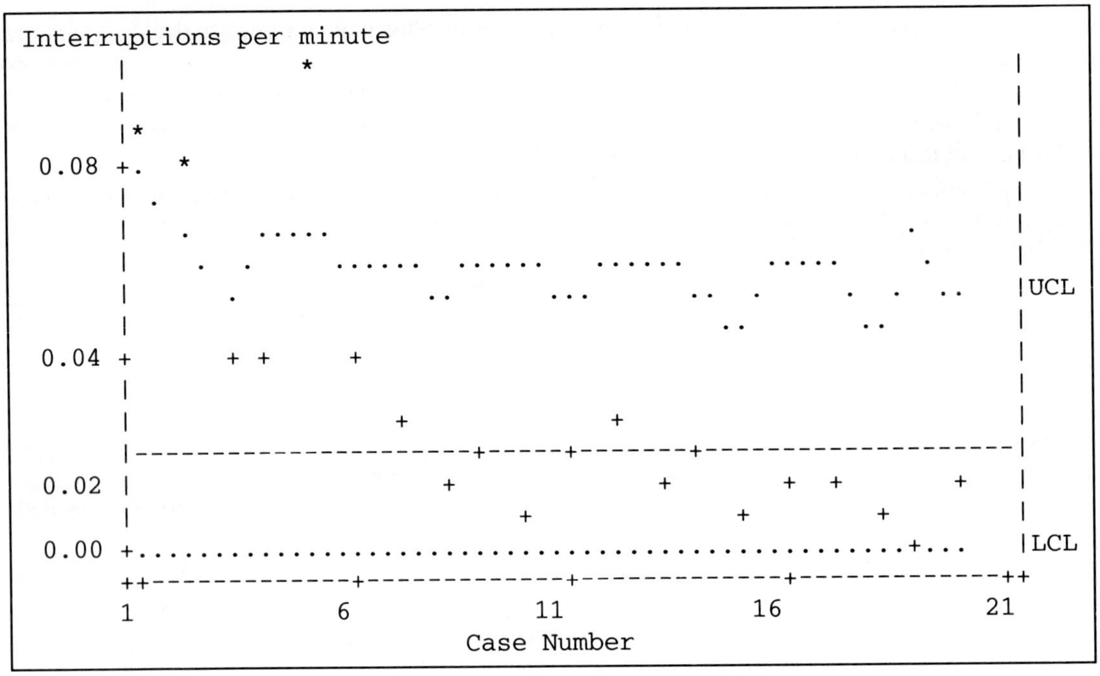

CASE **36**

VIOLENCE ON TV—II, INSTRUCTOR'S NOTE

Abstract: This case is about issues of methodology: parametric versus nonparametric tests, survey design, and unreliable test statistics, to name a few.

Level:	Intermediate.
Techniques:	Cross-tabulations, nonparametric methods.
Typical Class Discussion Time:	30–45 minutes.
Data Set Size:	94 cross-sectional observations, 8 variables.

GENERAL COMMENTS

This case calls for lots of careful thinking about what methodology should be applied to the data. Some students will have difficulty focusing on the analysis of data gathered in a way different from what they recommended in their responses to the first part of the case (and they may be right to so object). Others will apply parametric tests with no thought to appropriateness. Among those students who apply t tests of equality of means, some will test for equality of variances, and others won't.

Straightforward application of chi-square independence tests results in some expected frequencies less than five. Some students will report, and correct, this deficiency; others will not notice.

Some students will have important questions about the survey design. Others will ignore the survey and plunge right into data analysis.

You can use this case to uncover any or all of these issues. Careful organization and planning of the classroom discussion will be important to avoid turning it into a free-for-all.

QUESTIONS FOR ASSIGNMENT

1 What kinds of tests should be applied to these data to uncover differences in perceptions of violence?
2 Comment on the survey. Offer any suggestions for improving the survey design and implementation.

SUGGESTIONS FOR LEADING THE DISCUSSION

To begin this discussion, take the following descriptive statistics of VIOLENT categorized by AGE group, and say: "I was given the following information about our survey. Can you help me interpret this table?"

Expect a number of responses. Someone might say that the only two teenagers in the sample are both male and both rated TV violence as low (*not very violent* is 5). This can be used to foreshadow problems with a small sample size and chi-square tests. Someone might say that perception of violence increases with age group. You can then use this comment to uncover why comparison of means might not be appropriate here, and lead into nonparametric methods. Someone might say that it doesn't mean much of anything because the survey question about violence is so ambiguous, opening up a critique of the survey design. Let the discussion go where it will.

Breakdown of VIOLENT by AGE

	AGE	N	MEAN	MEDIAN	STDEV	MIN	MAX
VIOLENT	1	2	5.0000	5.0000	0.000	5.0	5.0
	2	36	2.917	3.000	1.228	1.0	5.0
	3	31	2.419	2.000	1.259	1.0	5.0
	4	19	1.947	2.000	1.026	1.0	5.0
	5	6	2.167	2.000	1.329	1.0	4.0

As you get into the hypothesis tests, stop the discussion and have the class decide on a significance level. (Some of the results differ qualitatively depending on the chosen significance level.) Later, if you want, return to this issue with something like: "How would your recommendations change if we had chosen a .10 significance level rather than a .05 significance level?" and, "How am I supposed to explain this discrepancy to the manager of KTDS?"

TYPICAL RESULTS AND APPROACHES

Sixty-five percent of the sample is female. Tidusville is a well-educated community, if the sample is representative, since 23% of the sample attend(ed) graduate school. It does appear that there are some concerns about violence on TV, as suggested by this frequency distribution of VIOLENT.

Frequency Distribution of VIOLENT

			CUMULATIVE	
RESPONSE	FREQ	PERCENT	FREQ	PERCENT
Much too violent	26	27.7	26	27.7
Somewhat too violent	22	23.4	48	51.1
Violent	21	22.3	69	73.4
A little violent	18	19.1	87	92.6
Not very violent	7	7.4	94	100.0
TOTAL	94	100.0		

Chi-square tests of independence between VIOLENT and the other variables result in some cells with expected values less than 5. Only seven respondents rated VIOLENT = 5; combining responses of 4 with responses of 5 for VIOLENT results in rejection of independence between VIOLENT and GENDER (P = 0.012), VIOLENT and CHILDREN (P = 0.045), and nonrejection of independence between VIOLENT and MARRIED (P = 0.135) at the 5% significance level. Further groupings are required for the remaining crosstabs to avoid expected cell frequencies of less than 5.

Parametric and nonparametric tests (two-sample t tests, rank sum tests, median tests) find differences in perceptions of violence by GENDER and no difference by marital status. The outcome of CHILDREN depends on which test is applied; the t and rank sum tests reject the hypothesis of no difference, but the median test fails to reject it at the 5% level.

Kruskall-Wallis finds differences in perceptions of violence by AGE group, education level (SCHOOL), and HOURS of TV watched at the 1% significance level. No difference arises by INCOME group.

The survey itself has some problems. Your students may note, as we did, the ambiguity of the VIOLENT question: How does "somewhat too violent" differ from "a little violent"? Is there a better way to ask this question of perceived violence? Single and divorced are lumped in the same category. Maybe these two groups have different perceptions of violence.

CASE **37**

COMMERCIAL PROPERTY, INSTRUCTOR'S NOTE

Abstract: We use this case to demonstrate the existence of, influence of, and solutions to multicollinearity.

Level:	Advanced.
Techniques:	Regression.
Typical Class Discussion Time:	45 minutes.
Data Set Size:	75 cross-sectional observations, 11 variables.

GENERAL COMMENTS

There is no relationship between the age of the property and selling price, which is somewhat surprising (although the mean year built is 1976 with a typical fluctuation of only 10 years). Location is relatively unimportant as measured by the variable TOWNSHIP. Year and month of the sale of the property are not significant (SYEAR and SMONTH). The primary use of the property and the quality of the construction can be important.

Overall size of the building (measured either by square feet, number of floors, or land size) is the important factor. The case is designed to concentrate on which of these three measures of size is best.

Certain data limitations exist. Of the 75 properties in the data set, only two are rated as GOOD in the quality of their construction and these two properties are both higher priced and larger than most others. There are only two hotels in the data set, both of which are large properties.

CLASS and DESIGN are very similar. Because of this, we concentrate on the CLASS variable (because it is easier to work with), but you might find some useful information in the DESIGN variable as well.

QUESTIONS FOR ASSIGNMENT

1 What are the primary determinants of selling prices of commercial property in Arapahoe County? What impact do these important factors have on selling price? Your answer should be a numerical one.

2 Are the three variables LANDSIZE, SQFEET, and FLOORS somehow behaviorally or statistically redundant to this problem? If so, what harm does this redundancy inflict? If this redundancy is a problem, how would you suggest solving it?

3 What is the importance of CLASS, DESIGN, location (TOWNSHIP), and quality of construction (QUALCON) on the selling price of property in Arapahoe County? Give a numerical answer.

4 Construct a model of selling prices of property in Arapahoe County, and show how this model might be used to predict other selling prices. How might your model be used to determine the accuracy of assessed values, if at all?

SUGGESTIONS FOR LEADING THE DISCUSSION

Begin the discussion by saying, "What is the sign of the relationship between PRICE and SQFEET?" Everyone should say this is positive. Then ask the same question about PRICE and LANDSIZE. If you are lucky, you will get some disagreement. Most students will say positive (based on the correlation or slope in the simple regression), while others will say negative (based on the sign of LANDSIZE in a multiple regression model). Focus the discussion on this discrepancy.

Another way to start the discussion is to ask how much you should raise the selling price of the property for an extra floor. If you're lucky, someone in class will say over a million dollars (referring to the simple regression of PRICE on FLOORS). Then ask how much a typical property sells for ($775,000) and ask someone to explain how an additional floor costs more than a million dollars whereas a typical building only sells for about three quarters of a million dollars. You can then talk about the relationship among FLOORS, SQFEET, and (maybe) LANDSIZE, and get the discussion of collinearity going that way.

TYPICAL RESULTS AND APPROACHES

The following paired correlation coefficients apply to these data.

	PRICE	LANDSIZE	SQFEET
LANDSIZE	.675		
SQFEET	.901	.871	
FLOORS	.676	.372	.628

The series of simple and multiple regression models below indicate classic symptoms of multicollinearity (*t* ratios are parenthesized).

Regressions of PRICE on Various Factors

Model	R^2
PRICE = -226359 + 51.3 SQFEET (17.8)	81.2%
PRICE = -128837 + 17.1 LANDSIZE (7.8)	45.6%
PRICE = -881639 + 1062192 FLOORS (7.8)	45.6%
PRICE = -57843 + 73.9 SQFEET (14.6) - 11.5 LANDSIZE (-5.1)	86.2%
PRICE = -185482 + 69.4 SQFEET (10.2) - 10.3 LANDSIZE (-4.1) + 99150 FLOORS (1.0)	86.4%

The wide swings in the regression coefficients, the negative sign on LANDSIZE in equations with SQFEET, and the insignificance of FLOOR when combined in an equation with SQFEET are all indications of multicollinearity. Since square footage is the strongest variable, as measured by the goodness-of-fit statistic, it makes some sense to use this quantity alone in subsequent analyses.

A second approach to solving the multicollinearity problem, and one that would immediately come to mind to the experts in this business, is to convert the dependent variable PRICE into price per square foot. This automatically eliminates the need to worry about size and permits students to concentrate on other aspects of the problem. Simple regressions of PRICE PER SQUARE FOOT on LANDSIZE and on FLOORS are statistically insignificant.

The following regression model might be used to price property in Arapahoe County, but note the standard error of the regression equation. The variables except SQFEET are binary variables created from the categories defined by CLASS and QUALCON.

Regression of PRICE Using Dummy Variables

```
PRICE= 3454999+41.4 SQFEET-1450274 MERCHAN-1225508 OFFICES-1551508 SPECIAL
                (12.49)       (-2.33)            (-2.10)            (-2.64)

          -1646422 WAREHOUS -1680954 MULTI -1683036 MANUFACT -2001648 FAIR
           (-2.77)              (-2.72)          (-2.58)             (-3.20)

              -2157370 AVERAGE
               (-3.63)

        s = 648243       R-sq = 87.5%      R-sq(adj) = 85.8%
```

CASE **38**

AIDS RATES, INSTRUCTOR'S NOTE

Abstract: An exercise in using the World Wide Web to extract and evaluate information. It also elicits discussion about intuitive impressions compared with objective reality and about dealing with data on sensitive issues. Forces consideration of the underlying data gathering process and the quality of the data obtained. A good case to assign early in a course.

Level:	Beginning.
Techniques:	Descriptive statistics.
Typical Class Discussion Time:	45 minutes.
Data Set Size:	No data set provided.

GENERAL COMMENTS

This case offers drill and practice in using the World Wide Web, and offers the students a chance to check their intuitions about rates with the more objective counts and rates provided by government agencies such as the Centers for Disease Control and Prevention. Neither formal technique nor business knowledge is required. Computing a few percentages and plotting or tabulating the results will handle the summary work here, but to answer the underlying question, the student must think hard about the underlying data-gathering process and the quality of the data. It's a good case to assign early in the semester.

QUESTIONS FOR ASSIGNMENT

1 What trends do you find cases of AIDS in the US?

2 How close were your intuitions (or those of the people in your survey) to the results in the data? How much variation was there in your survey results?

3 Provide an appropriate summary report of your survey results and those from the government data.

SUGGESTIONS FOR LEADING THE DISCUSSION

Data from the World Wide Web varies dramatically in quality. One way to begin discussion of this case is simply to ask for summaries of the data obtained, and then probe a bit for the quality of the data. The underlying issues are the same whether the data comes from the Web or from a book in the library, but custom suggests that people think harder about data from the latter, while data obtained electronically isn't viewed with as much suspicion as it probably should be.

Another possibility is to focus on the variation in survey results. In typical situations, when asked what the rate of AIDS is, "show-of-hands" surveys in class yield answers from 1 in 100 to 1 in 100,000,000. The mere range of variation is a key point here: it's why we need data rather than just intuition.

Finally, you can ask for appropriate graphs or other summaries to communicate the results.

TYPICAL RESULTS AND APPROACHES

Some students will confuse HIV with AIDS, and this may take some discussion. The fact that the definition of AIDS has changed over time can be used to focus attention on such issues in ascertaining trends. Data from various advocacy groups may be offered as results from the Web search. This can be used to direct attention to the problems of assessing data quality—at least in principle, we can find how CDC goes about its business. Other sources vary from the careful to the irresponsible. Most students will provide some sort of table or plot showing trends in the AIDS rates.

At this writing, the prevalence of AIDS is of the order of magnitude of 1 in 1000 in the USA. In parts of Africa it approaches one in three (but even there, more lives are lost to river blindness and malaria than to AIDS—if you start on this line of discussion, it leads to questions of individual fears versus population-based public health issues, and it can get complicated).

CASE **39**

SOFTWARE SERVICE CALLS, INSTRUCTOR'S NOTE

Abstract: A straightforward case for those with the discipline to explore it. Some simple scatterplots reveal most of what's going on, as long as you ask the right question (consider possible lagged variables as predictors). It's elementary, if you use informal techniques; intermediate, perhaps, if you use more formal time series or regression methods.

Level:	Beginning, intermediate.
Techniques:	Scatterplots, auto- and cross-correlation, regression.
Typical Class Discussion Time:	30 minutes.
Data Set Size:	18 time series observations, 4 variables.

GENERAL COMMENTS

This is a real situation, one in which a simple solution fits the needs of the business pretty well. The key is to consider whether lagged values of quantity sold, etc. can predict the number of service calls. Simple scatterplots then reveal the answer, though the cross-correlation function will do it, too.

QUESTIONS FOR ASSIGNMENT

1 What should Sam report to his supervisor?

If you want to provide more structure, you could use the following sequence, or something like it:

2 How variable is the number of service calls with which Sam's department must deal?
3 Using the data available, could Sam predict it any better? How? How much better?

If you want to give the whole thing away, you could ask:

4 Derive a regression model for SCALL by using the variables given and/or their lags.

SUGGESTIONS FOR LEADING THE DISCUSSION

The same questions given above work pretty well here. You should follow up on any models suggested, asking for justification. None of them are very complicated in this case. The key to this case is noticing that it's time-sequenced data, and you should consider lagged variables as predictors. Everything gets simpler after that, so we've labeled this beginning level, if you provide hints, or intermediate level, if you don't.

TYPICAL RESULTS AND APPROACHES

The volume of service calls doesn't appear to be related to the quantity sold or to the number of STRs or to MONTH in any useful way, but it does appear to be related to the quantity sold two months previously (Q lagged 2), as the cross-correlation plot that follows suggests. You might or might not want to display a plot like this, depending on how much time series material you plan to cover. The regression equation that comes from this approach (or from simply trying out a few different lags) is:

$$\text{SCALL} = 71.8791 + .53753(\text{Q, lagged 2 months})$$

which gives $R^2 = .79$ and a standard error of estimate of 6.54 (compared with the original standard deviation of 13.67). So Sam could be doing at least a little better in predicting than by using the average. Whether that can be translated into effective staffing patterns depends on other things, too, of course—how long does it take to train the staff on the product, how often is it practical to rearrange their schedules, and so forth—and this is perhaps worth bringing out in the discussion. Any equation like the one above is just a starting point for Sam, but hopefully it's a better starting point than he has now.

It may be worth pointing out that for complex software (the package is described as a "fairly large package for CAD . . .") it would be reasonable to expect that it could take two months for users to install and begin to use a new system. A wave of questions might be expected as the customers begin to use the product more intensively.

Cross-Correlation of SCALL and Q

```
              -1.0 -0.8 -0.6 -0.4 -0.2  0.0  0.2  0.4  0.6  0.8  1.0
LAG   CORR.     +----+----+----+----+----+----+----+----+----+----+
  -9  -0.028                            **
  -8  -0.071                           ***
  -7  -0.216                        ******
  -6  -0.053                            **
  -5  -0.181                        ******
  -4  -0.268                      ********
  -3   0.059                             **
  -2  -0.395                   ***********
  -1   0.287                             ********
   0   0.179                             *****
   1  -0.293                      ********
   2   0.826                             **********************
   3  -0.128                          ****
   4   0.134                             ****
   5   0.422                             ************
   6  -0.120                          ****
   7  -0.053                            **
   8   0.058                             **
   9  -0.026                            **
```

CASE **40**

TITANIC SHIPPING COMPANY, INSTRUCTOR'S NOTE

Abstract: A case using descriptive statistics, means and percentiles. Nothing fancy, but it corresponds to a real problem in practice. Apparently *average* response time doesn't mean much to the computer users, but something like the 75th percentile might.

Level:	Beginning.
Techniques:	Descriptive statistics, percentiles, scatterplots.
Typical Class Discussion Time:	30 minutes.
Data Set Size:	63 time-sequenced cross-sectional observations, 8 variables.

GENERAL COMMENTS

Simple descriptive statistics work fine here. Judging by the users' reactions over the last 13 weeks, it isn't very important to them to have the *average* response time less than 2 seconds (although they might be impressed by an average less than perhaps 1.5 seconds). Percentiles of the response time seem easier to relate to the users' views. For example, if the 90th percentile is less than about 3.3 seconds, the users rate the performance "OK," while as soon as it gets even slightly worse than that, they pan it.

This case combines some simple ideas about using percentiles rather than averages with some very real problems in service organizations: How can you relate users' often informal impressions of quality to something you can really measure and track?

QUESTIONS FOR ASSIGNMENT

1 How often during the past 13 weeks did the computer center meet its stated performance target of an average response time less than two seconds?

2 How often were the users satisfied with the response time?

3 Suggest an appropriate measure of aggregate response time for the computer center manager.

4 Using your proposed measure, suggest a standard level at which the users will be satisfied.

SUGGESTIONS FOR LEADING THE DISCUSSION

If you use the kind of assignment questions given above, most classes should be able to respond well to a general opening for the discussion. Something like, "Well, do the users have a point here?" or, "What do you think the manager should do?" might start things off. If you don't give assignment questions, you might use the assignment questions as discussion guides.

TYPICAL RESULTS AND APPROACHES

Various relationships become apparent through looking at some simple plots. Most students will compute something like POK (the fraction of users rating the service as "OK" in the plot below). They discover quickly that when response time (as measured by any of the various quantities in the data set) exceeds a certain point, POK plunges quickly to unacceptable levels. The point at which it begins its plunge is probably a good candidate for the manager's new proposed standard level of service. Students may also note that the performance deteriorates as the number of total users (NBAD + NOK) increases, which is not surprising.

The better measures of response time seem to be the percentiles, such as the 75th or 90th. The plot below gives the result for the 90th percentile. The decrease in approval is sharper for percentiles than for the average. Such a result is typical in response time studies: to know that (say) 9 times out of 10, the response time will be less than some amount is more important than to know the average. It's what you can *count* on that matters.

Titanic Shipping Company, Response Time Survey

```
POK: Fraction of Users Rating Response Time as OK
1.00+
    ******  ** *
    |       * *
    |
    |           *
0.67+
    |            *
    |            *
    |            *
    |
0.33+             *
    |              *
    |
    |               *
    |                          *     *     *
0.00+---------+--------***-*--****+-***-*-*****-*-****+
 2.49      3.38      4.27      5.16      6.05      6.94
  RESP90: 90th Percentile of Response Time (Seconds)
```

CASE **41**

CAMPAIGN STATISTICS, INSTRUCTOR'S NOTE

Abstract: A classic case of Simpson's paradox: within each income category, the effective tax rates actually went down, though in total, the rate went up. A good case to assign early, as it requires no special techniques and makes students think hard about ratios and their comparison.

Level:	Beginning.
Techniques:	Ratios, tabulations.
Typical Class Discussion Time:	20 minutes.
Data Set Size:	5 categories, 4 numbers in each.

GENERAL COMMENTS

For the total rate to go up when all the individual category rates went down, the mix had to shift. Presumably because of inflation, a higher percentage of people had incomes in the higher categories, which have higher tax rates. This is a good case to get students thinking about ratios and what goes into them. It sometimes provides a useful shock to those whose intuitions suggest such an anomaly can't happen.

QUESTIONS FOR ASSIGNMENT

Which category should the congressman cite in his upcoming campaign?

SUGGESTIONS FOR LEADING THE DISCUSSION

You can begin this one with the assignment question. Someone will normally point out that, well, er, actually *none* of the categories is helpful to the congressman's cause. That can lead into a discussion of whether it really is possible that the rates in *all* categories went down, while the aggregate rate went up. "How can that happen?" may serve to start the follow-on discussion. "How can we summarize all this?" can help to bring students to realize that there are really two forces at work here: the individual category rates and the mix between the categories. Probably any attempt to portray them both in the same table or chart will distort things, so students can learn not to oversimplify.

TYPICAL RESULTS AND APPROACHES

The table on the next page contains the actual rates, which reveal pretty much everything, as long as you're willing to look at the details. Most students have a hard time summarizing and portraying these data.

Total Income and Total Tax (in thousands of dollars)

Adjusted gross income	1974 income	1974 tax	1974 effective tax rate	1978 income	1978 tax	1978 effective tax rate
Under $5000	$ 41,651,643	$ 2,244,467	.054	$ 19,879,622	$ 689,318	.035
$ 5000 to $9999	146,400,740	13,646,348	.093	122,853,315	8,819,461	.072
$ 10,000 to $14,999	192,688,922	21,449,597	.111	171,858,024	17,155,758	.100
$ 15,000 to $99,999	470,010,790	75,038,230	.160	865,037,814	137,860,951	.159
$100,000 or more	29,427,152	11,311,672	.384	62,806,159	24,051,698	.383
Total	$880,179,247	$123,690,314	.141	$1,242,434,934	$188,577,186	.152

CASE **42**

HOSPITAL CHARGES, INSTRUCTOR'S NOTE

Abstract: What is the appropriate variable to be studied here—DAYS, CHRGS, or dollars per day? The results of the hypothesis tests depend on how your students answer this question. This is due, in part, to an outlier in the data set.

Level:	Beginning, intermediate.
Techniques:	Tests of hypotheses, confidence intervals.
Typical Class Discussion Time:	15 minutes.
Data Set Size:	289 cross-sectional observations, 4 variables.

GENERAL COMMENTS

Your students will find statistically significant differences between insurance carrier types and among physicians for DAYS and CHRGS. Charges per day ($/DAY below), which standardizes for the 14-day stay of the patient under Doctor 2's care, demonstrates no statistically significant difference between payor type or physicians in this data set.

Thus, the statistical evidence depends on whether the question involves dollars per stay in the hospital or dollars per day spent in the hospital.

QUESTIONS FOR ASSIGNMENT

1 Do charges incurred by a patient depend on the type of insurance the patient has? If so, how?
2 Do charges incurred by patients depend on which doctor treats them? If so, how?
3 What method would you recommend for estimating the total charges that will be incurred by a patient?

SUGGESTIONS FOR LEADING THE DISCUSSION

To begin the discussion, try this. Take to class an overhead of the printout below, and say: "Boy, am I steamed. I just got these data from the statistics department, and I'm going to fire Doctor 2!"

Students who spent a moment looking at the raw data before tackling the problem most likely anticipated the issue of $/stay versus $/day. The 14-day visit under Doctor 2 really stands out either by scanning the data or by examining a histogram. There is no right answer to which measure is best, but you should have a lively classroom discussion on the question of $/stay versus $/day.

Summary of PHYS by CHRGS

	PHYS	N	MEAN	STDEV
CHRGS	2	37	3489	2083
	4	21	2611	528
	6	25	2857	756
	7	40	2559	632
	10	54	2786	566
	11	30	3029	1147
	12	34	2787	785
	13	20	3125	1120
	14	28	2742	559

You can also have an interesting methodological discussion if you asked your students to create a model for forecasting charges incurred. The regression of CHRGS on DAYS gives different estimates from the simple average of $/DAY. Ask your students what their estimate would be of charges per patient and come prepared to play the devil's advocate role. Ask why the two methods provide such different answers.

TYPICAL RESULTS AND APPROACHES

Doctor 2 has the highest average number of days (2.7), but this physician has that one 14-day patient. On the other end of the spectrum, Physician 7 has patients who stay on average 1.97 days. Charges per day range from \$1260 (Doctor 11) to \$1491 (Doctor 13), but charges per stay range from \$2559 (Doctor 7) to \$3489 (Doctor 2).

Managed care patients (PAYOR = 1) incurred higher average charges (\$2966) per stay than did commercial insurance carrier patients (PAYOR = 0 at \$2714/stay), but on a per-day basis, the rank is reversed (PAYOR = 1 averages \$1357/day while PAYOR = 0 averages \$1434/day).

The *t* test for the difference between means by PAYOR rejects the equality of the means at the 5% significance level for DAYS and CHRGS (*t* ratios are 2.85 and 2.29 for DAYS and CHRGS, respectively). The same test fails to reject differences in means of $/DAY by PAYOR with a *t* ratio of −1.36.

The one-way ANOVA test for the equality of means by physician rejects the null hypothesis of equal means at the 5% significance level for DAYS and CHRGS, but not for $/DAY.

The regression equation of CHRGS on DAYS estimates a fixed fee of \$931 (the *y*-intercept) plus \$844 per day (the slope).

CASE **43**

COAL TESTING, INSTRUCTOR'S NOTE

Abstract: This case works well with two-way ANOVA. There are substantial effects due to laboratories, and basically none due to batches. There's enough data to check for interactions, of which there are a couple. The student can summarize and interpret the results in the original setting fairly quickly—as long as he or she understands what interaction means.

Level:	Intermediate.
Techniques:	Two-way ANOVA, multiple regression.
Typical Class Discussion Time:	30–45 minutes.
Data Set Size:	50 cross-sectional observations, 4 variables.

GENERAL COMMENTS

The abstract describes this one pretty well. You can do the arithmetic either by two-way ANOVA or by multiple regression with dummy variables, as you prefer. The main value of the case is that it makes students recognize an ANOVA problem, check for interactions, and interpret the results back in the original setting. Which laboratories are doing what? What should we conclude?

QUESTIONS FOR ASSIGNMENT

If you feel the need for more structure than the case provides, you might try assigning questions like:

1 Do there appear to be any substantial variations from laboratory to laboratory? From batch to batch?
2 What action do you recommend for the coal company?

SUGGESTIONS FOR LEADING THE DISCUSSION

We suggest starting with the assignment questions above. For each conclusion, ask for justification. At some point, ask the students to describe their conclusions in the form of a model. Those who have treated it as a problem in multiple regression will be a step closer to this than those who used a straight ANOVA approach, but encourage the latter to think in such terms, so that they understand the underlying model.

Once they've presented their models, be sure they've considered possible interactions. You should discuss their interpretation, too.

Finally, ask how we could best summarize our conclusions to present to the vice president, who never took a statistics course. Probably some sort of plot or table of means or maybe a short equation will result from this. Many students have difficulty with interactions, and this is a chance to make them connect the computer output with the original problem. Alternatively, show the scatterplot that follows and lead from that into a discussion of the ANOVA or regression printouts, showing how they offer a more detailed (refined?) description of the general pattern shown in the plot.

TYPICAL RESULTS AND APPROACHES

Two-way analyses of variance (or the equivalent regressions) without interactions show that Laboratory 3 is consistently giving readings lower than the others, and not much else is going on. The standard error of estimate is about 77, so it doesn't look as though successive readings would reliably fall within the plus or minus 100 range the national standard requires, even after removing laboratory and batch differences.

Including interactions gives a little more complicated picture. Laboratory 3 is still consistently low and Laboratories 1 and 2 are more or less consistent. Laboratories 4 and 5 are erratic: high on one batch, but low on the other, and more than just chance. Students might be encouraged to speculate on how this might happen—major changes in procedures between the processing of the two batches? The case provides little guidance here.

Even with interactions included, the standard error of estimate is about 50. Probably about on the standard, since about 95% of the repetitions would fall within the published plus or minus 100 range, but this is after removing all effects due to differences in laboratories and batches, so it's probably nothing to feel very good about. Indeed, the engineer who reported this to us said: "Quality control and round-robin testing of control samples have become standard operating procedures since . . . this data was collected. The ASTM changed several of its coal-testing specifications and standards as a result of this study."

Scatterplot of BTU versus LABNO

```
BTU
      +-----------------------------------------+
11800 +                             +          +|
      |                             +           |
      |+         +                  +          2|
      |                                        2|
      |2         +                              |
11700 +                                         |
      |3         +                              |
      |4         +                   2          |
      |          +         +                    |
      |          2         +         2          |
11600 +          +         2                   3|
      |          2         2         +         +|
      |                              +          |
      |                              +         +|
      |                                         |
11500 +                                         |
      |                    2                    |
      |                                         |
      |                                         |
      |                    2                    |
11400 +                                         |
      ++---------+---------+---------+---------++
       1         2         3         4         5
                         LABNO
```

CASE **44**

PERFORMANCE EVALUATIONS, INSTRUCTOR'S NOTE

Abstract: Lots of interesting issues arise in the case. As with many of our cases, though, there are questions that cannot or should not be addressed with this data set. The data are amenable to numerous statistical methodologies, including two-sample means tests, ANOVA, influential observations, and multiple regression.

Level:	Advanced.
Techniques:	Regression, tests of hypotheses.
Typical Class Discussion Time:	60 minutes.
Data Set Size:	144 cross-sectional observations, 10 variables.

GENERAL COMMENTS

This case uses a wide range of statistical methodologies and concepts. An influential observation is partly responsible for the perception that lag time relates to score. Analysis of variance and *t* tests of differences in means can be used to look at categorical properties of SCORE. Multicollinearity confounds the issue of a potential bias in evaluations against longer-time employees. This case is very comprehensive, and is one of the more difficult cases in this volume.

All things considered, we would have to conclude that the evaluation system is working relatively effectively. No really important differences between departments or grades or educational levels arise. Lower scores tend to be associated with employees who are no longer with the firm. We would be concerned if there was no statistically significant difference in scores between the current and former employees; that might suggest that some of the good employees are leaving the organization.

One recommendation to the firm might be to lighten up a bit on the three-month evaluations. One regression that we estimated suggests that three-month evaluations are lower on average by 13 points; this could be enough to throw an employee into the next lower raise category, all else equal.

QUESTIONS FOR ASSIGNMENT

The interesting questions that arise in the data set are the five questions listed in the case. One further issue, hinted at in the last paragraph of the case, is whether there is any difference in SCORE depending on whether the employee is still with the company at the end of the year. Additional questions for assignment might include the following.

1 Perform the hypothesis test that average scores do not differ by department. Do the same test for GRADE and TYPE.

2 Are there any differences between those employees who were still at XYZ at the end of the year, as opposed to those who left the company?

SUGGESTIONS FOR LEADING THE DISCUSSION

This case is so varied and comprehensive that there are numerous ways to lead a classroom discussion. One method, which is fairly straightforward, is to assign groups of students to provide answers to each of the five questions posed in the case. This might occur in informal responses from their seats, or you might also ask them for formal presentations. Because there are so many different ways to look at each of these issues, you are likely to get varied reactions from the others in the class.

Another fun method of doing this case is to split the class into two groups. One group is asked to present data supporting the five complaints listed in the case (this might be the employees' union representative, for example), and the other group is to be management determined to show the unbiasedness of the evaluation system. This approach can provide a great deal of entertainment, if everyone realizes it is all intended in a good spirit, while also reinforcing the notion that we can often see both sides of an issue in any one given data set.

TYPICAL RESULTS AND APPROACHES

The average score is 134 with a typical fluctuation of 25 points. Fifty percent of the observations are annual evaluations; 50% are three-month evaluations. The employee base is relatively new to the company. The average number of years working for the company at the time of the evaluation is 3.5 years, but there is much variability here: a typical fluctuation of 4 years around the average!

Fifty-one of the 144 employees have two years of education beyond high school, but have not yet earned a bachelor's degree; this is the largest group in terms of education. There is only one Ph.D. in the sample. Department 62 has the largest number of observations in this sample with 27 employees. Departments 70 and 40 have, respectively, only one and two employees each in the sample. They may be underrepresented here.

The average score drops by 18 points from those who currently work for the company to those who have left the company, and the difference is statistically significant. Annual evaluations are typically higher by 22.5 points over three-month evaluations, another statistically significant difference. Differences in mean scores by DEPT, GRADE, and EDUC are not statistically significant. The ANOVA results are suspect, though, because of small sample sizes in some categories.

Typical answers to the five specific issues in the case are suggested below:

a *Is there a relationship between LAG and SCORE?* This will take a little bit of thinking about what is being recorded in the variable LAG. If LAG is a negative number, the manager spoke to the employee that number of days after the due date. If LAG is positive, the manager completed the evaluation and spoke to the employee before the deadline. The correlation between SCORE and LAG is −0.15.

It appears, though, that the employees don't need to worry about this issue too much, since removal of an influential observation forces LAG to be statistically insignificant. The statistically significant, negative coefficient for LAG might be related to one influential observation. Observation 133 in the data set is probably wrong, and certainly odd, since it suggests that some employee who was not in the job at the end of the year, and who was having a three-month evaluation, had his or her evaluation 87 days before the due date, which was about three days after beginning the new job. Removing this observation from the data set in the second of the two regressions below causes LAG to become statistically insignificant. Here are two multiple regression models that show this point. The first regression below uses all 144 observations; the second excludes observation 133. The *t* ratios are in parentheses, and HERE? is a dummy variable indicating whether the employee was (HERE? = 1) or was not (HERE? = 0) employed at the end of the year. Some of the observations for LAG are large, because evaluations are sometimes done before an employee goes on vacation or on a lengthy business trip.

```
SCORE = 112 - 0.236 LAG - 1.83 EMP@EVAL + 3.77 POS@EVAL + 13.0 TYPE + 18.8 HERE?
              (-2.51)     (-2.53)          (3.74)           (3.03)       (3.75)

              s = 20.21         R-sq = 34.6%      R-sq(adj) = 32.2%

SCORE = 115 - 0.140 LAG - 1.77 EMP@EVAL + 3.61 POS@EVAL + 12.9 TYPE + 15.4 HERE?
              (-1.44)     (-2.52)          (3.67)           (3.09)       (3.07)

              s = 19.67         R-sq = 31.4%      R-sq(adj) = 28.9%
```

b* *Are there statistically significant differences in average scores between departments? No, according to the ANOVA results.

c* *Are scores being bumped up into the next highest raise category? Without knowing what the raise increments are, your students would have no way of knowing this. However, the dotplot of SCORES suggests some type of nonnormality is present in the scoring system.

Dotplot of SCORES

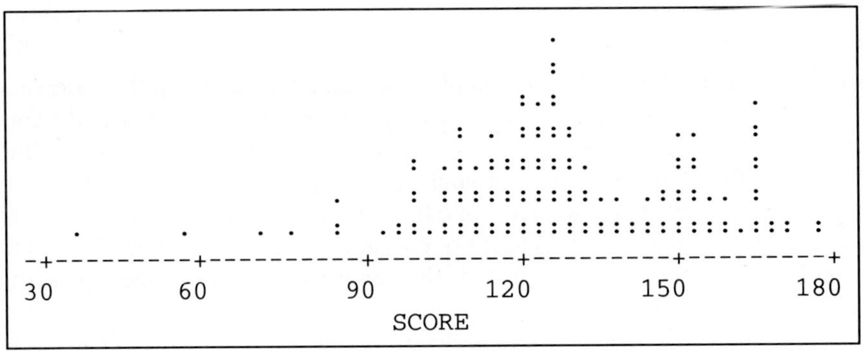

The relationship between annual scores and annual merit raises is as follows.

Summary of Percentage Raise by SCORE

SCORE	RAISE(%)
0 - 70	0.00
71 - 90	1.00
91 - 110	2.00
111 - 130	3.00
131 - 150	4.00
151 - 170	5.25
171 - 190	5.75
191 - 200	6.50

This scale has been used by the company since the appraisal system started. It's known by all employees, including the evaluators. The dotplot suggests there may be some "bumping up," particularly in the upper range.

d* *Can we determine whether three-month evaluations are unnecessary? No, because there are no data here about productivity or any other measure of performance. We do know, however, that three month evaluations are statistically lower than annual evaluations. The two sample *t* test of the difference between the two means for this problem generates a *t* ratio of 6.17.

e *Is the company trying to force early retirement by lowering scores of long-time employees?* The correlations between EMP@EOY and SCORE and EMP@EVAL and SCORE are 0.237 and 0.228, respectively. Simple regressions of SCORE on these two variables individually indicate that EMP@EOY and EMP@EVAL are statistically significant. However, when one of these variables is combined in a multiple regression model with POS@EVAL (shown above), EMP@EVAL becomes statistically significant and negative. One would suspect multicollinearity in this instance. There's probably a whole lot of unnecessary paranoia about this issue.

CASE **45**

AMCHEM'S TAX CREDIT, INSTRUCTOR'S NOTE

Abstract: A case for multiple regression, using a dummy variable for the special month. You get a good fit, and can answer the question pretty directly. Students may have more difficulty in translating the resulting regression equation into a defensible estimate for tax purposes, as they must make a couple of assumptions as well as understand the meaning of the regression equation.

Level:	Intermediate.
Techniques:	Multiple regression, dummy variables.
Typical Class Discussion Time:	45 minutes.
Data Set Size:	60 time series observations, 6 variables.

GENERAL COMMENTS

Most students will quickly seize on multiple regression as a way of attacking this problem, particularly if it's assigned in the middle of a course emphasizing it, but they may have difficulty in doing a complete job. The main difficulty comes in using the resulting equation to produce an estimate of production gas usage, which is not the dependent variable in the equation.

QUESTIONS FOR ASSIGNMENT

For a mature class:

1 As Harold, what do you do?

For a bit more structure:

2 Prepare an estimate of the amount of gas Amchem used for production purposes in the most recent year (year 5).

For less experienced classes:

3 Derive an appropriate model, predicting the total gas usage at Amchem based on the information given.
4 Interpret each element of your model in words.
5 Using your model, estimate the amount of gas used during year 5 for production purposes at Amchem.

SUGGESTIONS FOR LEADING THE DISCUSSION

Discussion can follow the same general lines as indicated by the questions above: For a mature class:

1 "Well, what should Harold do?" will work well if followed by requests to explain and defend the various choices.

For a bit more structure:

2 "What estimate of gas Amchem used for production purposes in year 5 should Harold use and how did you derive it?" will usually lead into a discussion of the various possible models and model selection issues.

For less experienced classes:

3 What model did you select?

4 Inquire about the meaning of each coefficient in the model, paying particular attention to the difference between gas per pound of production and total gas associated with chemical production, and so forth.

5 Using your model, estimate the amount of gas used during year 5 for production purposes at Amchem.

TYPICAL RESULTS AND APPROACHES

The most satisfactory approaches seem to come from regression models with GAS as the dependent variable. Some students will probably try creating dummy variables to allow for effects for each individual month, but they don't seem to explain much that isn't explained by DEGDAY, X, and Y. A dummy variable for the extended turnaround month of July of year 5 does seem to help a bit, though. A typical result is the model

$$\text{GAS} = 29856 + 20.6(\text{DEGDAY}) + 795.7(\text{X}) + 312.9(\text{Y}) - 21884(\text{if July of year 5}),$$

which has R^2 of .78 and a standard error of estimate of 6185. There is a hint of autocorrelation in the residuals (they have fewer runs than expected), but the Durbin-Watson statistic is acceptable and none of the autocorrelations is individually significant at conventional significance levels. Some relevant plots and printouts are attached for possible consideration as the basis for a more advanced discussion. Students might experiment with including lagged terms in the model. Those who do should be asked whether the improvements they obtain are worth the added complexity, particularly since the model has to be explained to the tax authorities. (Would an *auditor* find it plausible that March's production or gas usage even *predicted* April's gas usage, for example, much less *caused* it?)

The data show that Amchem produced 127.1 thousand-pound units of X and 167.2 thousand-pound units of Y during year 5, so from the model, 795.7(127.1)+ 312.9(167.2) = 153,450 cubic feet of gas are attributable to production of chemical groups X and Y.

Some students may advocate including the constant term (29856) or part of it in the estimate, on grounds that most of the overhead is associated with the production buildings, and it's in Amchem's interest to claim it. This can be defended, though it requires using information beyond what the data set gives. On the other hand, it would be fairly common for businesses to simply claim the credit and let the tax authorities dispute it if they want to. This may provide the basis for a discussion of how to incorporate the statistical results in the business situation from which they were drawn.

Regression of GAS on Other Amchem Variables

PREDICTOR VARIABLES	COEFFICIENT	STD ERROR	STUDENT'S T	P
CONSTANT	29855.8	2678.51	11.15	0.0000
DEGDAY	20.6234	2.06184	10.00	0.0000
X	795.714	237.144	3.36	0.0014
Y	312.906	100.385	3.12	0.0029
TURN	-21883.6	6561.68	-3.34	0.0015

R-SQUARED	0.7776	RESID. MEAN SQUARE (MSE)	3.825E+07
ADJUSTED R-SQUARED	0.7614	STANDARD DEVIATION	6184.75

Time Series Plot of Residuals of GAS on Other Amchem Variables

```
 RESGAS
         +-------------------------------------------------------------+
   13000 +                                                             |
         |                           7                                 |
         |                                       9                6 8  |
         |                                              6  9  23       |
         |                            8     4                          |
         |             3      01                                   7   |
    4000 +   3      0   45                   5     1                   |
         |  2                      5   9                               |
         | 1      8                 6           8        78            |
         |                 7 9          0         0             45     |
         |                                          2 45             9 |
         |     5     1                        6                        |
   -5000 +            2                  1 3                 1        0|
         |      6         6       4       2    7                       |
         |    4  7          8   23                   3                 |
         |         9                                                   |
         |                                                             |
         |                                                  0          |
  -14000 +                                                             |
         ++---------+---------+---------+---------+---------+---------++
          0        10        20        30        40        50        60
                                   Case Number
```

CASE **46**

NATURAL GAS PRICES, INSTRUCTOR'S NOTE

Abstract: This univariate time series has several distinct patterns. Natural gas prices decline steadily over the early part of the series, followed by a seasonal pattern. Near the end of the series is a notable spike. Students must wrestle with what to make of all of this. Students can update the data set by visiting the Energy Information Administration's Web site.

Level:	Intermediate, advanced.
Technique:	Time series, forecasting, regression.
Typical Class Discussion Time:	30–45 minutes.
Data Set Size:	163 monthly observations, 1 variable.

GENERAL COMMENTS

Wellhead natural gas prices in $/Mcf are provided from January 1984 to July 1997. Encourage your students to expand the time series and use the new data for out-of-sample model validation. (Expect that the given observations will have been revised by the Energy Information Administration by the time you do this case.)

Note that the case asks students to estimate prices for the first three months beyond the last month in the student's data set. Because the updated series is available on the Web, you can require them to forecast the three most recent months, whatever those might be at the time you assign this case.

Students will have a tendency to discard earlier data in the series on the grounds that the series has undergone an adjustment. Expect your students to derive a variety of forecasting models. The case lends itself to discussions of the competing forecasting methodologies.

QUESTIONS FOR ASSIGNMENT

1 Should any of the data be discarded? If so, how will you decide whether discarding data will improve the quality of your forecasts?

2 What model best forecasts the upcoming months? How did you decide what "best" means?

3 Should monthly variations in natural gas prices be incorporated in the model? If so, how?

SUGGESTIONS FOR LEADING THE DISCUSSION

Cases that require forecasting lend themselves to an in-class derivation of the root-mean-square error of prediction. For instance, if you assign question 2 above, students must decide what criterion will be used to choose the best forecasting model. Ask students what criterion they used. Once the class decides on some summary measure, such as the MSEP, ask everyone to calculate this quantity for their models. This way, you can have a basis of comparison for all models that might be brought up in the discussion.

TYPICAL RESULTS AND APPROACHES

Natural gas prices decline initially, and they display seasonal regularity for about five years, followed by a second period of declining prices. A spike in the winter of 1996–97 dominates the latter part of the time series.

Time Series Plot of Price

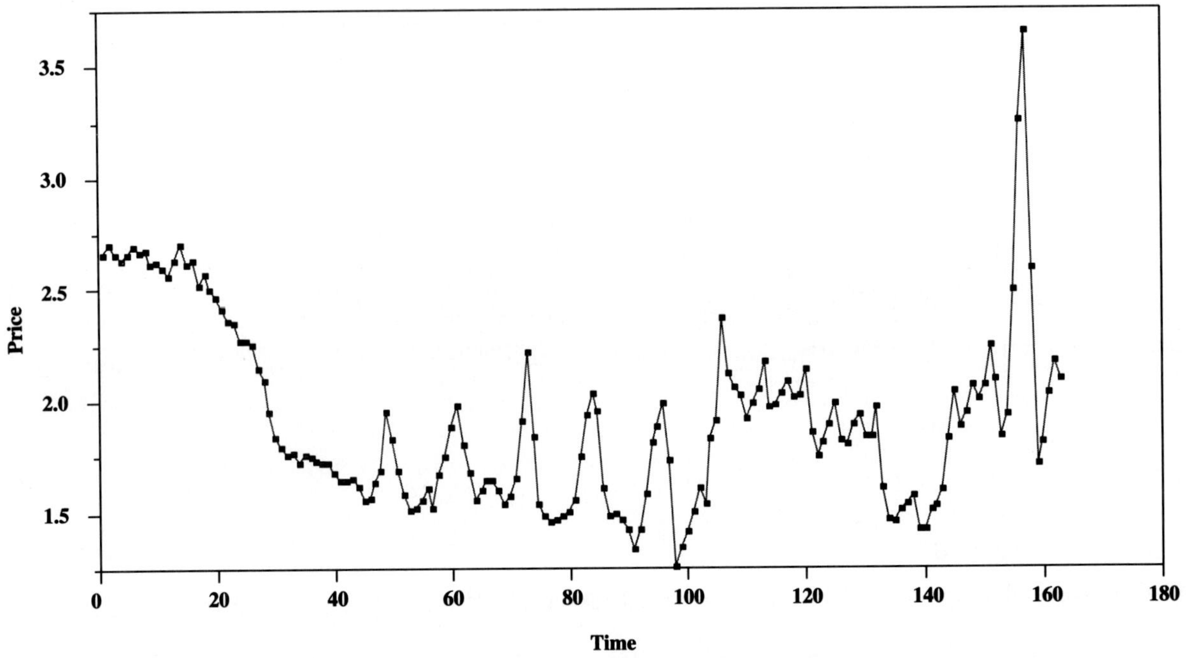

Unless they are familiar with ARIMA models, students will likely model the series using either a distributed lag regression model or one of the smoothing methods. ANOVA of Price by month fails to reject constant monthly averages over the complete time series.

Autoregressive Model of Price

$$\text{Price}_t = 0.18 + 1.2\ \text{Price}_{t-1} - 0.6\ \text{Price}_{t-2} + 0.2\ \text{Price}_{t-4}$$

R^2 = 84%
RMSE = 0.158

A single exponential smoothing model with constant 1.46 does as well as the moving average or double exponential smoothing models based on MAPE, MAD, and MSD:

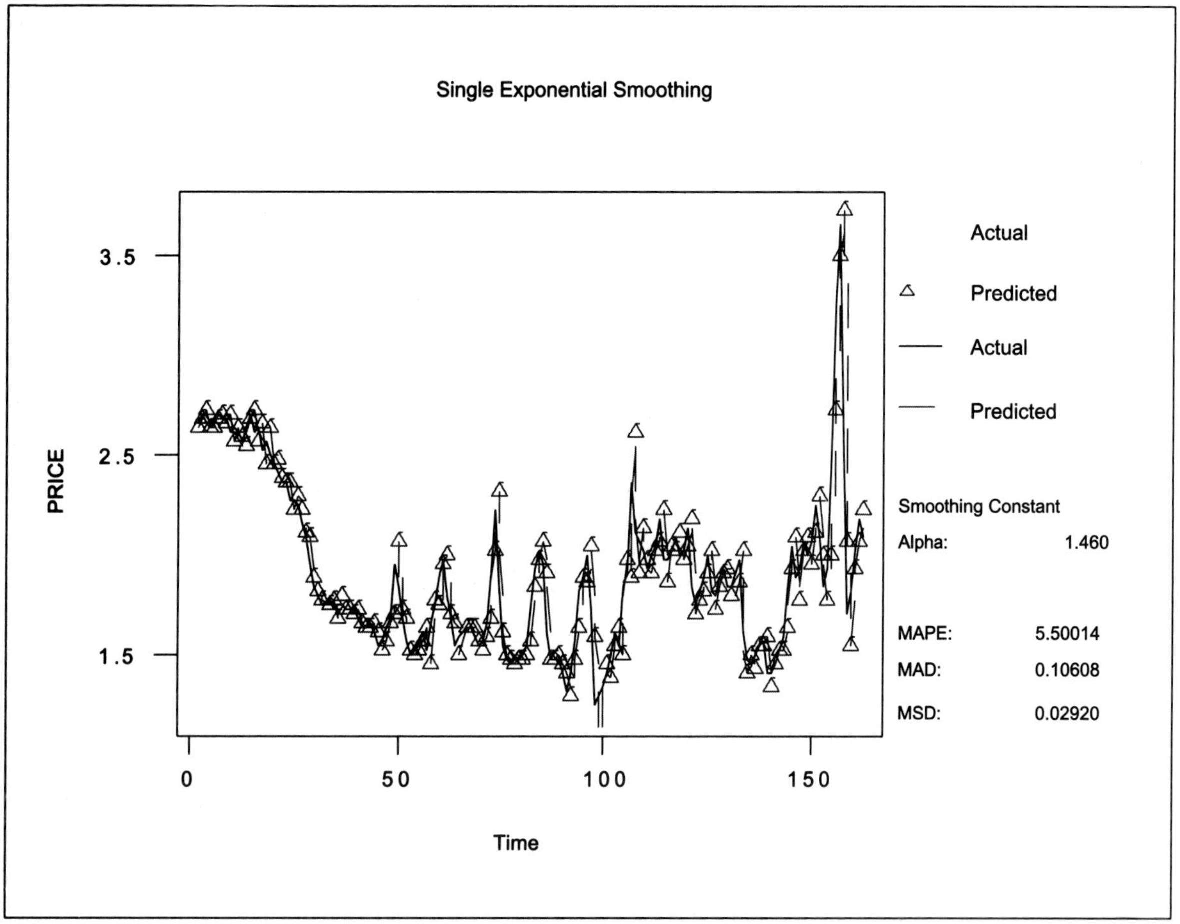
Single Exponential Smoothing
PRICE
3.5
2.5
1.5
0
50
100
150
Time
Actual
Predicted
Actual
Predicted
Smoothing Constant
Alpha: 1.460
MAPE: 5.50014
MAD: 0.10608
MSD: 0.02920

CASE **47**

WESTAR BEVERAGE SALES, INSTRUCTOR'S NOTE

Abstract: A case for multiple regression, advanced (if you give no guidance) and intermediate (if you give some hints). A key aspect is to consider the components of total sales separately rather than combined, as different factors are apparently influencing them.

Level:	Intermediate, advanced.
Techniques:	Multiple regression.
Typical Class Discussion Time:	45–60 minutes.
Data Set Size:	30 time series observations, 8 variables.

GENERAL COMMENTS

Underneath everything here, it's a fairly standard application of multiple regression, but students have often had difficulty sorting through the problem and discovering it. A key aspect of a good solution seems to be to treat the pub and the nightclub separately—different factors appear to influence them. When they're lumped together, you miss things and get some puzzling results.

QUESTIONS FOR ASSIGNMENT

For advanced students, we suggest no more than what's given in the case: let them figure out that regression can help. With this approach, we call the case advanced.

For less mature students (to make it an intermediate case), you might give a little structure:

1 What factors appear to influence beverage sales at the Westar?

2 Should Westar consider rehiring the piano player for the Pour Nous Lounge? How about happy hour food?

3 Can you suggest a new forecasting method for Jane to try out?

SUGGESTIONS FOR LEADING THE DISCUSSION

The best discussions of this case have come from general invitations to make recommendations, perhaps something like: "Well, did they do the right thing when they shot the piano player?" or, "Is there any hope that Jane can bring her department into the 20th century?"

You can follow up by requesting a specific model (which you can transcribe onto the blackboard). When several models have been proposed, you can switch to a discussion of the criteria that might be used to select among them. You can then compare and contrast the models based on those criteria. You could also use the regression printouts that follow as a basis for discussion, comparing the results of the two separate beverage regressions with the first one, which lumps all beverage sales together.

TYPICAL RESULTS AND APPROACHES

Students may treat beverage sales as a time series (which it is, of course), but the usual time series techniques don't seem to offer a lot in this case. Regression seems plausible, since we would expect beverage sales to be driven by some of the other independent variables. Regressing BEVTOT on the other independent variables (another obvious thing to do) yields the equation:

$$\text{BEVTOTAL} = 18250 + 2.58551\ \text{FOODNITE} + 29.2952\ \text{FOODPUB} + 525.451\ \text{MONTH} - 2.08149\ \text{ROOMS}$$

with an R^2 of .56 and standard error of estimate of 3397, if all variables with t statistics of 2 or more are included. This shows a time trend upward, but a puzzling negative relationship between ROOMS and BEVTOT, which Jane will certainly have a hard time persuading her colleagues to believe in. Admittedly, the coefficient of ROOMS is only marginally significant.

Running *separate* regressions for BEVPUB and BEVNITE yields more useful and interpretable results:

$$\text{BEVPUB} = 5877.2 + 2001.4\ \text{PIANO}$$

with $R^2 = .548$, $s = 940.5$ and:

$$\text{BEVNITE} = 10357 + 3.02(\text{FOODNITE}) + 26.8910(\text{FOODPUB}) + 587.9(\text{MONTH}) - 2.257(\text{ROOMS})$$

with $R^2 = .66$, $s = 3373$, where PIANO is a dummy variable equal to 1 for the first 13 months. The picture is a little different now: for the pub, neither ROOMS nor a time trend appear to be important, but the piano player was. For the night club, we still have the same puzzling negative relationship with room occupancy, a general upward trend, and promotional food appears to improve things. The effect of promotional food in the pub (FOODPUB) on BEVNITE is less plausible than the effect of FOODNITE, unless providing food in the pub convinces people to go to the night club. It is of marginal significance, in any case ($P = .0554$). The residuals from the regression for BEVPUB show no particular time series structure, confirming the decision to use regression. The residuals from the regression for BEVNITE do show some autocorrelation at lag 1, though whether they'd be judged significant based on this number of observations is a bit questionable. Students may wish to follow up on this if they have enough time series background.

A major lesson from this is the importance of not lumping things together too quickly (a mistake the case makes it easy to do since it conveniently provides the totals in the data set). Lumping all the beverage sales together hides the fact that different factors are influencing the different beverage sales.

The results prompted Jane Bartley to visit the Westar, where she discovered that, in fact, much of the business at the Pour Nous Lounge was from walk-in, local customers, more or less unrelated to the room occupancy at the hotel, but for whom the happy-hour ambiance and the piano player were important. Apparently this hadn't been understood at the more remote corporate headquarters, where the forecasting was done. The results for the night club remain a bit more puzzling in some ways. Students might want to investigate possible multicollinearity. They might also note that while PIANO and not FOODPUB appear important in predicting BEVPUB, PIANO and FOODPUB are highly correlated (and FOODPUB, though measured in dollars, might as well be a dummy variable, for practical purposes), so we may not be able to separate their effects.

The standard error of estimate is large enough here that individual monthly forecasts are probably not reliable. The main result of this analysis is to identify which *factors* seem to be important. Specific forecasting techniques will probably require more research.

Regression of BEVTOTAL on Other Factors

```
Dependent Variable:          BEVTOTAL---TOTAL BEVERAGE SALES
Independent Variable(s):     Constant
                             MONTH---MONTH NUMBER
                             FOODPUB---FOOD IN POUR NOUS
                             FOODNITE---FOOD IN GAIETE
                             ROOMS---ROOM OCCUPANCY
 2 observations omitted because of missing values
28 observations used in regression.
       REGRESSION COEFFICIENTS
IND. VBL.     COEFFICIENT      STD. ERROR     T
Constant      18249.76241      6993.97584   2.6
MONTH           525.45104       148.28616   3.5
FOODPUB          29.29519        13.41973   2.2
FOODNITE          2.58551         1.17482   2.2
ROOMS            -2.08149         1.03850  -2.0

ANALYSIS OF VARIANCE:
SOURCE          DF                SS                    MS
Regression       4      334641842.53894      83660460.63474
Error           23      265401511.56820      11539196.15514
Total           27      600043354.10714      22223827.92989
R-squared: 0.558 Std. error of estimate:         3396.93923
```

Regression of BEVPUB on PIANO

```
  Dependent Variable:          BEVPUB---SALES IN POUR NOUS PUB
  Independent Variable(s):     Constant
                               PIANO---
 2 observations omitted because of missing values
28 observations used in regression.

   REGRESSION COEFFICIENTS
IND. VBL.     COEFFICIENT      STD. ERROR     T
Constant           5877.2       242.82644  24.2
PIANO          2001.41538       356.37179   5.6

R-squared: 0.548 Std. error of estimate:         40.46276
```

Regression of BEVNITE on Other Factors

```
   Dependent Variable:        BEVNITE---SALES FOR GAIETE
   Independent Variable(s):   Constant
                              MONTH---MONTH NUMBER
                              FOODPUB---FOOD IN POUR NOUS
                              FOODNITE---FOOD IN GAIETE
                              ROOMS---ROOM OCCUPANCY
 2 observations omitted because of missing values
28 observations used in regression.
  REGRESSION COEFFICIENTS
IND. VBL.      COEFFICIENT       STD. ERROR     T
Constant       10357.07216       6944.41400   1.5
MONTH            587.92866        147.23535   4.0
FOODPUB           26.89097         13.32463   2.0
FOODNITE           3.02152          1.16649   2.6
ROOMS             -2.25687          1.03114  -2.2
R-squared: 0.664 Std. error of estimate:        72.86729
```

CASE **48**

PRICE-FIXING CASES, INSTRUCTOR'S NOTE

Abstract: This case can be used in a couple of ways: to focus attention on the assumptions underlying tests or perhaps to use a control chart. It can be assigned early in a course, and requires no specific business knowledge. It shows a decrease in the number of price-fixing cases, but by most measures, the differences are not particularly important.

Level:	Beginning.
Techniques:	Hypothesis testing, statistical process control, time series plots.
Typical Class Discussion Time:	30 minutes.
Data Set Size:	27 time series observations, 2 variables.

GENERAL COMMENTS

The number of indictments returned did go down in the Reagan years, but not to the point of being in and of itself very convincing. This can form the basis of a discussion of how the data can reveal only part of the full answer. To really determine if there was some sort of change would require more than just these data.

QUESTIONS FOR ASSIGNMENT

To emphasize the process control point of view, ask students to:

1 View the data as the results of a process and use the first 20 observations as their "base" period to construct an appropriate equation, graph, or other tool that describes the normal results of the process.
2 Use their tool to decide whether the results of 1981–1987 suggest that the process somehow changed.

To emphasize a hypothesis testing framework, ask students to:

3 Test the hypothesis that the number of indictments returned in the Reagan era did not differ from that of the previous 20 years. What assumptions are you making?

Either of these questions (particularly the second) makes the technical part of the case into little more than an arithmetic exercise, but leaves open for discussion the question of whether the assumptions underlying those techniques are satisfied, or how *well* they're satisfied. Better in our view to let the students decide the kind of technique that would provide the most insight and then carry it out, if you feel they can. The discussions of underlying assumptions are often the best part of the case, though, so be sure you allow them to develop.

SUGGESTIONS FOR LEADING THE DISCUSSION

The best discussions seem to come from this question in the case: "Do these data support the claim that the Reagan administration adopted a more lenient policy?" Usually the answers suggest further discussion topics such as, "What does it mean to *support* the claim?" or, "How big a decrease is significant?" You may wish to display a plot like the one that follows so that the class has something to focus on. The discussion can be an opportunity to let students wrestle with the issues of their own preferences or political views versus what the data have to say. People should defend their conclusions specifically (based on the data). The question in the case ("Do the data support . . .") can be interpreted in several ways, all legitimate, but it's easy to let the discussion become more political than statistical. In that case, the instructor may want to point out that perhaps the role of statistics in decision making isn't clear: should it be somehow pure and objective (probably ensuring its irrelevance) or just another way to present your side of things (probably ensuring it won't be trusted)? Students may also point out (correctly) that there's more to the question than these data can reveal.

TYPICAL RESULTS AND APPROACHES

Most students will plot the data over time (as they ought to). The last seven years are generally lower than the preceding ones. Most measures don't call this significant. A c-chart based on the first 20 years does not reveal anything as out of control. A t test on the differences between the means of the first 20 years and the last 7 years gives a t of about 1.3 (depending on the pooling assumptions made). If students use t tests or other tests, probe a bit to see if they have thought about whether the underlying assumptions about independence, sampling, and so on, have been satisfied. While some assumptions about successive years' data being independent within administrations can probably be defended, many others can't. Without some further external evidence, we wouldn't be convinced of a change.

Time Series Plot of CASES

```
CASES
   +-----------------------------------------------------------------+
15 +                                                                 |
   |                                                0                |
   |                                                                 |
   |                                                                 |
   |                            2           7                        |
   |                                      6                          |
10 +                                4  5                             |
   |  2                                       8           3  4       |
   |                  8      1                                       |
   |                                                                 |
   |               7                                                7|
   |                      0       3                                  |
 5 +        4   6                                9    1          6   |
   |1    3    5                                                      |
   |                                                    2            |
   |                                                                 |
   |                                                           5     |
   |                    9                                            |
 0 +                                                                 |
   ++------------+------------+------------+------------+-----------++
  1961         1966         1971         1976         1981         1986
                                 YEAR
```

CASE **49**

PROPERTY CRIMES, INSTRUCTOR'S NOTE

Abstract: This case is a fairly straightforward exercise in model building with multiple regression, although the statistical results might be unexpected. For example, students expect the unemployment rate to be an important indicator of property crimes—it is difficult to find any statistical relationship between these two variables here.

Level:	Intermediate.
Techniques:	Regression.
Typical Class Discussion Time:	30 minutes.
Data Set Size:	50 cross-sectional observations, 10 variables.

GENERAL COMMENTS

This model-building exercise is relatively straightforward, which should permit students to build some confidence in their abilities to undertake this type of exercise. We recommend that you make this assignment one of the first in the series of model-building exercises. Your students can concentrate on the basics such as interpreting the regression coefficients; there are few complicating factors such as categorical predictors or severe multicollinearity.

QUESTIONS FOR ASSIGNMENT

1 What are the primary determinants of property crime rates in the United States?

2 What would you like to know about property crime rates that cannot be answered by this data set?

3 What are the advantages and disadvantages of using data sets in which the observations are macro data rather than micro data? (Macro data are observations that are themselves calculated from other observations, such as state averages, as we have here. Micro data represent observations from the original entity of interest, such as individuals or households or firms.)

SUGGESTIONS FOR LEADING THE DISCUSSION

Because most students will end up with similar models, focus the discussion on their model-building *process.* Ask your students to carefully interpret their regression coefficients. More likely than not, they spent little time looking at those values during the model building. Before you leave the case discussion, make some mention of the disadvantages of performing regression on aggregate data.

TYPICAL RESULTS AND APPROACHES

The lowest crime rate is West Virginia's and the highest is Florida's. After using regression, these two states do not have large standardized residuals. This is a great example of how regression controls for other factors.

The typical crime rate per hundred thousand residents is 4559 with a typical fluctuation of 1232. Stepwise regression on this problem builds a model of CRIMES on URBAN, DROPOUT, and DENSITY, whether you use forward or backward elimination. In this regression, DENSITY has a statistically significant negative coefficient, which tends to contradict most people's expectations (the correlation between CRIMES and DENSITY is +0.06).

Two regression models are shown below with *t* ratios in parentheses.

```
CRIMES = 222 + 50.9 URBAN + 73.5 DROPOUT - 155 PUBAID       R-sq = 60.2%
                (6.35)        (4.04)          (-2.32)

CRIMES = - 1053 + 67.9 URBAN + 57.8 DROPOUT - 1.93 DENSITY   R-sq = 65.6%
                   (8.07)        (3.77)          (-3.67)
```

CASE **50**

ROCKVILLE PHYSICIANS' GROUP, INSTRUCTOR'S NOTE

Abstract: A case focusing on the implementation of statistical approaches in an organizational setting rather than the approaches and techniques themselves. The students get a general overview of the situation and the data set and must advise Dr. Eaton how best to summarize it. The organizational setting takes some attention here; it involves convincing people who aren't used to dealing with statistics in their day-to-day world to start using them. Follow-on cases (*Cases 50B, 50C* and *50D*) describe what actually happened.

Level:	Intermediate.
Techniques:	Descriptive statistics.
Typical Class Discussion Time:	45–60 minutes.
Data Set Size:	No data.

GENERAL COMMENTS

Like all communication, using statistics to influence an organization's behavior requires a sender and a receiver, both switched on and both tuned to the same frequency. Most statistics courses focus (understandably) on the sender's role: what should be computed and how it should be described. This case offers a change of pace. The students don't have the data themselves, only a brief description of their nature and a sample printout. In advising Dr. Eaton on how best to proceed with the results of the C-section study, they must therefore concentrate on the organizational situation, the physicians whom Dr. Eaton wants to convince to use statistics, and current trends in the practice of medicine—that is, on the receiver.

This case requires a sense for organizations and context. It probably works best for older students. It requires only basic statistical ideas (indeed, they don't even have to be carried out, only assessed), but a broader outlook than might be expected for younger students. Accordingly, we've labeled it intermediate.

QUESTIONS FOR ASSIGNMENT

1 What should Dr. Eaton consider as he prepares for the QAUM meeting?

2 What kinds of statistical summaries might be appropriate for Dr. Eaton's presentation at the QAUM meeting?

SUGGESTIONS FOR LEADING THE DISCUSSION

We suggest opening the discussion broadly: something like, "Well, how should Dr. Eaton proceed?" Topics usually mentioned include:

1 Dr. Eaton should keep it simple. Basic plots and charts showing frequency of C-sections, perhaps by doctor and by insurance class, gravida status, and the like.

2 Dr. Eaton should keep in mind that his goal is to enlist the physicians' support for quality measurement and monitoring in general at RPG, not just for the C-section study. He may anticipate a certain amount of resistance and should be careful not to let the meeting turn into a display of his own brilliance.[1]

3 There may be some discussion of whether tables or graphs are most effective.

4 Given the information in the case about the traditional way of running the QAUM meeting, some may suggest ways of running the meeting differently. This can lead to a good discussion of the process of change in organizations in general. Is it a good idea to sell the idea of change by changing the style of the meeting? Or would you do better to use the same familiar style (less threatening) and work up to the idea of change more slowly?

5 The question "Do you think Dr. Eaton should present separate results for each doctor? Should they be presented by name?" usually stimulates some disagreement. The answers are usually based on different underlying assumptions about the nature of RPG and the people involved, and the case doesn't really give information to confirm or deny most of them. On the other hand, the different assumptions are themselves revealing.

6 To answer another question that may come up: yes, this is a real situation, though the names of the individuals involved have been disguised.

The case has a number of technical terms. Students need to get used to dealing with such terms in analyzing data. Most of them are defined in the case, at least to some extent. Two that aren't are *parity* (the number of previous children born by the woman) and *gravida* (the number of previous pregnancies for the woman).

There are some serious limitations of the data. If the goal is to study reasons for C-sections, for example, such reasons are recorded only for cases in which C-sections are actually performed. The same condition might in principle have been present in cases in which no C-section was done, but it would not necessarily have been recorded in such a case.

As the discussion winds down, you can hand out *Case 50B,* which follows this note. It describes how Dr. Eaton actually proceeded. An assignment question for *Case 50B* can be, "What effect will Dr. Eaton's presentation have on the OBG physicians at RPG?"

Given 15 or 20 minutes to review *Case 50B,* most students will have something to say about the presentation. You may wish at this point to ask students to compare notes with their neighbors or work in small groups to prepare revised views. Some students may be critical of Dr. Eaton's presentation: it's busy, complicated, overwhelming, intimidating. Others will feel that he's done what has to be done: the physicians need to be awakened to the need for change, and some sort of shock is required. The use of sophisticated graphics was possible in 1989, but was nowhere near as widespread or taken-for-granted as it would be now, of course. Given the pace of technological change, students may not remember that far back.

You can then hand out *Cases 50C* and *50D,* either together or individually (if time permits), which describe briefly what followed. *Case 50C* confirms the suspicions of many students, namely, that Dr. Eaton's presentation didn't achieve its intended purpose. *Case 50D* suggests that matters may be more complicated than that, though, and that organizational change is a long-term matter, not determined in the first year. Whatever difficulties Dr. Eaton may have had with his presentation, he may have planted the seed of an idea, a seed that took a while to germinate.

TYPICAL RESULTS AND APPROACHES

Most of the results are described above under "Suggestions for Leading the Discussion." We use this case a few weeks into a course: long enough that students are used to computing things. This case serves as a reminder that the goal of statistics is never just to compute something: the organizational context has to be kept

[1]Dr. Eaton was not the sort of individual who would act in that manner, but the students have no way of knowing it. They *may* be able to note that he was at one point "one of them" and had practiced medicine at RPG for a long time, so his respect for his fellow physicians and their role shouldn't be seriously in question. In fact, he was highly respected by his colleagues. When he retired fully a couple of years after this case, they described his leaving as the "end of an era" in an advertisement in the local newspaper.

in mind continually. It also reminds those students who begin to feel frustrated about the computations that they are intended to solve real problems. Students who are stronger in the mathematical and computing aspects of things are often less sensitive to the organizational uses of the quantities computed. Indeed, they may never have considered them. They can learn from the others in the class, who may have more to offer in this case than in the more technical cases. It can be a good learning experience for all concerned: broadening the outlooks of the technically minded and empowering the mathophobes.

The case has a lot of ambiguity: the actual data are not provided, and many questions can't be answered without knowing more about the individual people involved than the case provides. On the other hand, that ambiguity leads to richer discussions, as different sets of assumptions will be brought to bear on it. Its primary purpose is to raise awareness, not to settle things.

CASE **50B**

ROCKVILLE PHYSICIANS' GROUP, PART B

The QAUM meeting at Rockville Physicians' Group (RPG) was scheduled for Tuesday, January 23, at 5:30 P.M. and as he reviewed the materials he had prepared for the meeting, Dr. John Eaton felt he had struck a reasonable balance between the quantitative-hard-sell and gentle-persuasion approaches to presenting the C-section information.

Two weeks before the meeting, a package of materials had been distributed to the OBG staff and a few other people. That package contained:

1 An agenda for the January 23 meeting.

2 A copy of "Continuous Improvement as an Ideal in Health Care," by Donald Berwick, from the *New England Journal of Medicine,* which participants were asked to read and be prepared to discuss.

3 Summaries of the C-section study. This consisted of several pieces: a general history and summary, a summary of the C-section data for all group patients, and a separate summary by individual doctor. Those verbal summaries were accompanied by about 40 pages of charts and tables produced by Lotus 1-2-3 and Harvard Graphics, depicting such things as means, variances, and distributions of various quantities of interest, as well as some charts designed to illustrate any differences in patient mix between the various doctors.

Extracts of the package of materials are presented on the next pages.

Dr. Eaton felt he had found reasonable compromises between the need to present the truth as the data revealed it and the natural reluctance of many physicians to trust numbers to tell them a whole lot about an intensely human activity such as medicine, particularly OBG, with so much individual variation. Still, it was the first such study focused on medical decision making they had run at RPG, and he couldn't help wondering how the meeting would go.

Excerpts from Dr. Eaton's Presentation

CESARIAN SECTION STUDY - RPG QAUM -
MATERIALS, METHODS AND PLAN OF THE STUDY

1. Data studied: This consists of all deliveries, vaginal and by C-section, from 10/1/88 to 9/30/89. It was entered into the auxiliary computer system by down-load from Cycare into records consisting of the following field for each transaction:

- Chart number
- Procedure Code
- Doctor number
- Date of Service
- Charge
- Financial class
- Diagnosis Code

In addition, charts were pulled and studied for all C-section cases and for a like number of vag delivery cases chosen by a random number technique. In all 354 charts, 178 C-section and 176 of 6448 vaginal deliveries, were pulled, data extracted and added to the data base from Cycare under the following fields:

- Maternal age
- Parity
- Gravida status
- Reason for C-section entered into two fields, "Reason_1" and "Reason_2" depending on what appeared to be the Immediate and Underlying reason for section.
- Potential reasons for section, coded as above, were recorded for cases where they were mentioned but a C-section was not done.
- Time in labor
- Apgar 1 and Apgar 5 scores
- Gestational age
- Birth weight
- Maternal death Y/N
- Fetal death Y/N

The abstracting of information from charts and keying it into the computer was done chiefly by Mary Spiegel, checking judgements with JE as needed. Most data was clearly presented in hospital labor records and consultation notes.

To deal with the reasons for C-section a two-digit code was devised such that when the first digit was a "1", the reason was applied to a pre-labor decision for section, and when it was a "2", to a decision made during labor. Reasons were almost always made clear in hospital admission and consultation notes. Listed below are the codes, the shortened labels used on graphs and what the code refers to.

Excerpts from Dr. Eaton's Presentation (*continued*)

Code # - Reasons for C-sect decision before onset labor:
11 "Prev" - previous C-section when this is a factor in C-section decision
12 "CPD" - CPD known in advance
13 "GYN" - variety of preexisting GYN problems, leiomyomas cervical disease, etc.
14 "Herpes"- active maternal herpes
15 "HBP" - high BP including pre-eclampsia
16 "Posn" - malpresentation, usually breech
17 "Fetal" - abnormal fetal status leading to C-section, including fetal distress before onset labor
18 "Steril" - the intent to tie tubes at surgery
19 "Pregn" - abnormalities of placenta, amniotic fluid, etc.

Code # - reasons for C-sect decision during labor:
21 "CPD" - becoming evident during labor. Some doctors might classify such cases as "failure to progress".
22 "Fail" - failure to progress, a mixed category that probably has elements of bony geometry, fetal position, uterine inertia, poor dilatation, poor response to Pit, etc.
23 "Memb" - premature rupture of membranes
24 "Distr" - fetal distress
25 "Wound"- disruption of prior C-section wound
26 "Posn" - Malposition, usually breech
27 "Bleed" - hemorrhage
28 "Praev" - placenta praevia
29 "Mechan" - mechanics of labor other than dystocia, such as prolapsed cord or arm, short cord, etc.

2. Data Handling: This material was manipulated using DbaseIII+, permitting various counts, sorts, filters and associations. Raw material from Dbase output was entered into small Lotus 123 spreadsheets for arithmetic calculations - sums, averages, breaking into ranges, calculation of confidence intervals, etc. Finally, the material in these spreadsheets was graphed using Harvard Graphics. When necessary (for data obtained from chart-review rather than down-load) data for vaginal delivery were adjusted upward from the study numbers to the full numbers by dividing by the percent studied.

3. Presentation: All of the information is presented in a series of graphs, each followed by a copy of the Lotus spreadsheet or, occasionally, by Dbase output sheets. The presentation is in two sections, an overview of RPG OB as a whole and observations on the differences among RPG Obstetricians. Each of these sections is introduced by a few pages of text with explanations of the graphs and comments on what I believe they say. It is sometimes difficult to make a graph transparent on first glance, partly due to limitations in labelling imposed by the software. Please note that the Y axis sometimes shows number.

Excerpts from Dr. Eaton's Presentation (*continued*)

```
"DELIV89.WK1"      JE
VAGINAL AND CESARIAN DELIVERIES,
( C-sections = CPT 59500 .and. charge >500. Vag deliv = 59410.)
                                                        95% confid level
  DOCTOR  # VAG DEL# C-SECT    TOTAL  % C-SECT    Upper      Lower
-----------------------------------------------------------------
     A        105        17       122     13.9%    20.0%      7.8%
     B        121        28       149     18.8%    25.1%     12.5%
     C        102        24       126     19.0%    25.9%     12.2%
     D         87        27       114     23.7%    31.1%     15.9%
     E        127        42       169     24.9%    31.4%     18.3%
     F         98        40       138     29.0%    36.9%     21.4%
 TOTAL        535       161       696
AVERAGE     107.0      32.2     139.2     23.1%    26.3%     20.0%
2 STD DEV    29.7      14.7      37.9      7.6%

BAYSIAN CALCULATION
Bayes theorum - formula = P(A|B) = P(B|A)*PA/PB
         A = having a C-section
         B = being a certain MD
         P(A|B) = probability of A if B is true = probability of having a C-sect
                  if B is the doctor = % Csect for that MD.
         P(B|A) = probability of being a certain MD if patient had a C-section
                = #Csect for that MD#/# all C-sect.
         PB = the probability of being that MD
            = # his/her deliv/total # deliveries.
         PA = probability of having a C-sect
            = average % C-section.
                                      vs direct calc
              A  P(A|B)    = 13.9%          13.9%
              B  P(A|B)    = 18.8%          18.8%
              C  P(A|B)    = 19.0%          19.0%
              D  P(A|B)    = 23.7%          23.7%
              E  P(A|B)    = 24.9%          24.9%
              F  P(A|B)    = 29.0%          29.0%

95% CONFIDENCE LIMITS FOR PERCENT C-SECTION     x = # C-sect
Formula: x/n +/- 1.96(sq rt((x/n*(1-x/n)))      n = # total deliv
                       Upper     Lower    Spread
              A        20.0%      7.8%     12.2%
              B        25.1%     12.5%     12.5%
              C        25.9%     12.2%     13.7%
              D        31.1%     15.9%     15.2%
              E        31.4%     18.3%     13.0%
              F        36.9%     21.4%     15.5%
          AVERAGE      26.3%     20.0%      6.3%
```

Excerpts from Dr. Eaton's Presentation (*continued*)

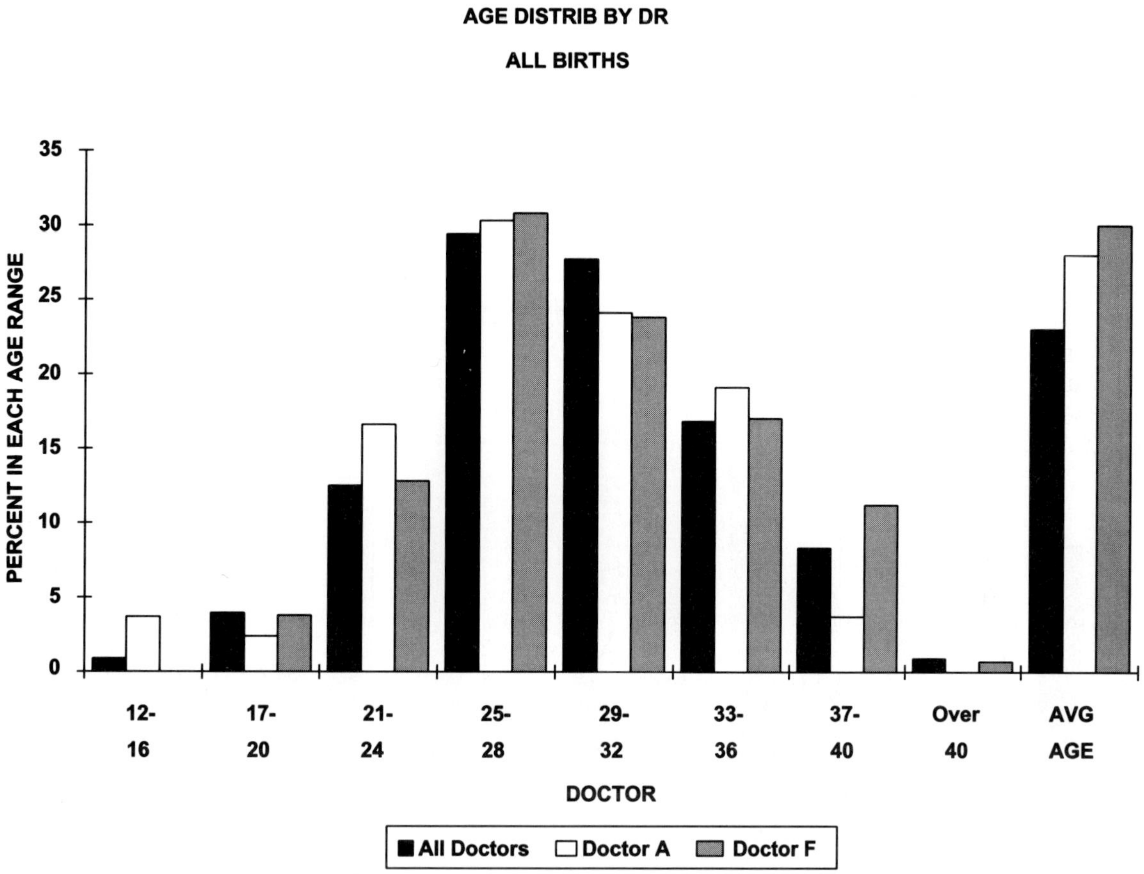

Excerpts from Dr. Eaton's Presentation (*continued*)

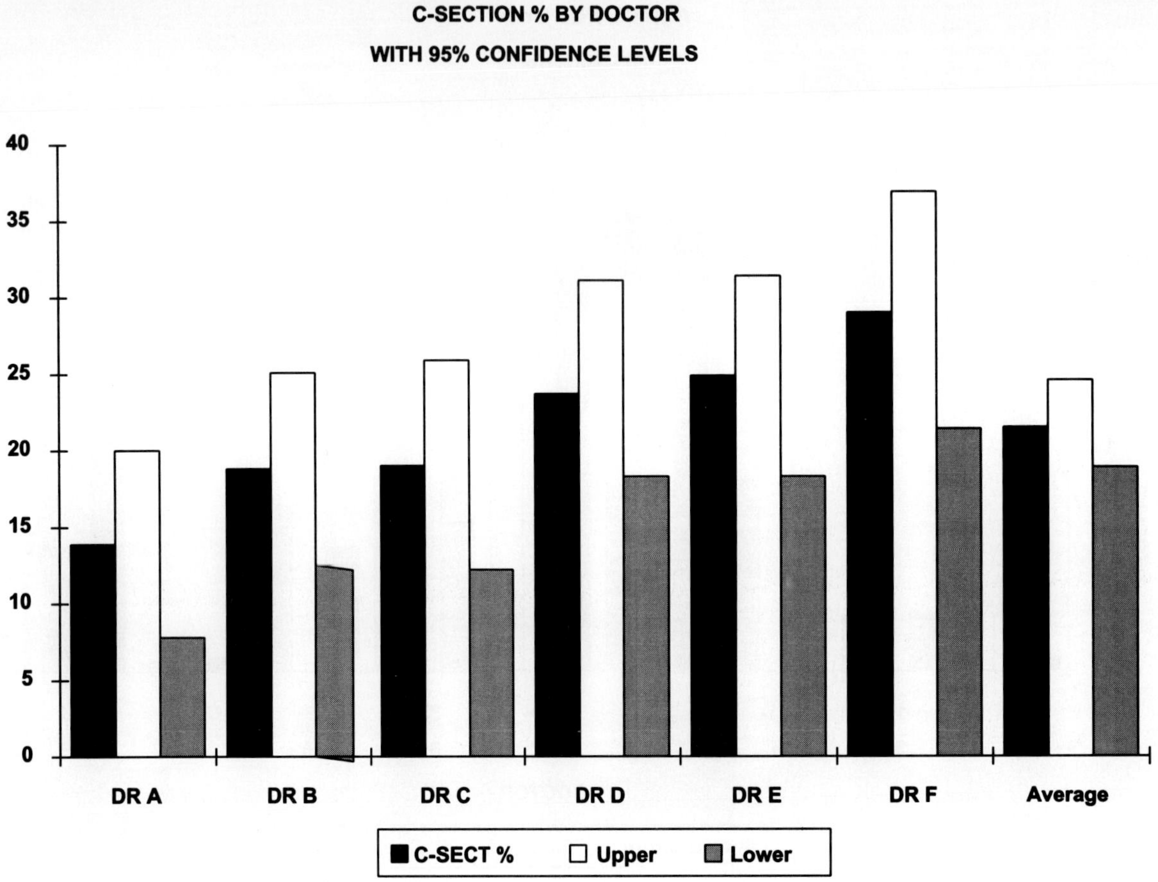

Excerpts from Dr. Eaton's Presentation (*concluded*)

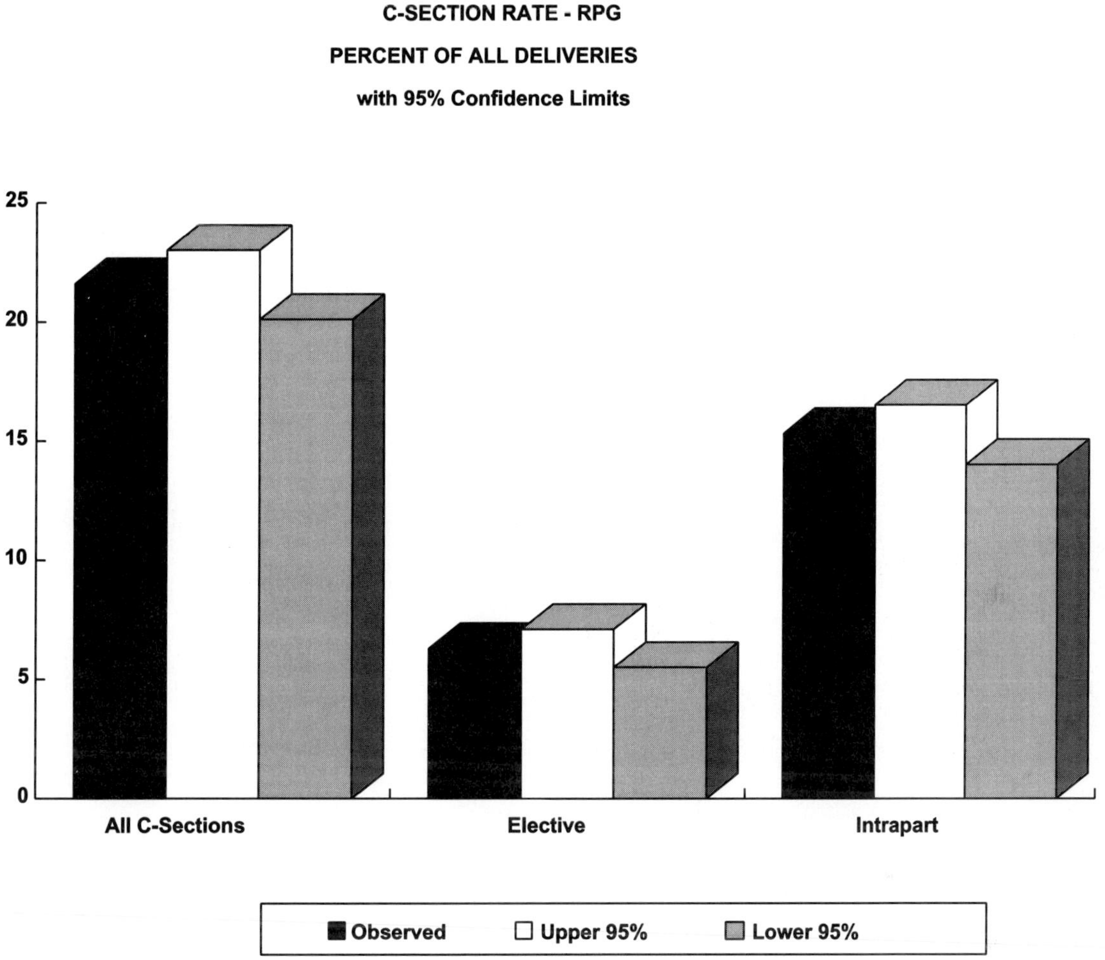

CASE **50C**

ROCKVILLE PHYSICIANS' GROUP, PART C

In a conversation in May (a year later), Dr. Eaton reflected on the C-section study and its aftermath.

I wanted them to do a follow-on sub-study, directed at the "failure to progress" diagnosis, which was a very common reason given for doing a C-section. What do we really mean by that? I wanted them to take notes on those cases as they happened and see what we could learn.

They decided consciously to drop the ball. They've done nothing with it at all. I did a study a year later, and the C-section rates were the same. Dr. F, who had the highest rate a year ago, still has the same rate, and Dr. A. (who had the lowest rate) actually raised his rate a little.

It's a matter of the psychology of physicians. Docs are focused on the *instance,* the particular case at hand. Dealing with overviews is foreign to them. Also, as a doc, you need to have very definite paradigms, approaches that you use, in order to deal with all the different situations you see in a typical day. When you ask them to consider maybe questioning that approach, you're fighting something which has become comfortable.

A couple of things are changing, though. When more young physicians start coming out of medical school with the notion that they are expected to monitor their own performance, look at the data themselves, not just expect someone else to do it, then some things will begin to get better.

The other thing is that pretty soon, the insurance companies will start forcing medical offices to do this sort of thing. They'll start awarding contracts based on whether or not the outfit has a quality monitoring program in place, or maybe use different reimbursement schedules. Once it starts hitting the docs in the wallet, they pay more attention.

I know some of my tables and charts were kind of busy, and the text and graphics were on separate pages. I've figured out how to combine text and graphics in the same report now, but I don't think graphics or the combination is enough. I'm committed to keep trying this kind of approach, and perhaps it will catch on. If not in OBG, then perhaps in another department.

CASE **50D**

ROCKVILLE PHYSICIANS' GROUP, PART D

In a follow-up conversation in November, Dr. Eaton reported that the C-section rate at RPG was now approaching 15% and that doctor F (who previous had a C-section rate of about 30%) now had a rate of about 20%.

CASE **51**

ABANDONED CALLS—I, INSTRUCTOR'S NOTE

Abstract: This case involves examination of two time series (daily and half-hour readings) of a service/queuing problem. Three solutions are under consideration by management: purchasing additional equipment, upgrading the existing equipment, and encouraging off-peak usage of the system. The data give no clear answer as to the preferred action, but provide significant insight to the problem. Students will likely propose additional solutions after looking at the data.

Level:	Intermediate.
Techniques:	Time series plots, ANOVA, correlation.
Typical Class Discussion Time:	45 minutes.
Data Set Size:	ABANDON1: 88 daily observations, 3 variables.
	ABANDON2: 240 half-hour observations, 3 variables.

GENERAL COMMENTS

This case requires students to advocate for one of three proposed solutions for achieving a target abandoned calls rate of 3% or lower. Two of the options, upgrading old modems and adding more modems, are feasible. However, since no cost information is given, it is unclear which of the two is preferred. Even so, the data suggest that either of these options may lower the abandonment rate.

The third option of encouraging off-peak usage appears viable. However, students who look beyond the numbers and into the practicality of this solution will argue against it. After all, the case suggests that getting the product to the customer in a timely fashion is important in this business.

Other options, unstated in the case, will be added to the list during the discussion. Some will suggest training the salespeople on use of the software. Others will recommend that the software be upgraded to eliminate the possibility of modem crashes resulting from operator error. Both are reasonable given the scenario in the case, but do not derive from examination of the data.

Some students will approach this case with control charts, without much thought about violation of the independence assumption.

QUESTIONS FOR ASSIGNMENT

1 What additional information, if any, might you need in order to decide which of the three options is preferred?

2 Are there any differences in CALLS, MODEMS, and HANGUPS from day to day or month to month? If so, what are these differences and how should they factor into your decision about process improvement?

3 Are there any differences in CALLS, MODEMS, and HANGUPS from hour to hour? If so, what information does the pattern add to your decision making?

4 Is the 3% target rate a reasonable one?

SUGGESTIONS FOR LEADING THE DISCUSSION

Try beginning the discussion by collecting students into three groups depending on their recommendation for process improvement. More than likely, a whole lot of students will be unwilling to commit themselves to only one group, but will argue the pros and cons of each option. Your job is to make sure that the resulting discussion is brought to closure in some clarified fashion. It will be useful to write the three options on the board, and list, for each option, what information argues in favor of (or against) that option and what additional information might be needed to complete the decision.

At some point, ask whether the 3% target rate is a reasonable one. This will provoke a flurry of responses. Although management has set the 3% rate (from where? based on what?), the target is ill defined. Do they want the target to be met always, or just on average? The answer will influence, to some extent, the proposed solution.

TYPICAL RESULTS AND APPROACHES

The time series plot of the 30-minute readings on incoming calls suggests a regular daily pattern; number of calls picks up during the early morning hours, peaking around 8:00 A.M. A second, but smaller, peak is observed during the afternoon hours. This pattern will lead some students to recommend that management encourage off-peak usage of the system.

Time Series Plot of CALLS: Half-hour Readings, Monday through Friday

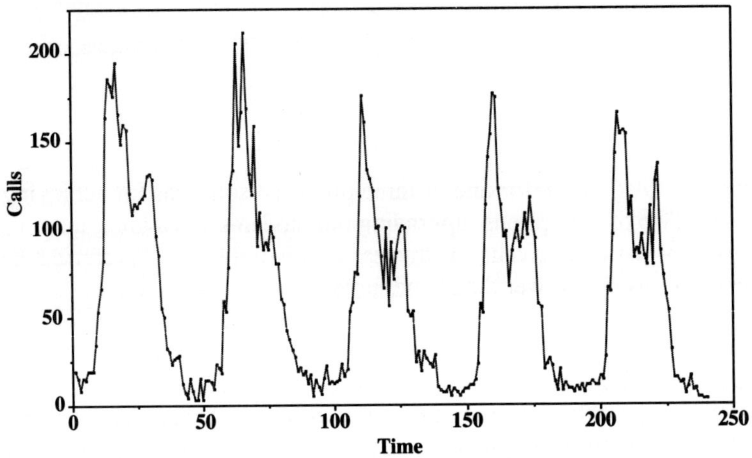

The daily data produces: an average number of incoming calls of 2846 per day with a standard deviation of 261, mean HANGUPS of 66 per day (standard deviation 122), and anywhere from 11 to 14 MODEMS used during this time period. The rate of abandoned calls (RATE = 100 × HANGUPS/CALLS) is 2.2% on average, less than the target of 3%. The average abandonment RATE is influenced by outliers; 19% of the days in the data set exceed the 3% target, the maximum daily RATE is 12.3%, and the median RATE is 0.25%.

There is no statistically significant difference from weekday to weekday for MODEMS, RATE, or HANGUPS. ANOVA indicates a higher average number of incoming calls on Mondays than on other days of the week.

The relationship between MODEMS and abandonment RATE is mixed. The early part of the time series shown below indicates that the target rate of 3% can be obtained even when as few as 12 modems are operable. During the middle and end of the data set, the abandonment rate can soar to 7% when the number of modems available drops to 13 or fewer. However, since the directive is to achieve a 3% or lower abandonment rate, most students will recommend upgrading or adding to the modem pool.

Daily Time Series Plots of Abandonment RATE, MODEM Availability, and Target Rate

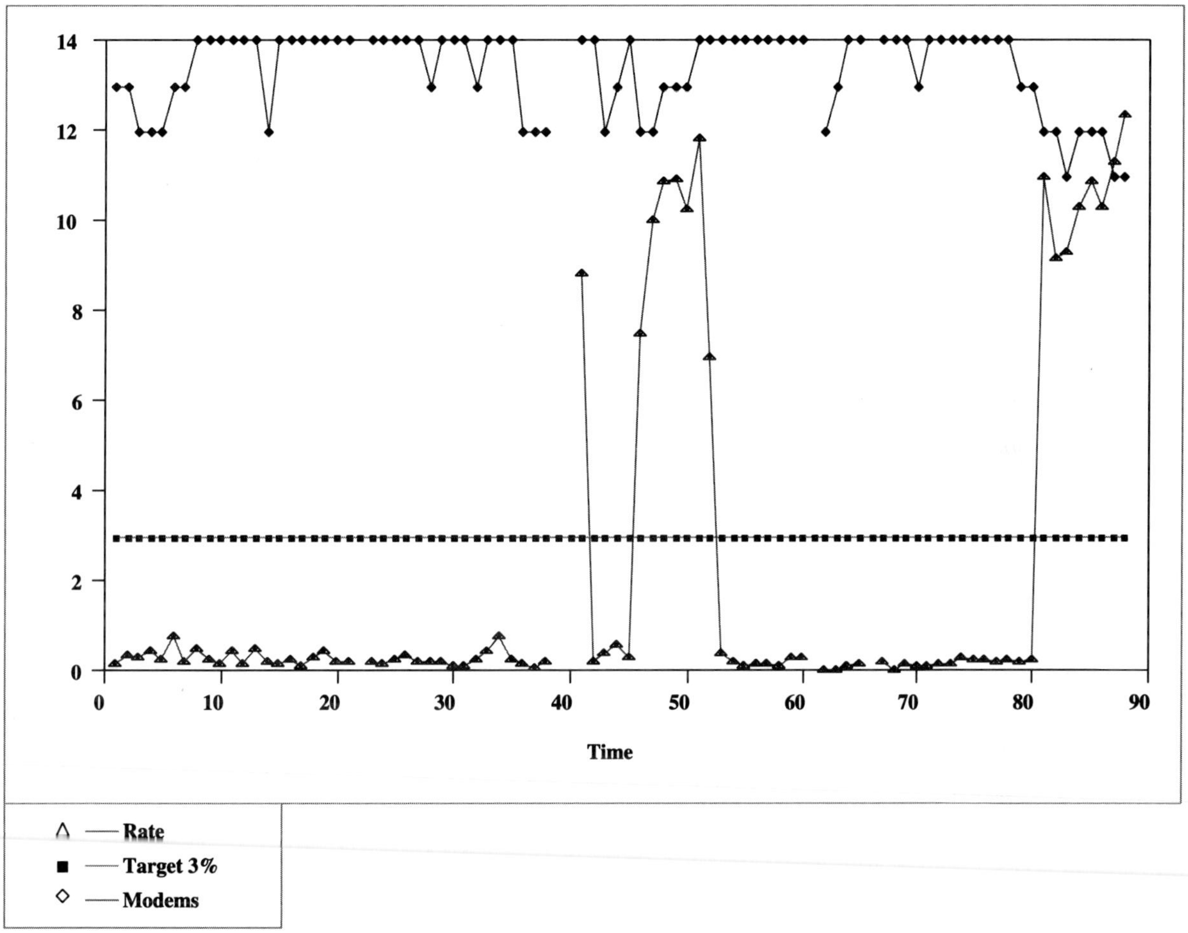

Some students will come to the same conclusion based on the correlation between HANGUPS and MODEMS (−48%) and/or RATE and MODEMS (−53%). The least squares fit of RATE versus MODEMS shows the influence of outliers on the relationship.

Least Squares Fit of Abandonment RATE versus MODEM Availability

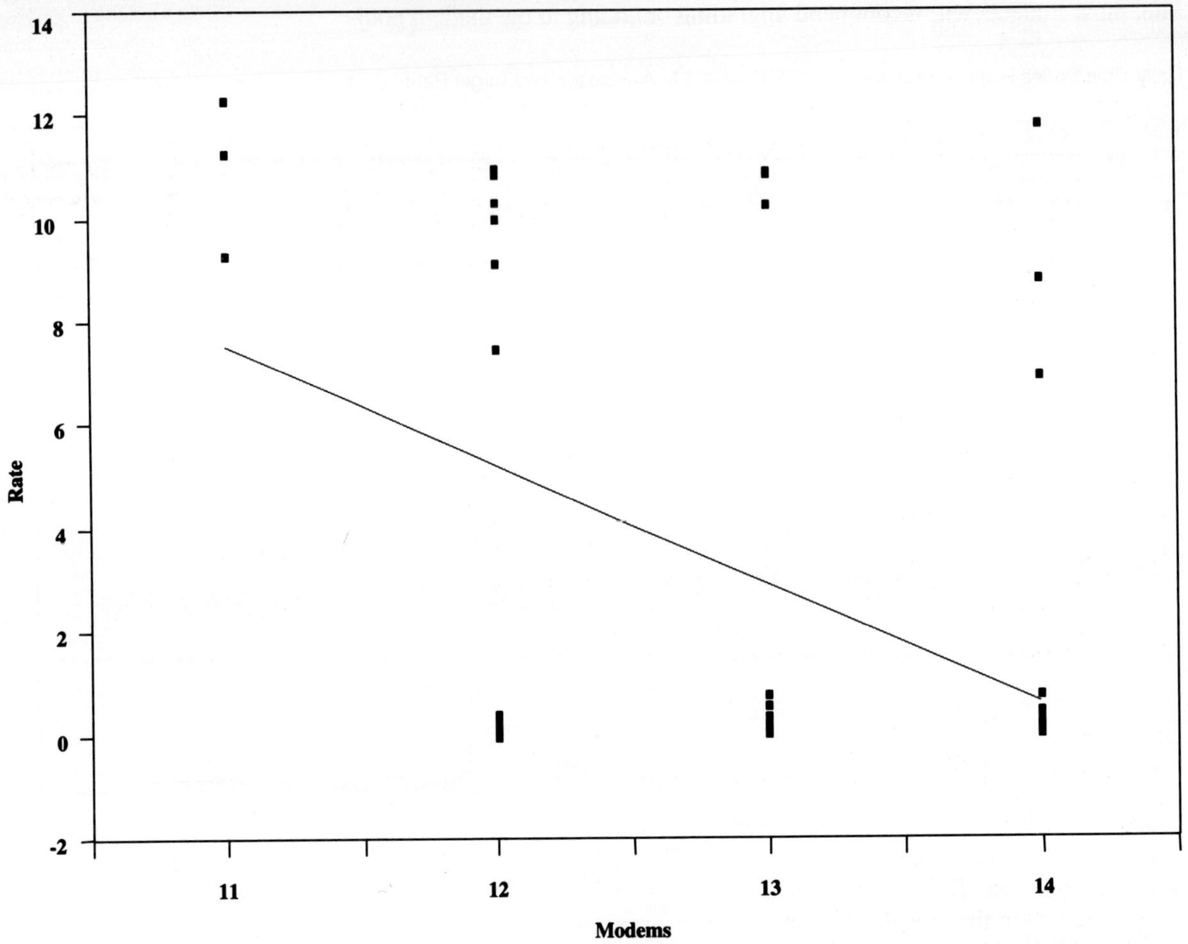

CASE **52**

ABANDONED CALLS—II, INSTRUCTOR'S NOTE

Abstract: This case is the second in a process improvement study. Part I of the series offered improvements that lowered the average abandoned-calls rate by almost 2%. This part of the case asks whether further improvements are possible, or even needed.

Level:	Beginning, intermediate.
Techniques:	Time series plots, ANOVA, nonparametric tests.
Typical Class Discussion Time:	30 minutes.
Data Set Size:	60 time series observations, 2 variables.

GENERAL COMMENTS

No fancy statistical techniques are needed here: standard parametric and/or nonparametric tests of differences in averages suffice. Sorting out what it all means, and how to convince the various interested parties, brings the excitement to the case discussion.

An unresolved issue from Part I of this series is management's directive about the target rate. In Part I, the desired rate was set at 3%, but unresolved were how this was to be measured (hourly, daily, weekly) and whether this meant on average. A clarification from on high appears in Part II—here, "abandoned calls are not to exceed 1% of all attempted calls" suggests that the rate can never exceed 1%. That's where the trouble lies.

The question stated in the case is whether adding a modem will lower the abandoned calls rates. The statistical tests say not, on average. However, the added modem is more likely to meet the "1% of all calls" directive from the company's CEO. Ask your students, either in their case write-ups or during a class discussion, to clarify this apparent contradiction. Doing so will help them understand several of the more puzzling aspects of statistical inference.

QUESTIONS FOR ASSIGNMENT

1 Test the hypothesis that the average rate of abandoned calls does not differ between the 13- and 14-modem populations. What assumptions (if any) underlie your test? Do the data appear to be consistent with these assumptions?

2 Test the hypothesis that the average rate of abandoned calls does not differ from weekday to weekday. What assumptions (if any) underlie your test?

3 Would control chart analysis help to answer the questions in this case? Why or why not?

SUGGESTIONS FOR LEADING THE DISCUSSION

As in Part I of this series, begin the discussion by telling students that they must decide whether to add the 14th modem. Unlike Part I, however, do not allow any fence-sitting. If you need to, give the undecided students a few minutes to review their analyses. Begin the discussion only when everyone is willing to sit in one camp or another.

Then let the two groups at each other. Your job is to play the devil's advocate. When the "13-modem" group presents their case, play the role of the salespeople's representative who wants to know why a difference of 0.15% should not be important to those out on the road selling the company's product. (The average RATE is 0.3% with 13 modems and 0.15% with 14 modems.) Force the students in this group to explain in nontechnical detail the notion of sampling error.

When the "14-modem" group recommends the additional modem, play the role of the CFO who is worried about "where this will all end." Ask for a guarantee that the 14th modem will meet the 1% goal (which, of course, they can't do). Then ask them to explain the difference in the recommendations between the two groups.

If students use control charts, highlight the various assumption violations that are likely to occur in this time series. Expect students to ask for more (or different) data; their discomfort is likely to revolve around accurately calculating small fractions, which is the crux of this case.

TYPICAL RESULTS AND APPROACHES

For this data set, the average abandonment rate (RATE = 100 × HANGUPS/CALLS) is 0.2%. A few outliers influence the average; the median RATE is zero.

Two-population and k-population tests of averages (or medians) from parametric or nonparametric methods all generate the same inferential outcome. There is no statistical difference between averages for the 13- or 14-modem populations for the variables CALLS, HANGUPS, or RATE. Similarly, there is no statistical difference between averages by day of the week for the same variables.

The sequence plot of RATE (below) demonstrates that the 1% maxim is violated most often when only 13 modems are available. Only the last of the four points exceeding the 1% ceiling occurs with 14 modems, and it just barely exceeds 1%.

Time Series Plot of Abandonment RATE and 1% Target Rate

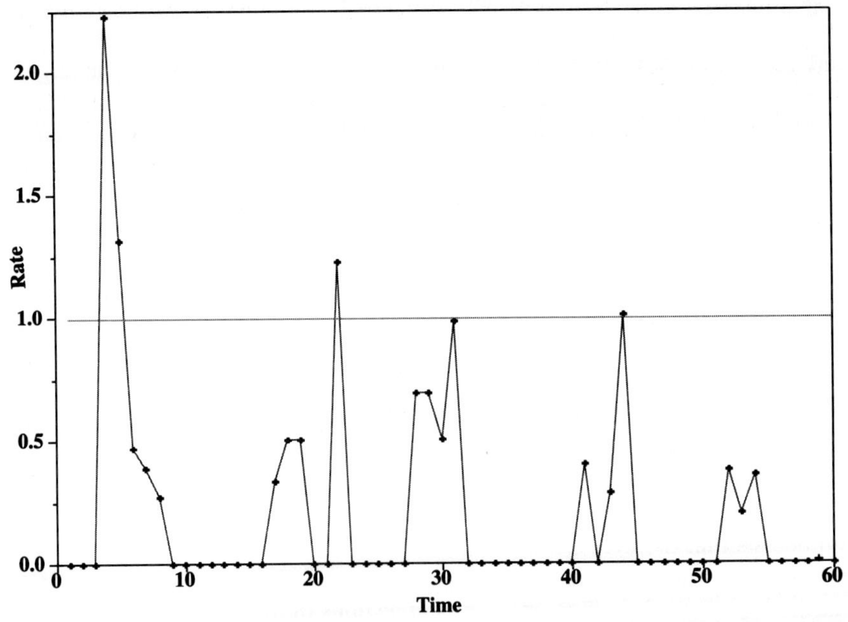

CASE **53**

EVERYBODY DOES IT, INSTRUCTOR'S NOTE

Abstract: A chance to think about procedures for obtaining needed data, in a situation in which people probably won't respond truthfully.

Level:	Intermediate.
Techniques:	Survey design.
Typical Class Discussion Time:	30–45 minutes.
Data Set Size:	No data.

GENERAL COMMENTS

Students will have to think a bit about this one. The problem is inherently difficult, and most statistical survey approaches will have many drawbacks. Problems of definition (What are we measuring? What's "inappropriate?" How much do we care?) rear their heads. The mechanics of dealing with the information (How can we identify all the files? Who should review them?) are daunting. It's an opportunity for students to explore an unfortunately real situation and the limitations of available information in a managerial context.

QUESTIONS FOR ASSIGNMENT

The case questions work well here, but you may wish to structure the problem a bit by asking other questions such as:

1 What measure of bad behavior is the appropriate one here? Percentage of employees? Percentage of system resources? (Measured how?) Over what time period?
2 What constraints must be placed on this inquiry? (Employees' legal rights, company policies, practical aspects—e.g., can we shut down the computers to examine them?)

For mature students, you might want to ask something like:

3 Should this problem be handled by IS at all? Why or why not?
4 Would it make more sense to ignore this particular problem, rehire the employee, and start a broader-based education program to control inappropriate use of computers? To change company policies to permit such use under certain conditions?

SUGGESTIONS FOR LEADING THE DISCUSSION

The case lends itself to a pretty general, free-wheeling discussion, and probably all that's necessary is to ask for recommendations, followed by reactions to them or counterproposals. It's helpful in this kind of discussion to write key points on the blackboard as they emerge (perhaps in such categories as "Constraints," "Objectives," "Methods," and/or "Strengths and Weaknesses"). At the end of the discussion, this list provides a ready-made summary.

TYPICAL RESULTS AND APPROACHES

There's probably no *good* approach here, and we favor the "punt" approach: forget it, and focus on developing policies and procedures to reduce this kind of behavior in the future. Cut your losses and rehire the employee: all the approaches we know of cost more than they're worth, and when all is done, you'll *still* have to rehire the employee and reeducate everybody anyway. If you feel you *must* gather the information, here are some approaches:

- Hiring an outside consultant to administer a survey probably provides the best guarantees of truthfulness, but even then, not all employees will be trusting.
- The mainframes could presumably be programmed to provide a sample of files to be reviewed for appropriateness, perhaps by the individuals' managers. The individuals and managers will presumably be unenthusiastic about this. How this could realistically be extended to all the personal computers is problematic.
- To be sure that inappropriate files aren't erased or disguised before the information is gathered, the inquiry should be more or less confidential. This raises concerns about privacy rights, and so forth. Alternatively you could announce the impending inquiry (and carry it out) not with a view towards getting good information, but hoping that employees *will* erase the offending files and stop engaging in this behavior.

In the actual situation on which this case is based, the IS division was directed to examine all file *names* and report anything that looked suspicious (BOWLING, BIORHYTHM, FLIGHTSIM, etc.) to IS management, which in turn was to speak to the manager of the individual employee involved. The IS division had no desire to be the bad guy in the eyes of the users (they already *were* for other reasons), but their protests were swept aside on policy grounds. A comedy of sorts ensued. One manager was duly notified that one of her employees had suspicious files labeled BIO1, BIO2, and the like. Investigation revealed that the names stood for "Basic Input/Output": the employee was a systems programmer, and the manager felt her time had been thoroughly wasted. The IS department employees gleefully renamed their own weekly status reports and other legitimate files with such names as GAMBLING, BIORHYTHMS, SOFTBALL, and so on, so that their *own* management would have to investigate them, only to find them innocent (at least innocent of *those* offenses). Finally, enough managers put pressure on Fred and the president, and the whole thing was dropped. Whether the original employee was rehired was not reported.

CASE **54**

MARRIAGE PREP, INSTRUCTOR'S NOTE

Abstract: A typical newspaper report. Students who are willing to talk back to the report a bit will become suspicious.

Level:	Beginning.
Techniques:	Ratios.
Typical Class Discussion Time:	15 minutes.
Data Set Size:	2 by 2 (or 2 by 3) cross-tabulation.

GENERAL COMMENTS

This case seems to be a simple chi-square (or test between two proportions) to those who are in the middle of learning statistical techniques. Some simple analysis of the data suggests, though, that maybe there isn't much to be learned from this report. It's a good case to use in raising awareness of the kinds of statistics with which we are regularly bombarded. It can also be handed out in class with no advance preparation, so it's useful in, say, the first class period.

QUESTION FOR ASSIGNMENT

The case questions here work pretty well. No further questions are really needed.

SUGGESTIONS FOR LEADING THE DISCUSSION

The assignment questions are fine here.

TYPICAL RESULTS AND APPROACHES

The usual conditions for applying chi-square tests or tests for the equality of two proportions don't really apply here, as not all cell sizes are greater than 5.

The main message, though, comes from some common sense and discipline. The article says that 4% of 64 couples were divorced in the treatment group, at least it *seems* to say that. How can this happen? Two couples out of 64 would be 3.125%, and 3 couples out of 64 would be 4.6875%, neither of which rounds to 4%. This is either simply wrong or there are some complicated issues that we don't know about. In either case, it's sloppy reporting. Similarly, you can't get 24% of 18, 14, or 32 using an integral number of people or couples.

Finally, the comparison group is specifically described as consisting of two subgroups, but only a combined percentage for these subgroups is reported. Why?

Beyond this, we don't know. Perhaps there are explanations for these discrepancies (e.g., maybe some of the original groups died, and were excluded from further analysis). The main point would seem to be that before we conclude anything from this article, we should probe a bit deeper.

CASE **55**

PREDICTING SNOWFALL, INSTRUCTOR'S NOTE

Abstract: An introduction to forecasting ideas, but with enough reality to require careful thought. Students will have to deal with missing data, choose a forecasting technique and criterion, and different months will probably require different techniques.

Level:	Intermediate.
Techniques:	Time series, forecasting.
Typical Class Discussion Time:	30–45 minutes.
Data Set Size:	60 time series observations, 10 variables.

GENERAL COMMENTS

Edgar Fielder (or was it Yogi Berra?) is quoted as saying something like "Prediction is hard . . . particularly the future!" Students may get a glimpse of the truth of this remark from this case. It lends itself to a variety of forecasting techniques, and as near as we've been able to make out, no one of them is uniformly better than another. The different months show different kinds of patterns. Students will have to deal with the amounts recorded as "trace" and with missing values, which will interfere with many built-in time series approaches. You can make the assignment simpler by asking for a forecast for January, which has no missing values.

QUESTION FOR ASSIGNMENT

We like the case question as given, leaving the students to look at various plots and to choose a few forecasting approaches to compare. You could also specify a few alternatives such as moving averages, simple exponential smoothing, random walk, and so forth, and ask that they be compared, if you wish to provide more structure.

SUGGESTIONS FOR LEADING THE DISCUSSION

Here's one approach:

- Before asking for forecasts, ask the class to suggest the criteria we should use to choose among alternative forecasts. A variety of measures of fit, absolute error, and the like, will usually be suggested, and they can be recorded on the blackboard.
- Probe these answers a bit, emphasizing particularly the issue of whether the method should be judged on the basis of how well it did on these data and how well the particular one-year forecast performs, as well as

on any external knowledge about how the process behaves. Nonnumeric criteria such as understandability should be considered here, too.

- Now solicit forecasts and the corresponding criteria, and record them on the board, too. Eventually you can decide the winner by class vote or other method based on the criteria listed. We've tried various schemes to motivate this; perhaps the best is to offer a (say) $25 check payable to the charity of the winner's choice. (You may wish to reserve the right to decide what is or is not a charity!)

This has the advantage of forcing students to think a bit about the alternative measures of fit, and their intents in using the forecasts (e.g., if you manage the ski area, does the June forecast really interest you much?), as well as the numerical techniques.

TYPICAL RESULTS AND APPROACHES

MINITAB's basic descriptive statistics for the various months are (with trace amounts of snow recoded as zero):

	N	N*	MEAN	MEDIAN	TRMEAN	STDEV	SEMEAN
SEPT	60	0	1.618	0.000	0.952	4.149	0.536
OCT	58	2	7.80	5.80	6.76	8.67	1.14
NOV	59	1	21.42	19.50	20.65	11.11	1.45
DEC	59	1	24.52	22.00	23.32	13.92	1.81
JAN	60	0	24.67	21.20	23.94	12.48	1.61
FEB	59	1	24.86	24.50	24.03	12.44	1.62
MAR	58	2	27.03	24.45	26.37	12.36	1.62
APR	59	1	14.97	14.00	14.66	9.51	1.24
MAY	59	1	4.084	2.000	3.093	6.662	0.867
JUNE	60	0	0.938	0.000	0.280	3.089	0.399

	MIN	MAX	Q1	Q3
SEPT	0.000	27.000	0.000	1.000
OCT	0.00	41.45	0.75	11.93
NOV	0.00	56.00	14.00	26.50
DEC	6.00	72.00	13.80	29.50
JAN	5.00	71.50	15.05	33.38
FEB	2.00	79.00	17.00	31.00
MAR	7.20	76.50	19.45	34.25
APR	0.00	36.00	6.70	22.00
MAY	0.000	41.000	0.000	5.500
JUNE	0.000	15.500	0.000	0.000

For most months, using the mean as a predictor works fine. The nonwinter months of September, October, May, and June are erratic, with occasional rare bursts of snow. The winter months display a more classical kind of time series structure.

For example, for January (the only winter month with no missing values), the basic time series plot and autocorrelation function are:

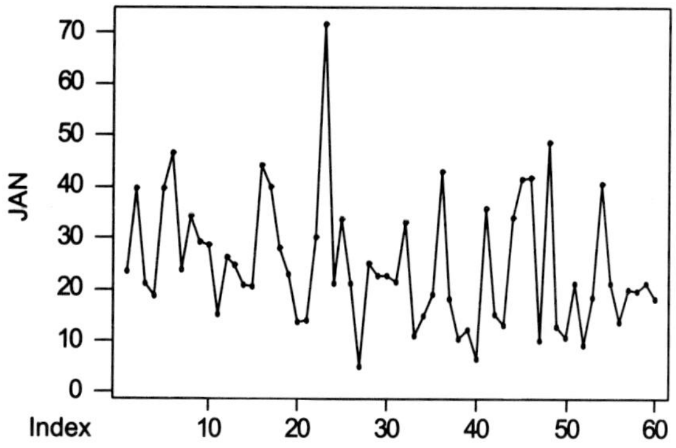

and

```
          -1.0 -0.8 -0.6 -0.4 -0.2  0.0  0.2  0.4  0.6  0.8  1.0
              +----+----+----+----+----+----+----+----+----+----+
  1   0.063                          XXX
  2  -0.081                        XXX
  3  -0.025                        XX
  4  -0.103                       XXXX
  5  -0.070                        XXX
  6   0.054                          XX
  7   0.143                          XXXXX
  8  -0.044                         XX
  9   0.107                          XXXX
 10   0.018                          X
 11  -0.101                       XXXX
 12  -0.020                         XX
 13   0.188                          XXXXXX
 14  -0.039                         XX
 15  -0.112                       XXXX
 16   0.003                          X
 17  -0.002                          X
```

The corresponding three-year moving average plot and forecast are:

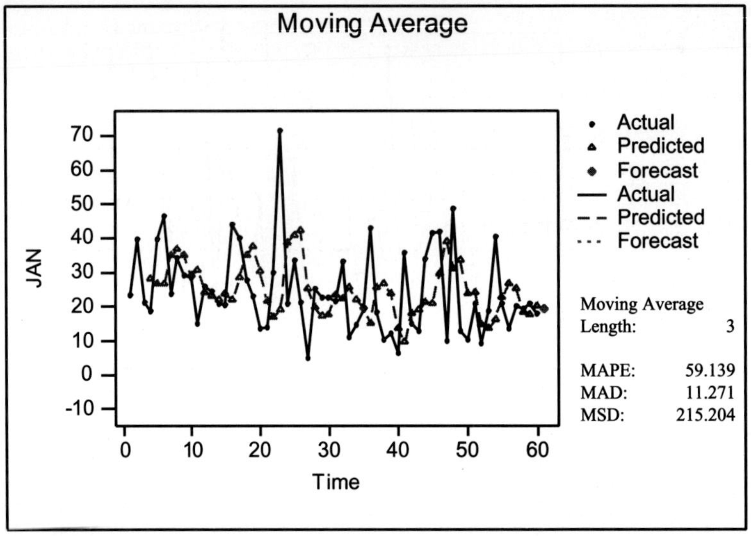

and

Period	Forecast	Lower	Upper
61	19.5	-9.25286	48.2529

A simple exponential smoothing approach yields:

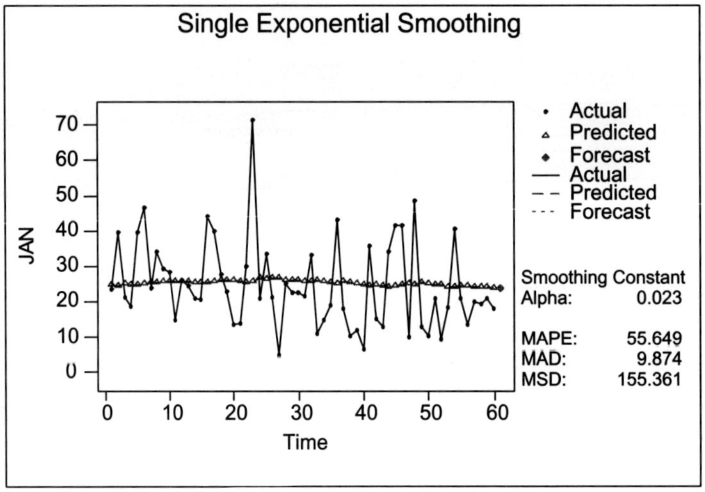

and

Period	Forecast	Lower	Upper
61	23.8217	-0.370628	48.014

This all nets out to:

Method	Forecast for period 61, January
Naive ($X_{t+1} = X_t$)	18.00
Average(X_{t+1} = average[$x_1, \ldots, x_t$])	24.67
Three-year Moving Average	19.50
Simple exponential smoothing ($\alpha = .023$)	23.82

The actual values for periods 61–63 were:

WINTER	SEPT	OCT	NOV	DEC	JAN	FEB	MAR	APR	MAY	JUNE
61.00	0.00	7.00	36.00	18.50	25.50	45.00	56.00	25.50	12.50	0.00
62.00	0.00	12.60	16.80	20.50	36.50	38.00	28.00	22.00	0.00	0.00
63.00	7.00	16.50	32.00	36.50	42.50	24.00	23.00	21.00	6.50	0.00

The average method gives the closest prediction in this particular case (January of period 61).

CASE **56**

CIGARETTE RANKINGS, INSTRUCTOR'S NOTE

Abstract: This case illustrates some simple market research ideas and can be used at a variety of levels of difficulty, from simple spreadsheet sorts up to multivariate cluster analyses.

Level:	Beginning, intermediate.
Techniques:	Sorting, averages, cluster analysis.
Typical Class Discussion Time:	30–45 minutes.
Data Set Size:	20 cross-sectional observations on 8 rankings.

GENERAL COMMENTS

This case illustrates a common issue in marketing research: it's not enough to know how your product is received; you must find out how it is received by different segments of the markets. It also illustrates how simple spreadsheet operations such as sorting can lead to insight.

QUESTION FOR ASSIGNMENT

The case questions work well here. To make the case easy, suggest that the students compute the average rank of each cigarette and sort the columns of the data matrix accordingly. The case then becomes mostly a drill-and-practice spreadsheet exercise. To make it more difficult, use only the questions given in the case.

For more advanced students, you might suggest they apply a clustering procedure, probably based on some sort of distance measure (depending on the computer facilities available and the methods you cover in your course), but much of the structure appears using nothing more than sorting.

SUGGESTIONS FOR LEADING THE DISCUSSION

Possible approaches here include:

- Play the marketing director's role, and ask, "Well, what does the market look like?" so that the students can structure the material for themselves.
- Ask specifically which cigarette is preferred. Follow up by asking, "Preferred *by whom?*" This leads into a discussion of whether there are meaningful groups in the data, and if so, whether we are interested in groups of cigarettes, groups of judges, or both.

TYPICAL RESULTS AND APPROACHES

The listing below shows the data with columns sorted by the average rank assigned by the 20 panels of judges.

PANEL	CIG1	CIG2	CIG5	CIG3	CIG4	CIG6	CIG8	CIG7
1	2	5	3	8	6	4	1	7
2	1	5	6	8	4	3	2	7
3	1	5	6	7	3	4	2	8
4	4	2	1	5	6	7	3	8
5	1	4	3	2	6	7	5	8
6	1	7	2	5	4	6	3	8
7	4	1	6	2	3	5	8	7
8	4	1	5	2	3	7	8	6
9	3	2	6	1	4	5	7	8
10	4	1	2	7	3	5	6	8
11	6	7	4	1	5	2	3	8
12	3	1	6	2	4	5	7	8
13	8	1	5	7	6	3	2	4
14	3	1	4	6	2	5	7	8
15	3	1	2	6	4	5	7	8
16	3	1	5	6	2	4	7	8
17	8	5	1	4	2	3	7	6
18	1	7	4	2	8	5	3	6
19	3	8	4	2	5	7	1	6
20	1	6	4	2	5	3	7	8
Average	3.2	3.55	3.95	4.25	4.25	4.75	4.8	7.25

Cigarette 7 is clearly in a (low) class by itself. Students will have to make some sort of decision about how much difference in (say) average rank is meaningful, but if we take a difference of 0.3 as meaningful, then the cigarettes would group into cigarettes 1; 2; 5, 3 and 4; 6 and 8; and 7. If we took a difference of 0.5 as important, the groups would be different, for example. If the data rows are sorted according to their ranking of cigarettes 1, 2, 5 and 3, we obtain the following listing.

PANEL	CIG1	CIG2	CIG5	CIG3	CIG4	CIG6	CIG8	CIG7
5	1	4	3	2	6	7	5	8
3	1	5	6	7	3	4	2	8
2	1	5	6	8	4	3	2	7
20	1	6	4	2	5	3	7	8
6	1	7	2	5	4	6	3	8
18	1	7	4	2	8	5	3	6
1	2	5	3	8	6	4	1	7
15	3	1	2	6	4	5	7	8
14	3	1	4	6	2	5	7	8
16	3	1	5	6	2	4	7	8
12	3	1	6	2	4	5	7	8
9	3	2	6	1	4	5	7	8
19	3	8	4	2	5	7	1	6
10	4	1	2	7	3	5	6	8
8	4	1	5	2	3	7	8	6
7	4	1	6	2	3	5	8	7
4	4	2	1	5	6	7	3	8
11	6	7	4	1	5	2	3	8
13	8	1	5	7	6	3	2	4
17	8	5	1	4	2	3	7	6
Average	3.2	3.55	3.95	4.25	4.25	4.75	4.8	7.25

CASE **57**

UTILITY BILLS, INSTRUCTOR'S NOTE

Abstract: This case involves forecasting several univariate time series for budgeting purposes. Messy data problems, such as missing observations, add to the challenge. This case is amenable to a wide range of introductory/intermediate forecasting methods, such as smoothing and regression techniques.

Level:	Intermediate.
Techniques:	Time series plots, regression, forecasting, control charts.
Typical Class Discussion Time:	60 minutes.
Data Set Size:	34 monthly observations, 8 variables.

GENERAL COMMENTS

This case permits an examination of several forecasting methods. One of the variables is best predicted by its historical average; others benefit from smoothing or regression methods. Several tangential issues will arise, though:

- Some students will decide to forecast the total utility bill, GAS($) + ELEC($) + WATER($) + PHONE($), rather than each individual series. In this instance, the seasonality in GAS($) will overshadow patterns in the other three variables.
- Some students will attempt to replace the missing observation for PHONE($) and the earlier missing observations for TEMP. Some may even go to the Web to find the TEMP observations—if so, give them kudos!
- Expect a debate on whether it is best to forecast dollar values or consumption values of the utility bills.

QUESTIONS FOR ASSIGNMENT

1 Do any of these variables appear to be random? If so, can you reasonably forecast future values from a random time series?

2 Given the objectives stated in the case, should you use quantity or dollar amount of the bill (or both) to prepare your forecasts?

3 What impact do the missing values for TEMP have on your analysis? How do you propose to deal with this?

4 Is there any reason, from Terry's point of view, to forecast total utility consumption rather than each utility separately? Is there any reason, statistically speaking, to forecast total rather than individual energy consumption?

SUGGESTIONS FOR LEADING THE DISCUSSION

We assign this case very early in the semester, long before students have been introduced to formal forecasting methods. We specifically tell them that they are to use whatever "seat-of-the-pants" approach they want to generate forecasts. They typically use simple graphics and descriptive statistics. We then return to the data set throughout the course as advanced methods are introduced. Because the students are familiar with the data, they readily see the advantages of the structured forecasting tools.

During the case discussion, we ask students to list their forecasts on the board along with the methodology used to generate those forecasts. We then supply them with the actual values for April through June of 1998 (below), and use classtime to introduce the MSEP. Given their forecasts, students are asked to calculate their MSEP, and we compare the values and underlying methodologies.

	Gas	Elec	Temp	Gas($)	Elec($)	Water	Water($)	Phone($)
April 98	50	483	50	29.87	40.45	3	10.20	35.22
May 98	23	357	61	17.92	31.80	3	10.20	40.76
June 98	22	464	73	17.46	39.01	4	11.54	38.84

If you have introduced regression as a forecasting tool, expect some students to report regression models in which the dollar amount of a bill is predicted by the quantity consumed. They'll get wonderfully impressive R-squared and *p*-values, yet only a few will recognize the implausibility of using such a model for out-of-sample forecasting.

Expect also that some students will choose to discard the first few observations for each variable on the grounds that those observations reflect various one-time costs such as down payments or installation fees.

TYPICAL RESULTS AND APPROACHES

The pattern in the monthly water bill, WATER($), is consistent with randomness according to the runs test ($p = 53\%$), and the Individuals and Moving Range charts indicate a stable process if the first month is discarded. This suggests forecasting WATER($) with its average: $9.29.

The PHONE($) variable is not consistent with randomness with a *p*-value of 0.009 for the runs test. (PHONE($) has one missing value in 1/97, which is replaced here by the average of the other January values in the data set: $35.26.) An MA(4), MA(5), or single exponential smoothing model (with smoothing parameter 0.205) all work well. The out-of-sample root mean square prediction errors for these models over April through June of 1998 are 2.64, 2.65, and 2.83, respectively. Within-sample, the single exponential model produces lower MAD (9.54) and MSD (178.61) than the other methods.

Single Exponential Smoothing Model of Phone($)

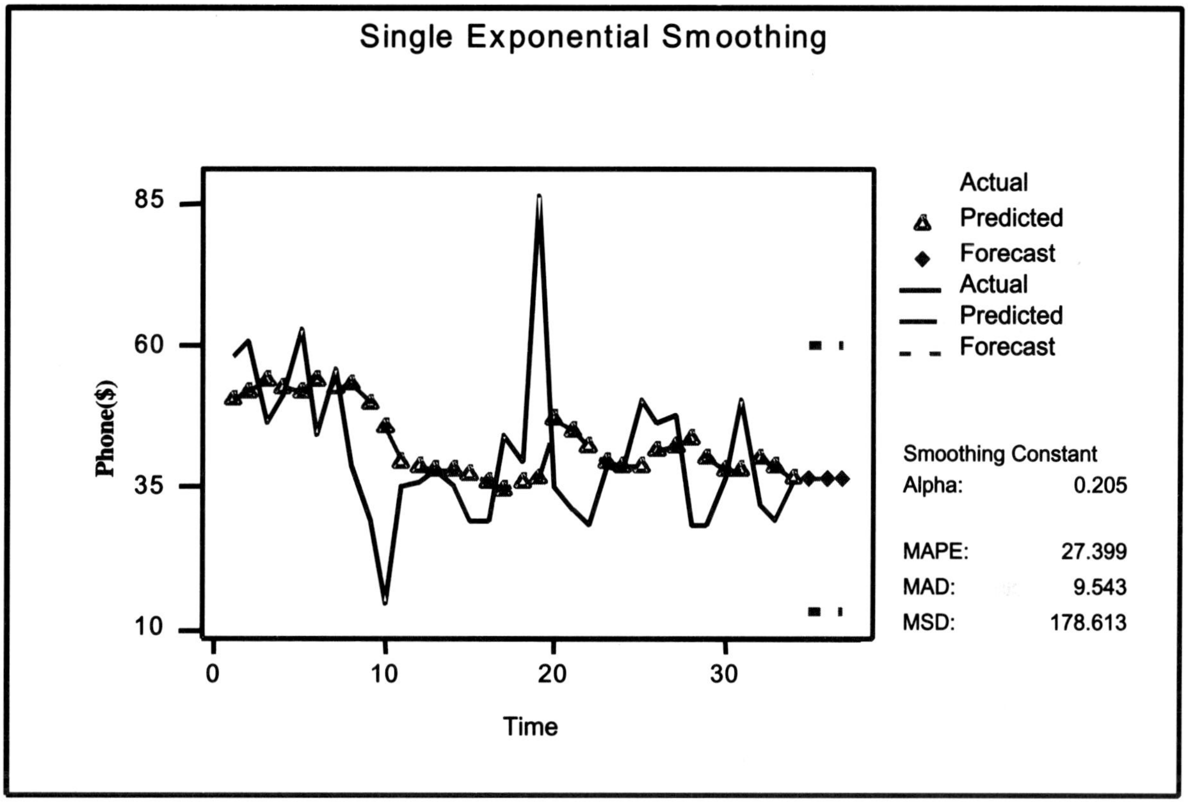

For more advanced treatments, the autocorrelation function of PHONE($) has a notable spike at lag 9.

ELEC($) does not appear to be a random series; its runs test *p*-value is 3.8%. The series is seasonal, and a regression equation using second- and third-quarter dummies produces ELEC($) = 39.1 − 5.76Quarter2 − 5.18Quarter3. Although both dummies are statistically significant at the 5% level, the R-squared of this model is only 20%. TEMP is only marginally useful as a predictor in regressions of ELEC($).

A single exponential smoothing model of ELEC($) produces the smallest within-sample prediction statistics among the rudimentary smoothing methods; a single exponential model with smoothing parameter 0.153 generates a MAD of 4.85 and MSD of 35.88. Regarding out-of-sample prediction accuracy, models using either the average of ELEC($) over the time series or the historical monthly averages to forecast April through June of 1998 produce the lowest RMSEP. These are 3.82 and 4.12, respectively.

The GAS($) variable is highly correlated with TEMP, and regression is helpful for modeling and forecasting GAS($):

GAS($)=102 - 1.2 Temp	R^2=90% s=7.2 RMSEP=9.4
GAS($)=39.8 + 25.5 Winter - 24.5 Summer	R^2=84% s=8.8 RMSEP=6.1

All coefficients in these regressions are significant at the 5% level. (The R-squared and s-values are not directly comparable, since the first equation has fewer observations, attributable to the missing values for TEMP. WINTER is defined as 1 for months December through February and 0 otherwise; SUMMER is 1 for May through August, 0 otherwise.) Based only on out-of-sample prediction accuracy, though, using past monthly averages as predictors of April through June of 1998 generates a RMSEP of 1.7.

Time Series Plot of GAS($)

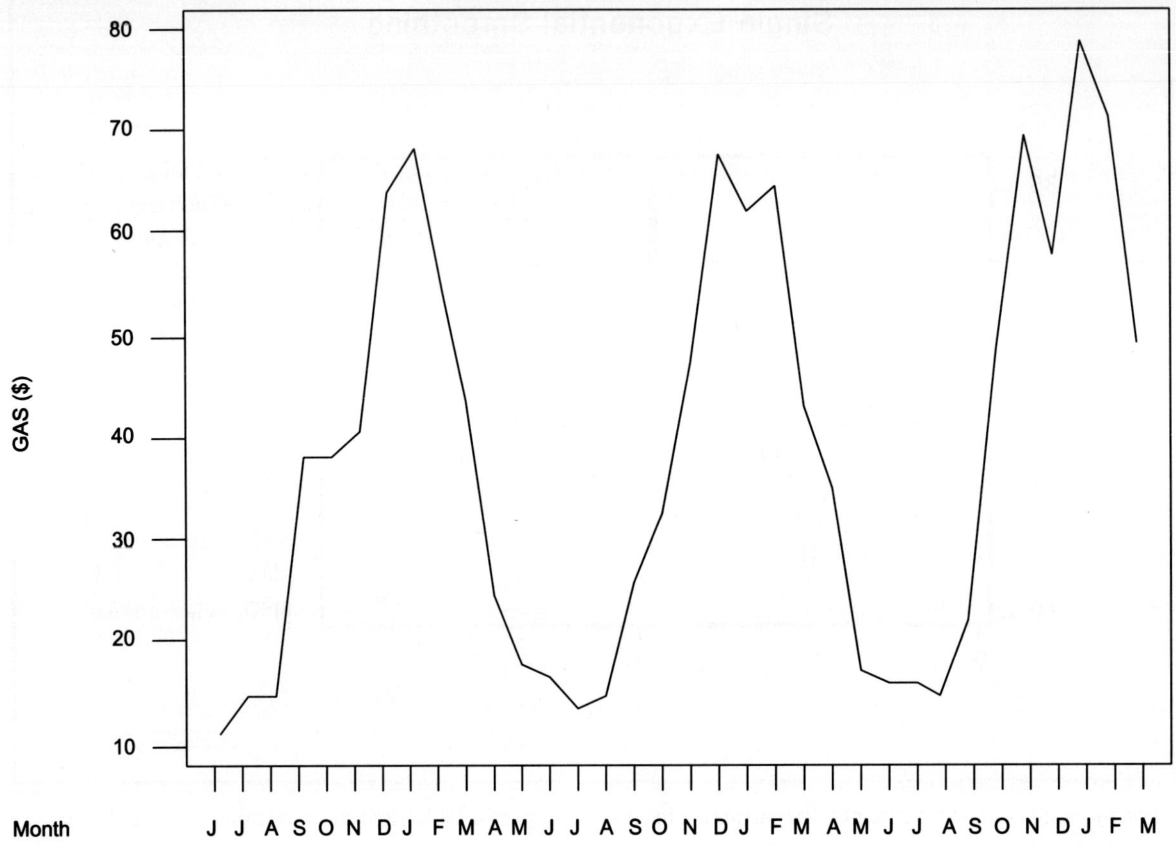

CASE **58**

JOB HOPPING AT CHIPS, INC., INSTRUCTOR'S NOTE

Abstract: The response variable in this case is binary and the predictors are either nominal or metric. Two-population means tests and chi-square tests can be used to answer one of the assigned questions. Because so many of the predictors are statistically insignificant, though, students will torture the data in search of statistical significance while losing sight of the important managerial question: what does all this mean?

Level:	Beginning, intermediate.
Techniques:	Cross-tabulations, Chi-square tests, two-population hypothesis tests.
Typical Class Discussion Time:	30 minutes.
Data Set Size:	71 cross-sectional observations, 13 variables.

GENERAL COMMENTS

Students are asked to prepare a profile of a job-hopping employee. The data really don't much help with this task. Of the 12 potential indicators of turnover, only 1 has strong statistical significance: whether the employee is full-time or part-time. Experience, previous job-hopping activity, education, and technical skills have little to do with turnover. From a managerial point of view, that's a nice outcome: if the company wants to minimize turnover, it should offer full-time instead of part-time employment opportunities.

Although that's the statistical answer, it may not be the appropriate managerial decision. The case states that part-time people are not paid benefits, which begs the question of whether losing an employee to a competitor after training is more expensive than the costs associated with full-time employment. Information from the human resources department is needed to address this larger issue.

QUESTIONS FOR ASSIGNMENT

1 Which of the potential indicators of job-hopping are nominal? Ordinal? Continuous? How does this categorization influence the choice of statistical tests you might undertake?

2 Which of the potential indicators of job-hopping are statistically significant?

3 Should the sample size influence which statistical tests you undertake? If, so, how and why?

SUGGESTIONS FOR LEADING THE DISCUSSION

There are several ways to focus the class discussion. If you prefer to emphasize the managerial decision-making process, begin the discussion by asking students to provide a profile of a typical job hopper. Before letting the discussion drift toward technical details, ask what you, as the head of human resources, should do to cut down on job hopping.

If you would rather emphasize hypothesis-testing methods, ask students how many different statistical techniques they used in preparing their case. Chi-square and *t* tests, and small sample versus asymptotic tests, should be expected. Play the devil's advocate, and ask students why they chose to use the test that they did. If you take this approach to the discussion, though, ensure that you don't leave the case without addressing the question of how all this information helps Chips to minimize job hopping.

TYPICAL RESULTS AND APPROACHES

Of the twelve possible predictors of KEEPER, only two have hope of demonstrating statistical significance: years of programming experience (YRS_PRG) and full-time employment status (FULLTIME). The two-sample *t* tests for the metric predictors generate *p*-values of: 18% for YRS_DP, 7% for YRS_PRG, 27% for JOBS_DP, 13% for JOBS_PRG, 13% for JOBS_CON, 73% for COLLEGE, 56% for GRAD, 26% for GENER3, 96% for GENER4, and 11% for DATABASE.

Means and Standard Deviations (parenthesized) by KEEPER

	YRS_DP	YRS_PRG	JOBS_DP	JOBS_PRG	JOBS_CON
KEEPER=0	10.03 (6.0)	8.84 (6.0)	4.03 (2.4)	3.73 (2.5)	0.54 (1.0)
KEEPER=1	12.06 (6.7)	11.71 (7.0)	4.91 (4.0)	4.94 (3.9)	1.09 (1.8)

	COLLEGE	GRAD	GENER3	GENER4	DATABASE
KEEPER=0	3.03 (1.6)	0.46 (1.0)	2.30 (1.1)	0.78 (0.9)	0.84 (0.8)
KEEPER=1	3.15 (1.3)	0.62 (1.2)	2.56 (0.9)	0.79 (0.8)	1.18 (1.0)

The chi-square test of independence between KEEPER and ENDUSER is 0.04 (*p*-value = 84%); the same test for KEEPER and FULLTIME is 10.43 with *p*-value of 0.1%.

Cross-tabulations of KEEPER and ENDUSER and KEEPER and FULLTIME

ENDUSER

KEEPER	Count Expected	0	1	
	0	15 14.5915	22 22.4085	37
	1	13 13.4085	21 20.5915	34
		28	43	71

FULLTIME

KEEPER	Count Expected	0	1	
	0	30 23.4507	7 13.5493	37
	1	15 21.5493	19 12.4507	34
		45	26	71

CASE **59**

MEDICATION ERRORS, INSTRUCTOR'S NOTE

Abstract: An easy case for Pareto charts, though some of the questions can't be answered on the basis of the data given.

Level:	Beginning.
Techniques:	Tables, Pareto Charts.
Typical Class Discussion Time:	30 minutes.
Data Set Size:	13 departments, 16 error types, 3 calendar quarters' data.

GENERAL COMMENTS

Nothing very complicated is required here, though students will have to wrestle a bit with which question is to be answered. We don't have any data on workload or patient-days or staffing over the three quarters, so in that sense we can't really tell if things are getting better or worse. On the other hand, we *can* determine trends about which errors are more common, either in the aggregate or within specific departments, though without further data, we probably can't get much insight into *why*.

QUESTIONS FOR ASSIGNMENT

You can make this case a bit simpler (in terms of computer manipulations) by asking students to focus on one or more specific departments (or on the grand total), but otherwise the administrative interests listed in the case form a good basis. Some possible questions include:

1 Which departments seem to have the most errors? Why? Has this changed over the time period involved?
2 Which types of errors are most common? Why? Does it differ for different departments?
3 Are we getting better or worse? Why do you say so?
4 Which types of error and which departments (if any) should we focus on improving in the next few months?

Several of these questions are unanswerable from the data given, but recognizing that is part of data analysis.

SUGGESTIONS FOR LEADING THE DISCUSSION

We like to start with, "Well, are we getting any better?" because it forces a discussion of the quality of the data *before* discussing the results. Once people begin to understand that without some sort of measure, say, of how many *total* medications were administered (not just those leading to errors), the number of errors is hard to interpret in terms of "better" or "worse."

Many students will want to say that therefore, we can learn *nothing* from the data, and this is an opportunity to make them be more specific, to frame the exercise in terms of "which questions we *can* and *can't* we answer," rather than just concluding that the data are either "adequate" or "inadequate" in general.

Basically, the questions we *can* answer relate to numbers of errors of different types and different departments, not those relating to *why.* In that sense, the data might provide some preliminary guidance or focus on certain areas, but probably not for any firm conclusions.

You can finish off the discussion by asking for recommendations. If your students have enough background, you might also raise the issue of whether some errors are more serious than others, whether some errors might just be tolerated, or whether in the spirit of TQM all errors should be viewed as opportunities for improvement.

TYPICAL RESULTS AND APPROACHES

Pareto charts for the grand totals across all departments for the first three quarters follow. Error types 104 (medication omitted), 108 (wrong dose given), and 106 (dose not given close to correct time) account for about half the errors in all three quarters. Error types 105 (duplicate or extra medication) and 109 (wrong medication) also contribute. Type 106 is hard to interpret, and we would need to know how "close" is close (15 minutes? Three hours?) to make conclusions about it. Type 104 is always most common, and that might be scary.

Pareto Chart for Error Type, First Quarter, 1995, All Departments

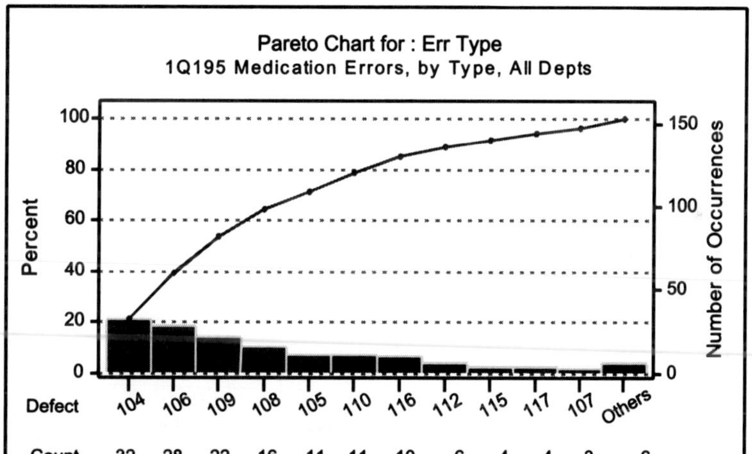

Count	32	28	22	16	11	11	10	6	4	4	3	6
Percent	20.9	18.3	14.4	10.5	7.2	7.2	6.5	3.9	2.6	2.6	2.0	3.9
Cum %	20.9	39.2	53.6	64.1	71.2	78.4	85.0	88.9	91.5	94.1	96.1	100.0

Pareto Chart for Error Type, Second Quarter, 1995, All Departments

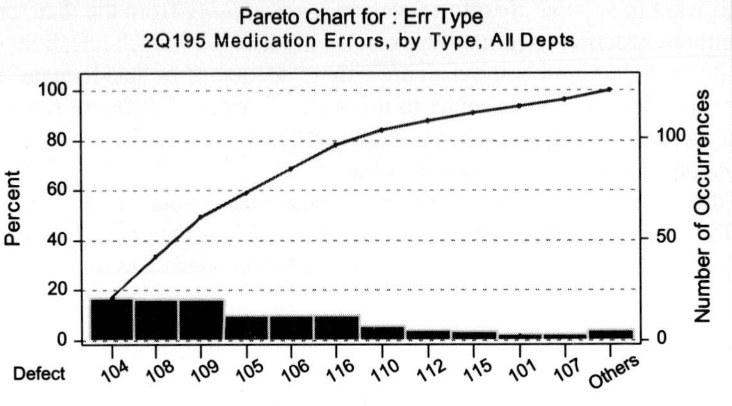

Count	21	20	20	12	12	12	7	5	4	3	3	5
Percent	16.9	16.1	16.1	9.7	9.7	9.7	5.6	4.0	3.2	2.4	2.4	4.0
Cum%	16.9	33.1	49.2	58.9	68.5	78.2	83.9	87.9	91.1	93.5	96.0	100.0

Pareto Chart for Error Type, Third Quarter, 1995, All Departments

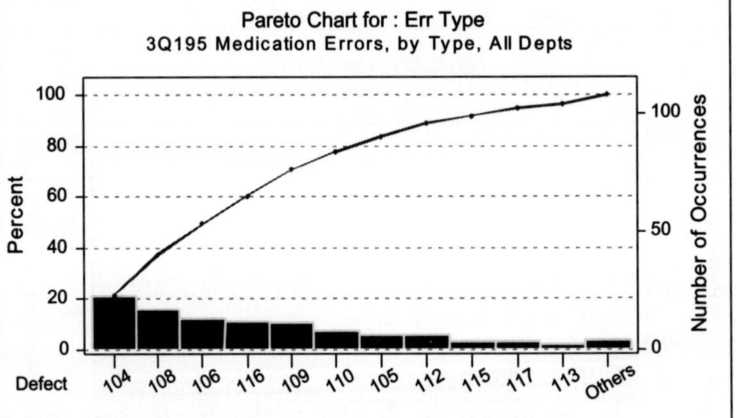

Count	23	17	13	12	11	8	6	6	3	3	2	4
Percent	21.3	15.7	12.0	11.1	10.2	7.4	5.6	5.6	2.8	2.8	1.9	3.7
Cum%	21.3	37.0	49.1	60.2	70.4	77.8	83.3	88.9	91.7	94.4	96.3	100.0

Table 4 probably needs no chart for its interpretation. The medical and OB departments have the most errors. Since for all we know, they have the most patients and/or the most medications to administer, it's hard to know if this makes us happy or not. On the other hand, if it were our job to follow up and investigate medication errors, those are the departments where we would apparently spend most of our time. This kind of result (that the data answer some questions but not all) is helpful in persuading beginning students to get beyond sheer memorization of statistical techniques.

CASE **60**

CLINICAL TREATMENT OF DEPRESSION, INSTRUCTOR'S NOTE

Abstract: This case can be used to demonstrate a number of statistical methodologies: two-sample means tests, chi-squares tests, regression, ANOVA, and ANCOVA. Virtually all levels of students will be able to do something interesting with the data. However, a careful critique of the experimental design is needed to arrive at meaningful conclusions.

Level:	Intermediate, advanced.
Techniques:	Hypothesis tests, cross-tabulations, regression, ANOVA, ANCOVA.
Typical Class Discussion Time:	45–60 minutes.
Data Set Size:	343 cross-sectional observations, 6 variables.

GENERAL COMMENTS

This case poses several questions: one about treatment modality and the other about cost prediction. Students are likely to miss the boat on the first question. In order to determine from these data whether the two types of depression respond differently to different treatment types, it must be assumed that treatment modality can be loosely measured by physician. Since Jones uses both medication and psychotherapy, and Smith relies on medication, this might be a reasonable assumption. Unfortunately, a confounding factor intervenes: the physicians' bedside manner. This part of the case provides a wonderful example of a flawed experimental design.

Several statistically significant length-of-stay prediction models can be devised, although the R-squared is disappointingly low in even the best models.

QUESTIONS FOR ASSIGNMENT

1 What statistical method should be used to answer the first question in the case?

2 Critique the experimental design described in the case.

3 Can one or more of the variables in the data set predict length of stay? If so, prepare a detailed description of how these predictions are to be derived.

SUGGESTIONS FOR LEADING THE DISCUSSION

It's fun to begin this case discussion by saying that you intend to terminate one of the two doctors and ask which one it should be. Expect a number of different responses, depending on which statistical methodology the student used. Some will choose Jones based on a simple comparison of the average lengths of stay (6.4 for Jones and 5.8 for Smith). If two-sample means tests were used, students may conclude that the data don't provide an answer. Others may recommend termination of Smith based on an ANCOVA model. Because so many of these variables interact in various ways, it's difficult to flush out the important factors. Your job is to ensure cohesion by the end of the discussion. Pair recommendations with methods on the board, and discover with the students why the answers differ. Some students will rightly argue that firing anyone based on these data alone is ill advised.

If students attempt advanced methods such as ANOVA with interactions or ANCOVA, do not leave the discussion until each of the coefficients from their models is interpreted in the context of the problem at hand. Doing so provides students with an invaluable lesson on, "What does all this mean?"

A less controversial way to begin this discussion is to ask whether the students were able to develop a method to predict patient costs. Regardless of how you begin the discussion, expect a sometimes-heated debate about cost and quality trade-offs in medical care treatment.

TYPICAL RESULTS AND APPROACHES

The average patient age is 41 years. The average length of stay is six days. Seventy-eight % of the patients are female. Dr. Smith treated 56% of the sample. The sample contains approximately equal numbers of major depressive disorder and bipolar disorder cases. Eighteen % of patients in this sample were treated on an outpatient basis. Dr. Jones treated no male patients on an outpatient basis. The distribution of age of patients with major depressive disorder is bimodal, with peaks at 40 and 70 years of age.

Among the eight different two-population means tests that can be performed on these variables, four render statistically significant differences at the 5% significance level: average length of stay (LOS) for inpatient versus outpatient status ($t = 4.5$); average length of stay for depression versus bipolar ($t = 3.0$); average age for inpatient versus outpatient status ($t = 3.2$); and average age for depression versus bipolar ($t = 3.9$). According to the two-population means tests, there is no statistically significant difference between average age of men and women, or between treating physician; there is also no statistically significant difference in length of stay between men and women or physician.

Two-way cross-tabulations of the four binary variables indicate statistical significance between GENDER and PHY only (Pearson's chi-squared = 40.7; $P < 0.001$):

Cross-Tabulation of GENDER and PHY

		Phy		
	Count Expected	0	1	
Gender	0	9 33.449	68 43.551	77
	1	140 115.551	126 150.449	266
		149	194	343

ANOVA on LOS using all nominal predictors and interactive terms produces significance for: I/P, the GENDER * Md/Bi interaction, and the Phys * Md/Bi interaction (the latter at the 10% level):

ANOVA for LOS

Term	Estimate	Std Error	t Ratio	Prob>|t|
Intercept	5.507	0.739	7.46	<.0001
Gender[0-1]	0.162	0.753	0.22	0.8298
Phy [0-1]	0.604	0.703	0.86	0.3912
I/P[0-1]	1.294	0.574	2.25	0.0250
Md/Bi [0-1]	0.049	0.505	0.10	0.9220
Gender[0-1]*Phy[0-1]	0.687	0.591	1.16	0.2460
Gender[0-1]*I/P[0-1]	-0.146	0.596	-0.25	0.8066
Gender[0-1]*Md/Bi[0-1]	1.084	0.429	2.53	0.0119
Phy [0-1]*I/P[0-1]	0.192	0.481	0.40	0.6901
Phy [0-1]*Md/Bi[0-1]	0.670	0.353	1.90	0.0588
I/ P[0-1]*Md/Bi[0-1]	-0.704	0.429	-1.64	0.1017

Students will find a number of potential ANCOVA models of LOS. An illustrative model is shown below. However, with an R-squared value of 11% and s = 5.9 (the average LOS is six), the predictive ability of the model is open to discussion.

ANCOVA of LOS

```
Los = - 2.14 + 0.174 Age - 0.120 Age*Phy + 4.22 Phy + 2.76 Md/Bi
        (0.15)  (<0.00)      (0.01)             (.025)     (<0.00)

     s = 5.919   R-sq = 11.3%   R-sq(adj) = 10.2%
```

CASE **61**

DEER-TRAIN COLLISIONS IN EASTERN HOKKAIDO, INSTRUCTOR'S NOTE

Abstract: A case rich with possibilities for descriptive statistics and plots. The size of the data set means that the logistics of editing it to fit into student versions of statistics packages may take some time, and to answer questions about time trends requires some careful selection of data points, but simple histograms and plots give most of the information.

Level:	Beginning, intermediate.
Techniques:	Histograms, scatterplots.
Typical Class Discussion Time:	30 minutes.
Data Set Size:	696 cross-sectional observations, 16 variables.

GENERAL COMMENTS

Students may feel frustrated by the foreign setting of the case and the unfamiliar issues it deals with, but as the case points out, real managers do have to be prepared for these aspects of their jobs. It's a good case for either a very general exploratory exercise or for specific questions designed to elicit specific techniques.

QUESTIONS FOR ASSIGNMENT

Here are some possibilities for general explorations:

1 Prepare a two-page report summarizing what the data have to say about the questions given in the case.
2 Prepare (and presumably deliver) a five-minute summary of your answers and recommendations to railway management.

For more specific exercises, you might try:

3 Prepare histograms for YR, MO, HR, and PLACE. Interpret them in the context of the problem.
4 Prepare scatterplots of combinations of these variables, and interpret them in the context of the problem.
5 Can you tell from these data whether the number of deer has been increasing or whether perhaps some other aspect of railway operation has led to the change in the number of collisions? If so, how?

In either case, it helps to focus discussion on the interpretation of the results.

SUGGESTIONS FOR LEADING THE DISCUSSION

The assignment questions may serve as the organizing themes for the discussion here. Either have the students present their results to stimulate discussion, or proceed systematically through the histograms and plots, asking for reactions and interpretations.

TYPICAL RESULTS AND APPROACHES

Students with restricted software will often have difficulty in deciding which variables and/or observations to include in their data sets. We coded MO into a general time variable, including the effects day, for example, and did the same with year and time of day. The resulting histograms follow.

Histogram of YR, Complete Data Set

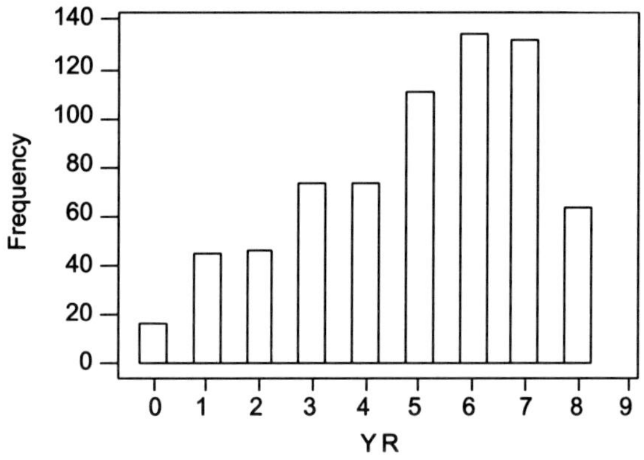

Histogram of MO, Complete Data Set

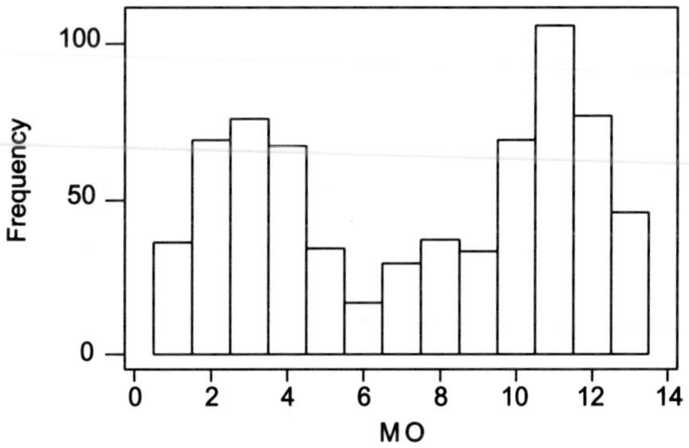

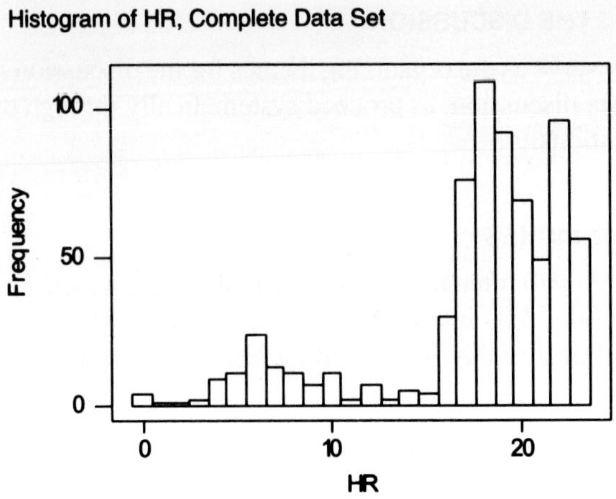

From these, we see that collisions have indeed been increasing over the years, that summer months have fewer collisions, and hours of darkness provide the overwhelming majority of collisions. The histogram of PLACE below shows that there are specific spots on the line, notably between Kushiro and Nemuro, where most of the collisions occur.

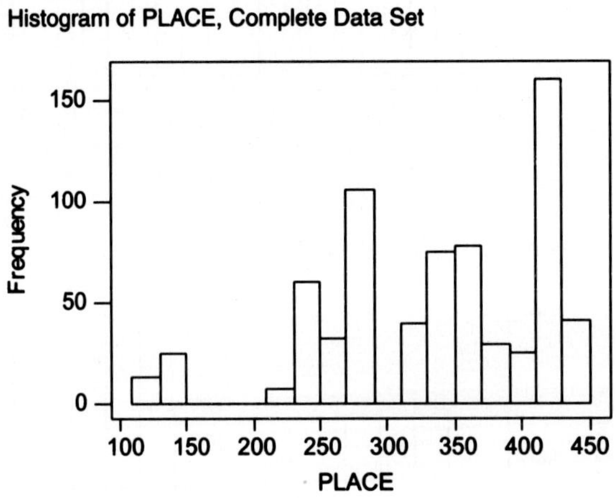

To attack the question of whether the number of deer have been increasing is more tricky. We have data only about the numbers of collisions, not the number of deer, and not even the total number of trains running on those tracks. On the other hand we are told in the case that certain trains did run regularly during those periods, so if we restrict attention to collisions involving those trains, we have at least partially compensated for any change in schedules or general amount of traffic.

For these trains, we obtain the plots that follow:

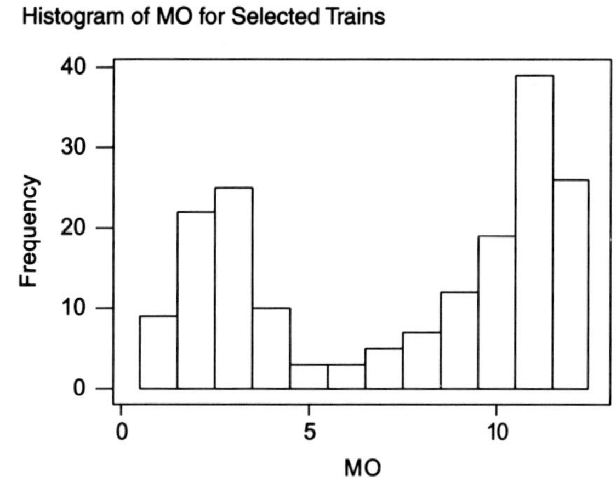

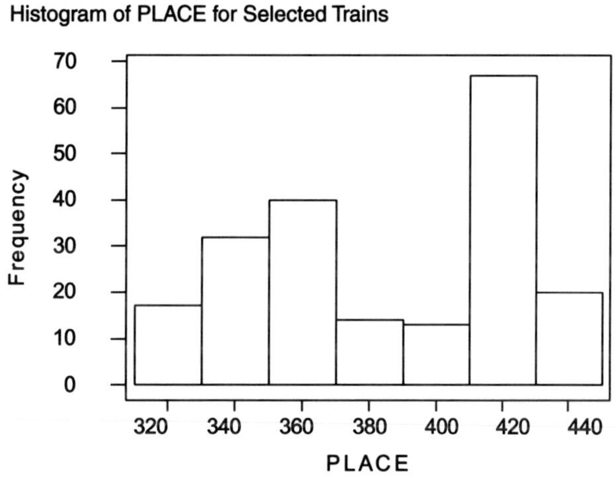

These histograms show pretty much the same patterns as the complete data sets (considering that the trains selected ran only between Kushiro and Nemuro). On the other hand, the histogram of YR, which follows, shows that before 1991 there were no collisions at all on the Kushiro–Nemuro stretch, while afterwards it's about 60 per year or more. Since we're told the number of trains was constant, perhaps this increase is a proxy for an increased number of deer in the area of the railway line.

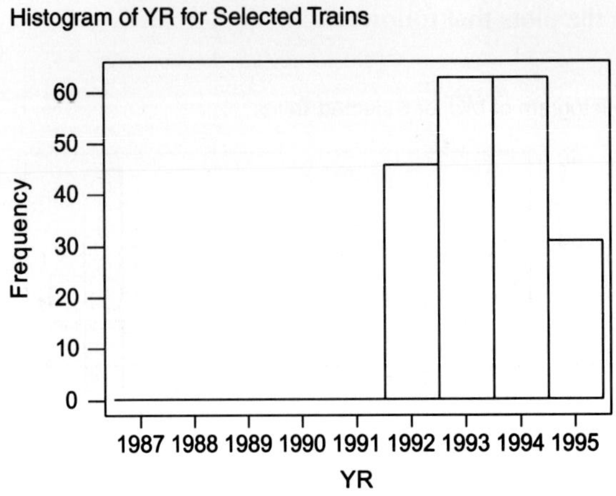

Again for these selected trains, the plot of HR versus PLACE (below) may serve as a vehicle for a good class exercise: ask students to explain why the points fall on straight line segments, or why some lines slope upwards and some down (among other things, apparently the trains run on time). The corresponding plot for MO shows no particular evidence of seasonal effect in the location of the accidents. Many other simple plots and statistics may also be presented.

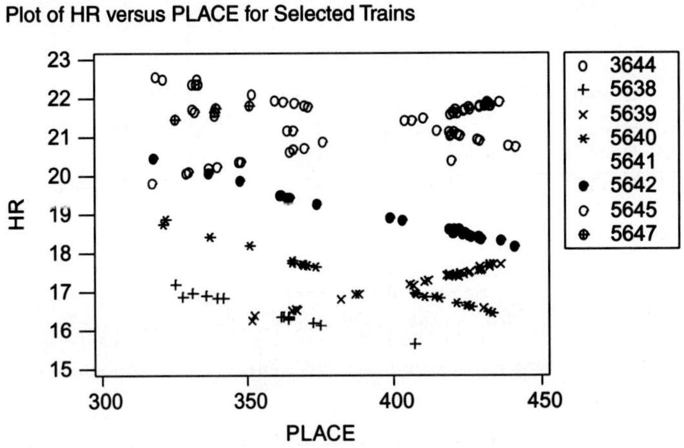

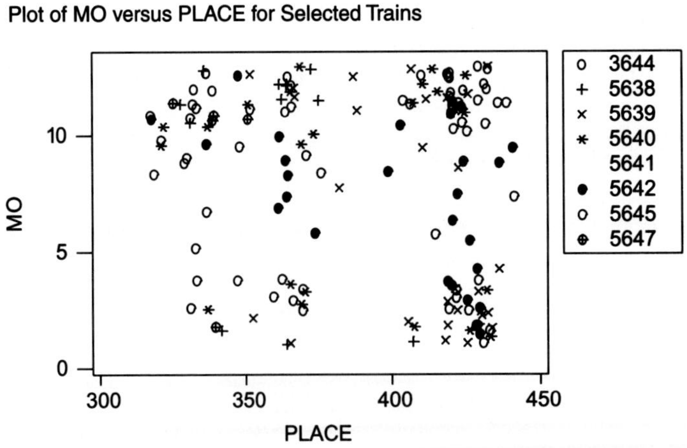

CASE **62**

PUBLIC DEBT IN GERMANY, INSTRUCTOR'S NOTE

Abstract: An easy case, good for getting students used to case discussions. Simple ratios or percentages are revealing. Better students will probably have legitimate questions about the definitions of the underlying data, as well as the quality of the data from the former East Germany. With luck, the subject matter may increase student awareness of some current global issues.

Level:	Beginning.
Techniques:	Ratios, plots, tables.
Typical Class Discussion Time:	15–30 minutes.
Data Set Size:	16 observations, 3 variables.

GENERAL COMMENTS

Simple ratios are all that is required to address this case, but students will need to decide which ratios are meaningful, and display them in some meaningful form. The case can be used on the very first day of class to help students get used to the idea that cases will be discussed in class, or to the idea that statistics is the process of learning from data and communicating what's been learned, not just doing calculations.

QUESTION FOR ASSIGNMENT

The case question is fine to begin with. You may wish to specifically ask students to prepare a one-minute summary for presentation (or a one-page summary). For the first day of class, you may break the students into groups and ask each group to report back in about 30 minutes with a quick one-minute summary.

SUGGESTIONS FOR LEADING THE DISCUSSION

If you've asked for presentations, you can ask for comments or questions about the presentations given, choice of summary technique, and so forth. At the beginning of the course, the answers can be used as the basis of a mini-lecture showing how the issues raised by the discussion will be addressed by later topics in the course.

TYPICAL RESULTS AND APPROACHES

The former eastern provinces had dramatically less debt in 1991 than did the western states, though we might question exactly what "public debt" means for a country only 1 year out of a 50-year history of communism.

We might also question the data. Some simple descriptive statistics follow for 1991 and 1996, separately for the eastern and western states.

	EAST	N	MEAN	MEDIAN	TRMEAN	STDEV	SEMEAN
DEBT1991	0	10	9682	8032	8823	5230	1654
	1	5	768.2	748.0	768.2	69.9	31.3
DEBT1996	0	10	11623	9940	10887	5680	1796
	1	5	7732	7837	7732	1248	558

	EAST	MIN	MAX	Q1	Q3
DEBT1991	0	3900	22334	7053	12005
	1	684.0	861.0	708.5	838.0
DEBT1996	0	4584	24547	8469	14983
	1	5834	8869	6575	8836

By 1996, things are not so dramatically different, as the following dotplots show

```
EAST
0
                   .  .   :.:     .   .                 .
          +---------+---------+---------+---------+---------+---------DEBT1991
EAST       .
1          ::
          +---------+---------+---------+---------+---------+---------DEBT1991
EAST
0
                   .   .    ..: .      .   .                .
          +---------+---------+---------+---------+---------+---------DEBT1996
EAST
1
                      .  .. :
          +---------+---------+---------+---------+---------+---------DEBT1996
          0       5000     10000     15000     20000     25000
```

In terms of fraction of increase (the data are already on a per-person basis) we obtain the statistics below. The western states have risen by anywhere from about 5% to 43%, while the eastern states have risen by anywhere from 780% to 1200% (both on a per-person basis). This may serve as a jumping-off point for discussion about whether the base year (1991) offers comparable starting points for comparing east and west (most conclude that it probably doesn't).

	EAST	N	MEAN	MEDIAN	TRMEAN	STDEV	SEMEAN
1996/1991	0	10	1.2173	1.2027	1.2103	0.1044	0.0330
	1	5	10.098	10.697	10.098	1.651	0.738

	EAST	MIN	MAX	Q1	Q3
1996/1991	0	1.0561	1.4341	1.1563	1.2762
	1	7.799	12.010	8.451	11.446

CASE **63**

JOBS IN DENVER, INSTRUCTOR'S NOTE

Abstract: Beginning statistics students can explore this case using sequence plots to describe trends and seasonality in the seven monthly variables. Intermediate methods of smoothing, regression, or ARIMA may also be applied to the data. Changes in definitions interfere with getting a clear picture of the trends in two of the variables.

Level:	Intermediate, advanced.
Techniques:	Time series methods, forecasting, autoregression, regression.
Typical Class Discussion Time:	45–60 minutes.
Data Set Size:	94 monthly observations, 7 variables.

GENERAL COMMENTS

This case is amenable to a number of forecasting methodologies. Because most of the series are quite regular, this is a good case to build students' confidence in forecasting techniques. A major redefinition of two of the series complicates matters. Even so, autoregressive models do a pretty good job of providing forecasts in these cases.

As with many government data series, these are constantly undergoing revisions. Students can find the actual values for November and December of 1997 and January 1998 (which they are asked to forecast in the case) by going to the Web. By doing so, though, they will find that the data we gave them through October 1997 has been revised. For those of you who prefer multistage cases, have the students undertake the case as written, then send them to the Web to get updated data and redo their analyses. The URL for the website of the revised data is http://governor.state.co.us/gov_dir/labor_dir/lmi/ali/98dnvrws.htm.

QUESTIONS FOR ASSIGNMENT

1 Describe any anomalies you see in the data. To what do you attribute these oddities?

2 Describe any regular patterns you see in the data. To what do you attribute these regularities?

3 Is it preferable to forecast each of the individual variables (Build, Make, TCU, Services, Trade, and FIRE), or just total Denver employment? Why?

SUGGESTIONS FOR LEADING THE DISCUSSION

Begin the discussion by asking for a description of each of the individual time series. This serves as a warm-up for a subsequent discussion of the three-month out-of-sample predictions. Ask students to write their forecasting models on the board. Compare the models, and forecasts generated from them, as a basis for the discussion.

TYPICAL RESULTS AND APPROACHES

Services and Trade jobs are the largest components of total jobs in Denver, comprising 29% and 24.4% of the total, respectively. The third largest group (15.8%) is the "other" category (mining, government, educational, and other employees) hidden in the Total variable. Although construction jobs account for the smallest piece of total jobs on average, their percentage of the total increases over the time series from 3.6% in January 1990 to 6% by October of 1997.

Services, Trade, and Build demonstrate clear seasonal patterns. Services, Trade, Build, and FIRE are trending upward. Perhaps because of a change in industry definitions, it's difficult to determine the recent trends in TCU and Make.

Five major changes in industry definitions occur over this time period:

- Make between December 1990 and January 1991, and between December 1994 and January 1995.
- Services between September and October 1991.
- TCU between December 1994 and January 1995, and between December 1995 and January 1996.

This information is hidden in a footnote to the original data table, and the students have no way of knowing about it. Although the data source indicates that each of these is a major change in definition, the only notable influence is on Make and TCU between December 1994 and January 1995. This break point is shown on the sequence plot of Make and TCU below.

Sequence Plots of Make and TCU

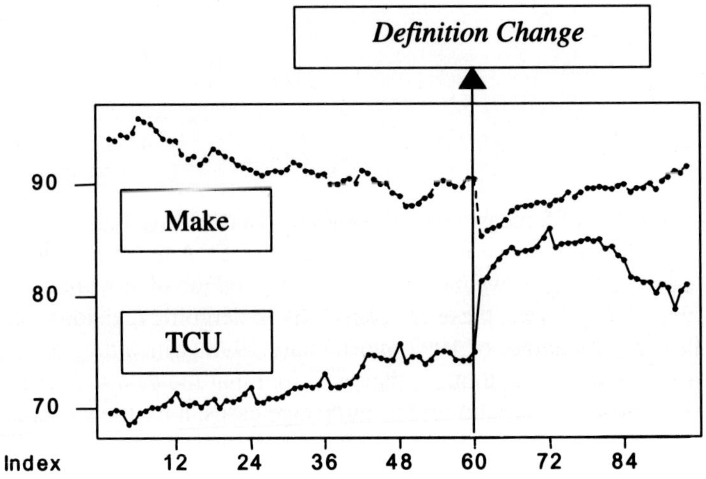

Sequence Plots of FIRE and Build

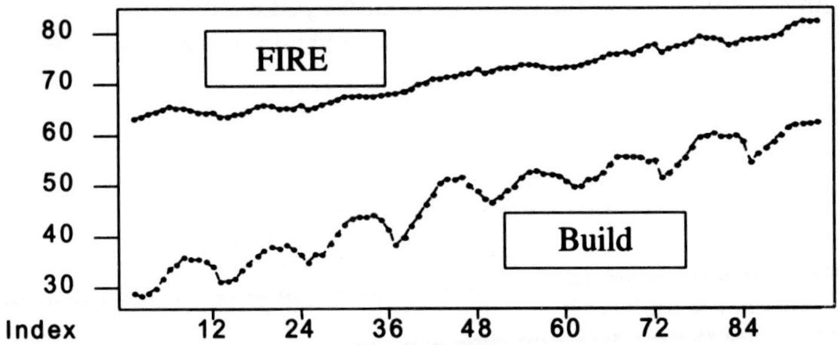

Sequence Plots of Services and Trade

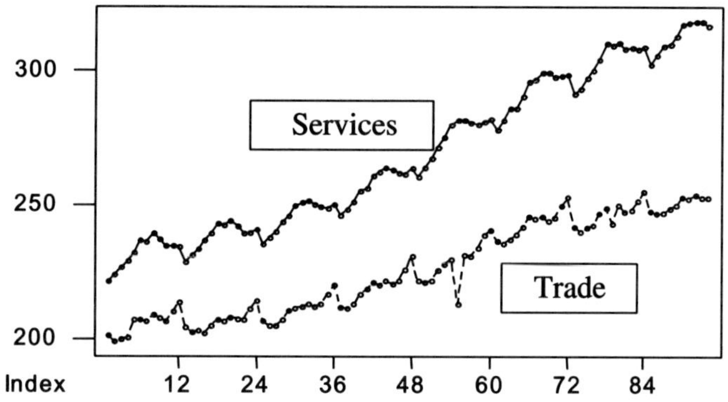

Total employment in Denver is extremely regular, with peaks in December, troughs in January, and a minor drop in employment from June to July each year:

Sequence Plot of Total Denver Employment

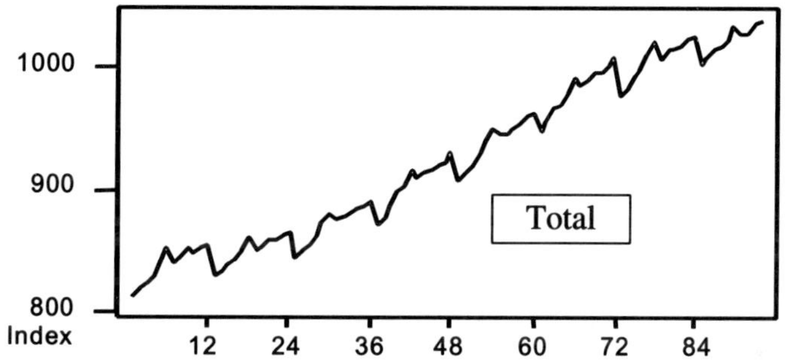

Regression models can be used for predicting future values of FIRE, Build, Services, Trade, and Total by modeling the upward trends and monthly seasonality in these variables. On the next page is a compendium of suggested models. The variable January is 1 if that month is January (0 otherwise), and so forth. All coefficients in the table are statistically significant at the 5% level:

Forecasting Models for FIRE, Build, Services, Trade, and Total

	Response Variable				
	FIRE	Build	Services	Trade	Total
Constant	61.816	29.973	218.236	198.162	815.409
Time	0.204	0.341	1.048	0.621	2.424
January		-2.575	-5.395	-3.369	-19.996
February		-2.341		-5.303	-16.045
March		-1.670		-5.062	-11.831
April				-3.933	-7.880
May					
June	0.822	2.019	5.577		
July		2.778	5.029	-3.697	
August		2.949	5.293		
September		2.332			
October		2.191			
November		1.471			
s	1.0	1.5	4.5	4.5	9.7
R-sq (adj)	97.1%	97.8%	97.7%	93.7%	97.9%

Predicting TCU and Make is problematic because of the discontinuity in the time series. One approach might be to combine TCU and Make into one variable and proceed with regression modeling. Smoothing is another obvious candidate for these two models. Alternatively, the AR(1) models for TCU and Make (*p*-values parenthesized) are:

AR(1) Models for TCU and Make

$$TCU_t = 1.33 + \underset{(<0.001)}{0.984}\ TCU_{t-1} \qquad s = 0.918 \quad R\text{-sq} = 97.4\%$$

$$Make_t = 5.74 + \underset{(<0.001)}{0.936}\ Make_{t-1} \qquad s = 0.692 \quad R\text{-sq} = 89.9\%$$

CASE **64**

THE ADVERTISING EXPERIMENT, INSTRUCTOR'S NOTE

Abstract: This little case can be very puzzling. Most of the ratios cited in the case can be computed one way or another, and all are vaguely plausible. Attempts to run regression models lead to some of the numbers, too, and multicollinearity rears its head. Students must either explain in words the board of directors can understand what multicollinearity means and how it limits our ability to make conclusions, or find a way to avoid it.

Level:	Intermediate.
Techniques:	Computation of ratios, multiple regression.
Typical Class Discussion Time:	30–45 minutes.
Data Set Size:	10 observations, 6 variables.

GENERAL COMMENTS

This case can be used as a take-home exam reasonably well. The amount of computation involved isn't large—a couple of regressions and/or some manual computation or spreadsheet work are all that is needed, but clear thinking is required to state and interpret the conclusions.

QUESTIONS FOR ASSIGNMENT

You could ask for a three-minute oral presentation of these data if you feel the class is ready for that, or you could ask something like: "Prepare a one-page summary of these data and your conclusions for the board meeting next week. Be prepared to answer their questions about the data."

SUGGESTIONS FOR LEADING THE DISCUSSION

If you ask for presentations, that will take care of starting the discussion. Otherwise you may have to ask for conclusions, and the bases for those conclusions. If not all of the numbers in the case have been presented or defended, you might play the role of the spokesperson for that number on the board: "I know you've worked hard on this, but my people have told me the appropriate conclusion is X [whatever conclusion you hope to bring out]. I wonder if you could help me reconcile the differences between your conclusions and theirs" or something similar.

TYPICAL RESULTS AND APPROACHES

Here are some derivations of the numbers cited in the case:

"It brought in about \$500 for each \$1000 spent." This comes from a regression model, SALES = 632085 + 522.5(ADEXP). It omits all other possible influences and gives an $R^2 = .033$ and $s =$ \$171,323. We probably shouldn't place too much faith in it.

"Every ad dollar returned more than \$1.30 in sales." This comes from a regression model, SALES = −4847 + 397.6(VISITS) + 1350(ADEXP). This is a good model ($R^2 = .998$, $s =$ \$8180.9), and if accompanied by a plausible interpretation (see below), it's probably the best explanation. It does point out that a high R^2 like this one might still be accompanied by a standard error of estimate that doesn't appear small on an absolute scale.

"On a per salesperson basis, sales were actually lower in regions with the higher ad budgets." This is true: total sales in the original region were 3,617,780 and in the special regions, 3,560,009. The total number of sales folk in the two groups of regions was 24 in each case. True, but an incomplete explanation; it takes no account of any other factors that might have influenced things.

"Every dollar spent in advertising resulted in average sales anywhere from \$3 to more than \$6." This is also true. Sales per ad dollar, computed within each region, vary from 2.626 to 8.311. The same criticisms as above apply to this approach.

What seems to have happened is that for reasons unknown to us, the salespeople in the regions with the increased ad dollars made fewer visits. Maybe the extra dollars made them cocky and/or lazy. Or maybe the salespeople in the other regions felt underprivileged and decided to show the upstarts what ordinary legwork could accomplish. We don't know, but the regression model that allows both VISITS and ADEXP to have effects seems more plausible.

There are any number of difficulties with the experiment as an experiment, of course, and students (particularly those who are having a hard time with the analysis) may take refuge in them. The issues include: Are the regions really comparable? What about data from the year before? and so on. These are perfectly legitimate issues. On the other hand, the situation described is plausible, too: right or wrong, ill-designed experiment or not, others will have made conclusions based on the data, and students will need to be able to examine others' conclusions, assess them, and explain their own views.

CASE **65**

STAFFING THE PHONE DESK, INSTRUCTOR'S NOTE

Abstract: Descriptive statistics and time series plots are sufficient to address the issues in this case. The subtleties involve questions about which statistics and plots should be used. Students are also asked to ponder whether the data are representative of some larger process.

Level:	Intermediate, advanced.
Techniques:	Graphs, time series plots, ANOVA.
Typical Class Discussion Time:	45 minutes.
Data Set Sizes:	260 hourly observations, 3 variables, and 53 weekly observations, 3 variables.

GENERAL COMMENTS

This case involves using time series data to staff a help desk. Two scenarios are posed: that the data are representative, and that the data may not be representative but future data will be available. Students are asked to prepare management for either outcome. Students are forced to think early on about when it is reasonable to generalize sample results to a larger population.

If the data are representative, the hourly data indicate that calls peak during the midafternoon hours, with afternoon call activity about five times heavier than at start of day. Again, if the data are representative, income tax preparation season coincides with a higher volume of call activity. Staffing assignments can be made proportional to these patterns. However, the case states that management wants to have calls covered and to have minimal idle time. Some students may use averages to construct their staffing schedule, while others will use maximums (or perhaps even third quartiles).

Data may not be representative because the month of March 1996 may not be typical of other months, the year may not be typical, and so on. Your more inquisitive students will recommend that management use company records or other documentation to understand whether any usual activities may have been going on during this time period. The second question in the case hints that subsequent data might be studied using control chart methods. On a practical level, though, those methods won't be of much help in making real time changes in staffing.

Students will need to know the average length of an incoming call in order to determine whether the existing staffing needs are sufficient.

QUESTIONS FOR ASSIGNMENT

1 Construct a weekly staffing schedule for the 50 employees. Describe how you arrived at your schedule and any reservations you might have about it.

2 Construct a yearly staffing schedule for the 50 employees. Describe how you arrived at your schedule and contingency plans if you think them necessary.

3 Write a report to management indicating how to use new data that will be collected in the future.

SUGGESTIONS FOR LEADING THE DISCUSSION

We begin the case discussion by simply posing the first question stated in the case. Unless you assigned questions 2 and 3 above, students will most likely not have constructed detailed staffing schedules. You can use classtime to challenge them to do so. First, have them discuss the historical patterns in the data. Then put on the board an hourly time grid and ask them to fill in recommended numbers of staff members for each hour. Have them work in groups. Although you could ask them to do the same thing for a monthly schedule, it will likely eat up a lot of class time, and they will have gotten the message by then anyway.

Before leaving the discussion, make sure that you touch on the issues of representativeness.

TYPICAL RESULTS AND APPROACHES

Average number of calls peaks between 3 and 4 P.M., with a mysterious dip between 4 P.M. and 5 P.M. The hour from 8 A.M. to 9 A.M. is the slowest time on average. Variability in calls received increases throughout the day as well, peaking at 3 P.M.

Descriptive Statistics for CALLS Grouped by HOUR

HOUR	N	MEAN	MEDIAN	STDEV	MIN	MAX
8	20	22.35	22.00	5.10	14	37
9	20	51.25	49.00	8.66	36	69
10	20	69.45	66.50	11.19	47	92
11	20	93.25	89.50	16.78	71	143
12	20	88.75	86.00	21.17	46	128
13	20	96.40	92.50	14.86	74	131
14	20	105.25	108.00	17.34	67	127
15	20	114.05	111.00	13.58	94	145
16	20	88.15	90.00	24.01	43	131
17	20	105.05	105.00	32.69	11	159
18	20	96.00	83.50	31.67	49	172
19	20	75.15	65.00	29.62	25	144
20	20	58.90	55.50	21.94	25	101
ALL	260	81.85	84.00	32.09	11	172

ANOVA of CALLS by HOUR rejects equivalence of means ($F = 30.4$, $p < 0.0001$, or Welch's ANOVA with unequal variances: $F = 142.1$, $p < 0.0001$):

Scatterplot of CALLS against HOUR

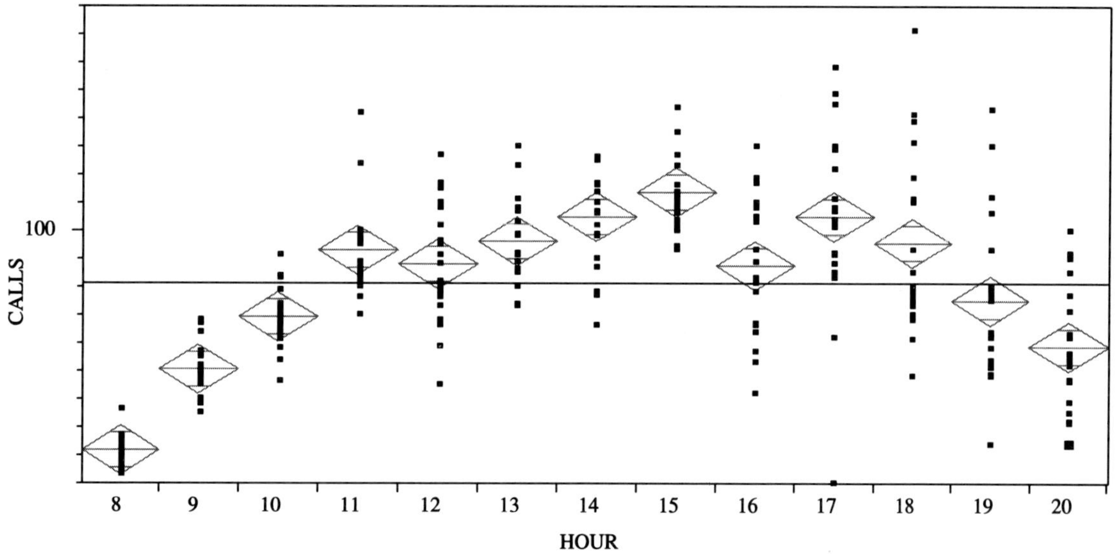

Understanding monthly patterns poses a challenge because of the way the data were recorded. The data were first collected on a weekly basis and then aggregated to monthly data. However, because not all months have exactly 28 days, the number of days in each MONTH (as shown in the data set) need not be the number of days in that calendar month. Similarly, because there are 53 weeks of data, it appears that the data begin the week that included March 1, 1995 (a Wednesday), and end the week that includes Thursday, February 29.

Your students will likely say that the "month" of January has the highest number of calls, and June the lowest, on average:

Descriptive Statistics for AVGCALLS Grouped by MONTH

	MONTH	N	MEAN	MEDIAN	STDEV	MIN	MAX
	1	5	2247	2214	422	1713	2895
1996	2	4	2018	1962	277	1764	2383
1995	3	5	1723	1604	351	1428	2321
	4	4	1703	1633	217	1536	2009
	5	5	1397.6	1347	188.9	1247	1712
	6	4	1348	1225	263	1199	1741
	7	4	1750	1676	303	1475	2171
	8	5	1403	1384	143.9	1279	1642
	9	4	1444	1354	292	1205	1866
	10	5	1639	1626	280	1240	1945
	11	4	1513	1492	76.4	1452	1616
	12	4	1329	1363	380	839	1749
	ALL	53	1631	1561	375.6	839	2895

For an overall picture of call activity, it's probably sufficient to talk about monthly trends, but for scheduling purposes, a week-to-week view of the data will be more helpful:

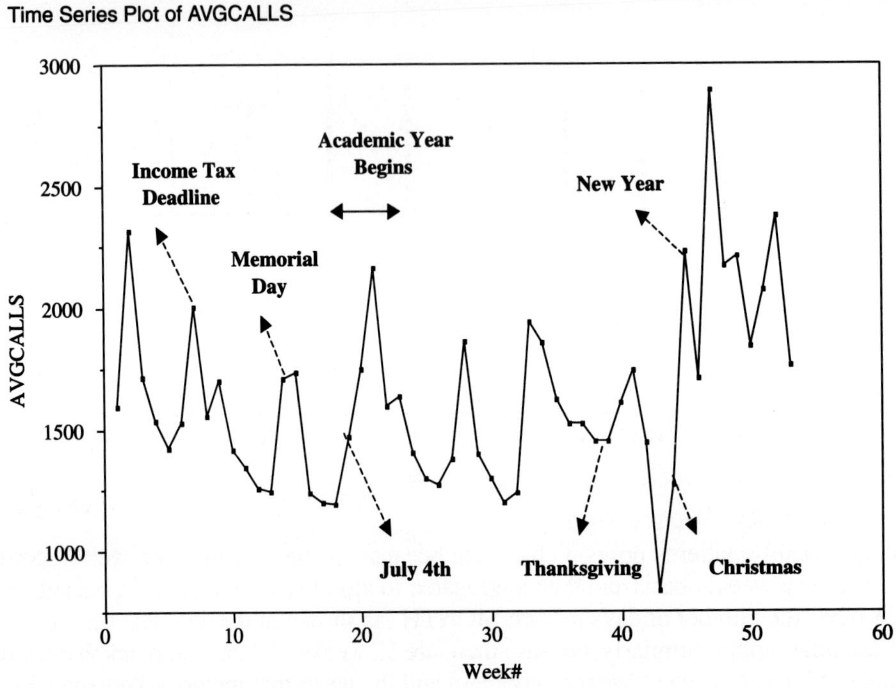

Given that TIAA/CREF is a retirement fund for academicians, it appears that tax season and the beginning of an academic year are the busiest times at the call desk. The *p*-value for the runs test of AVGCALLS is 5.5%, but the individuals and moving range control charts indicate special cause variability at week 47 (tax season *and* the beginning of a new academic semester?).

CASE **66**

HOME EQUITY LOANS, INSTRUCTOR'S NOTE

Abstract: Inferential methods applied to the data overturn initial expectations based on descriptive statistics alone. Students might create new variables from the data set to resolve this apparent dilemma; accounting students will likely suggest several useful transformations of the data. Different inferential methods can produce different results, providing an opportunity to explore with your students various assumptions that underlie the competing techniques.

Level:	Intermediate.
Techniques:	Descriptive statistics, two-population hypothesis tests, nonparametric methods.
Typical Class Discussion Time:	30–45 minutes.
Data Set Size:	59 cross-sectional observations, 13 variables.

GENERAL COMMENTS

Assign this case at the beginning of an introductory statistics class, and revisit it when inferential methods are introduced later in the course. When students first see the data set, they will base recommendations on numerical differences. When they examine the data again using inferential methods, they will change their minds about certain variables. By looking at the data set twice, students must come to grips with the difference between numerical and statistical differences.

Because many variables are interrelated, the case foreshadows the need for multivariate methods.

QUESTIONS FOR ASSIGNMENT

1 Test the null hypothesis that net worth does not differ on average between applicants accepted and rejected for a loan. What assumptions underlie your statistical test?

2 Repeat question 1 for all the other variables in the data set.

3 What follow-up study would you like to do, if any? What is the need for the additional study, and how you would go about doing it?

4 Are there any outliers in the data set that influence your recommendations? If so, how do they influence your recommendations, and what do you propose to do about them?

SUGGESTIONS FOR LEADING THE DISCUSSION

Students who examine the data set without inferential methods will come to class with myriad recommendations. Differences arise because of differing perceptions of how big numerical differences have to be before they are important. Contradictory recommendations will also arise as some students, but not others, sort the data looking for lurking or confounding variables. The discussion of "who sorted what and why" can go on indefinitely; be prepared to put closure to these discussions or you'll never get anywhere.

For example, start the discussion by asking whether there seems to be a gender bias in the decision to award a loan. Those who look at proportions may agree; those who look at gender and income will disagree.

End the discussion when the class has come to a consensus on which of the variables in the data set influences loan selection. Keep the list, assign the case again after inferential methods are known, and then begin the second discussion by writing the list from the first discussion on the board. Ask if they've changed their minds, and if so, why. Expect responses along the lines of, "Because I now know that these two averages are statistically different." Press them to express what that means. Continue to push this point until light bulbs go on around the room.

TYPICAL RESULTS AND APPROACHES

Relative to applicants rejected for a loan, those awarded: had higher average incomes (by $25, 357), higher average net worth (by $104, 207), less total debt on average (by $16,714), less revolving debt on average (by $1,640), and higher average equity ($18,031); had 4.5 fewer outstanding delinquent accounts on average; had worked an average of 3.7 more years at their job at the time of the application; and had been a customer of the bank for 3.4 years longer on average. Those awarded loans asked for about $3,460 more on average than those rejected. There is little difference between accepted and rejected applicants based on age or years of residence at the time of the application. Twenty percent of the applicants were women.

Acceptance or rejection of a loan is not statistically determined (at the 5% level) by most variables in the data set. These variables are Revolve, Tot Debt, Age, Equity, Loan Amt, Reside, and Year Job. This is true regardless of which statistical method (parametric or nonparametric) or measure of location is used. Because students will be surprised to discover that so many intuitively appealing variables are not statistically significant, some will create new variables from the data set. Two that we've seen most often are total debt less the requested amount of the loan and the leverage ratio (total debt/income). Of these two, only the leverage ratio is a statistically significant predictor of the Accept? result. In the following table, L Ratio is the leverage ratio, Debt-Amt is total debt minus the requested loan amount, and t ratios presume equal variances.

Descriptive Statistics for Metric Variables by Accept?

	Accept?	N	MEAN	MEDIAN	STDEV	t-ratio (p-value)
Age	0	11	39.64	39.00	8.26	-0.16
	1	48	40.12	39.00	9.09	(87.1%)
Income	0	11	57238	45000	31127	-1.77
	1	48	82595	67200	45007	(8.2%)
Del Acct	0	11	5.36	4.00	7.16	4.24
	1	48	0.792	0.000	1.304	(<0.1%)
Equity	0	11	60910	53660	32475	-1.13
	1	48	78941	66125	50266	(26.2%)
Loan Amt	0	11	15.77	15.00	7.05	-1.00
	1	48	19.23	16.35	10.88	(32.1%)
Reside	0	11	3.956	3.000	3.008	-0.21
	1	48	4.172	3.000	3.174	(83.8%)
Year Job	0	11	4.39	2.50	6.00	-1.41
	1	48	8.10	5.00	8.26	(16.6%)
Revolve	0	11	10221	7705	9726	0.62
	1	48	8581	6362	7449	(53.7%)
Tot Debt	0	11	182564	175317	82918	0.55
	1	48	165850	141633	92405	(58.4%)
NetWorth	0	11	70172	81383	47028	-1.93
	1	48	174379	115528	176638	(5.9%)
Customer	0	11	2.105	0.500	2.502	-2.24
	1	48	5.493	4.375	4.854	(2.9%)
L Ratio	0	11	3.563	3.383	1.160	5.25
	1	48	2.118	2.104	0.731	(<0.1%)
Debt-Amt	0	11	182548	175307	82916	0.55
	1	48	165831	141615	92403	(58.0%)

Only two variables are statistically significant predictors of Accept? regardless of method used: L Ratio and Customer. Three variables are statistically significant at the 5% level for some tests but not for others as shown below (*p*-values are parenthesized):

Parametric and Nonparametric Hypothesis Tests for Income, Del Accts, and NetWorth

	Income	Del Accts	NetWorth
t ratio (equal variances)	-1.77 (8.2%)	4.24 (<.0001)	-1.93 (5.9%)
t ratio (unequal variances)	-2.22 (3.8%)	2.11 (6.1%)	-3.57 (<0.1%)
Mann-Whitney (χ_1^2)	2.93 (8.9%)	8.27 (0.4%)	5.92 (1.5%)
Median test (χ_1^2)	0.07 (78.7%)	3.40 (6.5%)	2.55 (11.1%)
Van der Waerden test (χ_1^2)	4.24 (4.0%)	10.28 (0.1%)	6.43 (1.1%)

Accept? and Gender are statistically independent if you are willing to overlook small cell frequencies:

Crosstabulation of Accept? and Gender

ROWS: Accept?   COLUMNS: Gender

	0	1	ALL
0	4	7	11
	36.36	63.64	100.00
	33.33	14.89	18.64
	6.78	11.86	18.64
	4	7	11
1	8	40	48
	16.67	83.33	100.00
	66.67	85.11	81.36
	13.56	67.80	81.36
	8	40	48
ALL	12	47	59
	20.34	79.66	100.00
	100.00	100.00	100.00
	20.34	79.66	100.00
	12	47	59

CHI-SQUARE = 2.143 WITH D.F. = 1
p-value = 0.143

CELL CONTENTS --
COUNT
% OF ROW
% OF COL
% OF TBL
COUNT

CASE **67**

GAS STATIONS, INSTRUCTOR'S NOTE

Abstract: ANOVA answers the questions posed in the case. Two anomalies of the data set provide interesting twists: an unbalanced design and one miscoded observation.

Level:	Intermediate.
Techniques:	Two-population hypothesis tests, ANOVA.
Typical Class Discussion Time:	30 minutes.
Data Set Size:	192 cross-sectional observations (one miscoded), 4 variables.

GENERAL COMMENTS

We give you the data set exactly as it was given to us: observation 59 must be miscoded, since LOCATION cannot take on the value of zero. You may either let students discover this for themselves and make appropriate adjustments, or let them know about the miscoded observation before they tackle the case. If you've been sending the message all along that they are responsible for ensuring reasonability of the data and if your group is fairly mature, let them discover and act on their own. Undergraduates may become annoyed if you don't confess beforehand, or at least give a not-so-subtle hint.

Location has an influence on sales; presence of a car wash does not. However, only 6 of the 192 (or 191) observations have a car wash, begging the question of how certain we should be about the results.

Stations open longer sell more gas. This self-evident outcome will induce some students to create a new variable: gallons sold per hour per day. Results differ only slightly with this response variable used.

QUESTIONS FOR ASSIGNMENT

1 What statistical technique should be used to determine whether the presence of a car wash influences fuel sales? What assumptions underlie your test?

2 What statistical technique should be used to determine whether location has an influence on fuel sales? Is this the same technique you chose in question 1 above? Why or why not?

3 The data set contains few observations with car washes. Does this bother you? If so, why? Would you recommend collecting a new data set with equal, or approximately equal, numbers of stations with and without car washes? Why or why not?

SUGGESTIONS FOR LEADING THE DISCUSSION

The questions posed in the case should be used to open the discussion: how do fuel sales differ by location? Existence of a car wash? Hours of operation? After coming to some consensus on the answers, use class time to ascertain what assumptions underlie the statistical results and whether students tested for the validity of these assumptions in this case. Some students may create a new variable GALLONS/HOURS to control for stations open longer. Not much different arises, qualitatively, by using this response variable in place of GALLONS.

TYPICAL RESULTS AND APPROACHES

The results given here are based on a data set without miscoded observation 59. The average number of gallons sold in 1997 was 161,437. Ninety % of the stations are open 24 hours per day. Only 3% have carwashes. Fifty-one % of the stations are located in a metropolitan area but not along a major highway (LOCATION = 2); 6% are in rural areas off major highways (LOCATION = 3). The remainder, 43%, are located along a major highway (LOCATION = 1).

There is no statistically significant difference in average gallons sold between stations with or without carwashes at the 5% significance level. The *t* ratio assuming equal variances is –0.678; the *t* ratio assuming unequal variances is −0.665.

There are, of course, more gallons sold on average when stations are open 24 hours a day. The difference, 57,266 gallons per station, is statistically significant at the 5% significance level. There is no statistically significant difference between gallons sold per hour per day (GALLONS/HOURS) between the 24-hour and 18-hour stations using the *t* test with equal variances. There is a difference if unequal variances, the most likely scenario, is assumed; the *t* ratio in this case is 2.05 (*p*-value = 4.8%).

ANOVA of gallons sold by location is unambiguous. Stations located along a highway sell more fuel on average than those not located on a highway, whether metropolitan or rural. This result is robust in the face of nonnormality or heteroskedasticity.

ANOVA of GALLONS by LOCATION

```
ANALYSIS OF VARIANCE ON GALLONS
SOURCE      DF         SS         MS         F        p
LOCATION     2  1.658E+11  8.291E+10     20.78    0.000
ERROR      188  7.500E+11  3.989E+09
TOTAL      190  9.158E+11
                                      INDIVIDUAL 95% CI'S FOR MEAN
                                      BASED ON POOLED STDEV
 LEVEL       N      MEAN      STDEV   -------+---------+---------+---------
     1      83    194686      84314                                (---*---)
     2      97    137816      39854                  (--*---)
     3      11    118853      38231   (----------*----------)
                                      -------+---------+---------+---------
```

CASE **68**

ATTITUDES TOWARD NUCLEAR POWER, INSTRUCTOR'S NOTE

Abstract: A straightforward application of contingency table thinking, but issues of nonresponse and interpretation must be dealt with.

Level	Beginning.
Techniques:	Contingency tables, chi-square.
Typical Class Discussion Time:	15–30 minutes.
Data Set Size:	Various numbers of rows, 4 or 5 categories of responses, based on over 4000 individuals.

GENERAL COMMENTS

This case may illustrate some of the issues in social surveys. The hard work, of course, has been done: the survey is already designed and the data are gathered. On samples of this size, the chi-square test reveals most of what's there.

QUESTION FOR ASSIGNMENT

The case question is fine to begin with. You may wish to specifically ask students to prepare a one-minute summary for presentation (or a one-page summary).

SUGGESTIONS FOR LEADING THE DISCUSSION

Probably most students will apply chi-square or something equivalent to these data, although if it is early in the course, they may have difficulty with some of the logistics of extracting the data and putting into whatever format is required for their program. Discussion might best be focused on the *interpretation* of the results. Once a significant chi-square is discovered, make sure the students return to the cross-tabulation and interpret the results in words.

TYPICAL RESULTS AND APPROACHES

Various MINITAB printouts of the different chi-square calculations follow. Except for the first printout, they use only the response categories of 1 through 4. In some cases (e.g., see the calculations for urban versus provincial) the results are different if the *no response* category is included. Students should be encouraged to reflect out loud on what this might mean or what follow-up action might be taken. In each case, the expected counts are printed beneath the observed counts.

Chi-Square Printout for Location by Response, Including "No Response"

```
             Resp1     Resp2     Resp3     Resp4    NoResp     Total
Urban          465      2255       700       470       110      4000
            466.21   2236.95    706.59    461.93    128.31

Provincial      80       360       126        70        40       676
             78.79    378.05    119.41     78.07     21.69

Total          545      2615       826       540       150      4676

ChiSq =      0.003 +   0.146 +   0.061 +   0.141 +   2.614 +
             0.019 +   0.861 +   0.363 +   0.834 +  15.468 =  20.510
df = 4
```

Chi-Square Printout for Location by Response

```
             Resp1     Resp2     Resp3     Resp4     Total
Urban          465      2255       700       470      3890
            468.42   2247.54    709.93    464.12

Provincial      80       360       126        70       636
             76.58    367.46    116.07     75.88

Total          545      2615       826       540      4526

ChiSq =      0.025 +   0.025 +   0.139 +   0.075 +
             0.152 +   0.152 +   0.849 +   0.456 +  1.872
df = 3
```

Chi-Square Printout for Sex by Response

```
             Resp1     Resp2     Resp3     Resp4     Total
Male           335      1298       330       204      2167
            260.94   1252.03    395.48    258.55

Female         210      1317       496       336      2359
            284.06   1362.97    430.52    281.45

Total          545      2615       826       540      4526

ChiSq =     21.020 +   1.688 +  10.842 +  11.508 +
            19.309 +   1.550 +   9.959 +  10.571 +  86.446
df = 3
```

Chi-Square Printout for Age by Response

	Resp1	Resp2	Resp3	Resp4	Total
18-24	74	323	166	63	626
	75.38	361.69	114.25	74.69	
25-29	22	273	93	74	462
	55.63	266.93	84.32	55.12	
30-34	62	272	87	85	506
	60.93	292.35	92.35	60.37	
35-39	52	246	92	52	442
	53.22	255.38	80.67	52.74	
40-44	78	344	105	64	591
	71.17	341.46	107.86	70.51	
45-49	66	325	78	48	517
	62.25	298.71	94.35	61.68	
50-54	64	234	71	45	414
	49.85	239.20	75.56	49.39	
55-59	24	227	53	38	342
	41.18	197.60	62.42	40.80	
60-64	40	181	47	56	324
	39.01	187.20	59.13	38.66	
65-79	63	190	34	15	302
	36.37	174.49	55.12	36.03	
Total	545	2615	826	540	4526

```
ChiSq =  0.025 +  4.138 + 23.445 +  1.829 +
        20.332 +  0.138 +  0.895 +  6.466 +
         0.019 +  1.417 +  0.309 + 10.047 +
         0.028 +  0.344 +  1.593 +  0.010 +
         0.656 +  0.019 +  0.076 +  0.602 +
         0.225 +  2.314 +  2.834 +  3.036 +
         4.015 +  0.113 +  0.275 +  0.391 +
         7.169 +  4.375 +  1.420 +  0.193 +
         0.025 +  0.205 +  2.488 +  7.781 +
        19.508 +  1.379 +  8.090 + 12.276 = 150.500
df = 27
```

Chi-Square Printout for Education by Response

	Resp1	Resp2	Resp3	Resp4	Total
Elem	83	395	118	54	650
	78.52	376.12	118.72	76.64	
High	302	1348	489	249	2388
	288.47	1381.80	436.16	281.57	
Univ	158	858	214	227	1457
	176.01	843.08	266.12	171.79	
Total	543	2601	821	530	4495

```
ChiSq =  0.256 +  0.948 +  0.004 +  6.688 +
         0.634 +  0.827 +  6.401 +  3.767 +
         1.842 +  0.264 + 10.207 + 17.741 = 49.579
df = 6
```

Chi-Square Printout for Occupation by Response

	Resp1	Resp2	Resp3	Resp4	Total
Empl	255	1094	374	219	1942
	234.99	1125.36	351.84	229.81	
Self-Emp	72	357	49	78	556
	67.28	322.19	100.73	65.80	
Fam. Bus.	36	91	21	15	163
	19.72	94.46	29.53	19.29	
Homemaker	116	778	258	185	1337
	161.78	774.77	242.23	158.22	
Student	30	125	70	26	251
	30.37	145.45	45.47	29.70	
None	36	165	44	10	255
	30.86	147.77	46.20	30.18	
Total	545	2610	816	533	4504

```
ChiSq =  1.704 +  0.874 +  1.396 +  0.509 +
         0.331 +  3.760 + 26.567 +  2.263 +
        13.432 +  0.126 +  2.465 +  0.954 +
        12.956 +  0.013 +  1.027 +  4.533 +
         0.005 +  2.875 + 13.228 +  0.462 +
         0.858 +  2.009 +  0.105 + 13.490 = 105.942
df = 15
```

Chi-Square Printout for Household Type by Response

```
           Resp1     Resp2     Resp3     Resp4     Total
One-pers      15       114        40        10       179
           21.11    103.53     33.29     21.07

One-gener     81       435       122        71       709
           83.62    410.06    131.86     83.46

Two-gener    333      1605       461       343      2742
          323.39   1585.87    509.97    322.77

Three-gener 91         396       197        95       779
           91.88    450.54    144.88     91.70

Total        520      2550       820       519      4409

ChiSq =    1.769 +   1.059 +   1.352 +   5.817 +
           0.082 +   1.517 +   0.738 +   1.860 +
           0.285 +   0.231 +   4.702 +   1.268 +
           0.008 +   6.603 +  18.749 +   0.119 = 46.159
df = 9
```

CASE **69**

ALICIA'S UNION DATA, INSTRUCTOR'S NOTE

Abstract: This case offers a misleading correlation. Usually, students can guess at the underlying explanation from some simple plots, but it can also be addressed by using partial correlation. This all leads to a discussion of the ethics of summarizing data.

Level:	Intermediate.
Techniques:	Descriptive statistics, partial correlation.
Typical Class Discussion Time:	30–45 minutes.
Data Set Size:	50 cross-sectional observations, 3 variables.

GENERAL COMMENTS

The technical part of this case is pretty straightforward, and that's intended, because its main purpose is to lead into a discussion of the underlying ethical question: should the boss cite the correlation of −.9 that Alicia mentioned? It's a true number, as far as it goes, and it might influence a few people to see things his way, but other numbers and explanations are much more plausible and defensible. Of course, it's not likely that they'll see that in the 10:00 A.M. meeting, and his career may be on the line if the union election goes the wrong way.

QUESTIONS FOR ASSIGNMENT

We recommend using the statement in the case, and little else. For a good discussion, you want a wide variety of approaches to appear in the discussion, so don't structure it too much. Let the ethical issue emerge from the discussions.

If you want to focus the debate specifically on the ethics issue, you could assign the "Ethical Guidelines for Statistical Practice: Report of the Ad Hoc Committee on Professional Ethics" as reading to go with the case. It's in *The American Statistician* 37, no. 1 (February 1983), pp. 5–6. You could even ask them, "If the boss cites the −0.9 figure, would he be violating any of the ethical guidelines? Which ones?" but we prefer to let students wrestle with the issues before discussing the guidelines.

SUGGESTIONS FOR LEADING THE DISCUSSION

Start off by playing the part of the boss: "Well, what do you have for me? We're late as it is, and I have to get to that 10 o'clock meeting. . ."

Follow that up by asking different class members for their recommendations. Approaches will differ. Some will feel that the boss should use the data (it's the other side's problem to attack it if they don't like it). Others will feel that's somehow not right, because it's a misleading summary or because the other side won't have a chance to examine the data before the deciding vote at 12:00 P.M., or just for some vague hard-to-articulate reason. Some will have specifically analyzed the data and discovered alternative analyses; others will not. It can be a good exercise for those who haven't probed a bit deeper to discover how misleading the surface calculations might have been.

At some point, you can then turn the discussion to the issue of the principles that should be used to decide about what should or shouldn't be presented. When that has been discussed for a while, you can point out the ethical guidelines mentioned above, if you haven't assigned them already. We prefer to assign the guidelines for reading after discussing the case, as the need for such guidelines is generally more apparent and they have more impact. Would-be lawyers may be interested to learn that statisticians may "accept no contingency-fee arrangements," and must "apply statistical procedures without concern for a favorable outcome." Students may wish to compare the latter canon with the widespread use of statistics in advertising. Depending on the students' backgrounds, it can also lead into a discussion of methods of seeking truth or justice. Statistical methods are not really designed for adversarial or advocacy processes. Einstein's remark "Subtle is the Lord, but not malicious" is as good a statement of this difference as any.

TYPICAL RESULTS AND APPROACHES

The correlation between UNION and LIFEINS is -0.968, so Alicia is right, strictly speaking. On the other hand, plots quickly reveal that both UNION and LIFEINS are tightly related to AVESAL: firms with high average salaries have low union representation, and life insurance is directly related to salary (indeed, it's often a multiple of the salary). The salaries seem to be the driving force here. A partial correlation of UNION and LIFEINS after removing the effects of AVESAL is $-.134$ (which wouldn't be considered significant at conventional significance levels).

So what should the boss say? Most, but not all, feel it isn't in his long-term interest to cite the $-.9$ figure, either because it's unethical, or because the other side must inevitably find out about his cheating sometime, and then they'll really be out to get him. Some say the data can be interpreted in many ways and that he probably has other arguments he can make to bolster his case. For example, he could try to focus attention on salaries as the underlying issue. Life insurance will take care of itself. If he decides to use the $-.9$ figure, he is violating at least these guidelines:

1 He's not documenting known inaccuracies in the data (or at least in his analysis).
2 He's clearly choosing the result with regard for a favorable outcome.
3 Since the decision is to be made within a hour or two, he isn't really making the data available for others to look at.

Given that he's under a lot of pressure for a favorable outcome, some students (particularly younger students) may feel these guidelines shouldn't dominate his thinking. Some may point out (correctly) that "the selection of a statistical procedure may to some extent be a matter of judgment" (also from the ethical guidelines). The correlation of $-.9$ represents one such procedure, and the boss is entitled to have that one considered. This is probably the best specific argument for using the number, but most folks, on balance, feel the boss shouldn't use it.

One possible ending for this session is simply to mention that real statistics happens in tough situations, not just at easy, free-of-pressure, calm, reflective times. Sooner or later, ethical issues like this will have to be confronted, and it will be painful. The more you've thought ahead of time about what the principles or guidelines ought to be, the less painful those confrontations will be. They'll never be easy.

CASE **70**

HEART SURGERY, INSTRUCTOR'S NOTE

Abstract: This is an intermediate or advanced case, depending on how much you ask the students to do with it. It offers room for practice with many tools, as well as an attempt to interpret the results in a real setting.

Level:	Intermediate, advanced.
Techniques:	Tables, regression.
Typical Class Discussion Time:	45 minutes.
Data Set Size:	378 cross-sectional observations, 11 variables.

GENERAL COMMENTS

This case can be used twice in a course, once using descriptive statistics and cross-tabulations, and once using more advanced tools such as multiple regression. The first time, it illustrates the need for regression. The second time, it illustrates some of the complexities in dealing with unbalanced data with multiple categorical factors.

QUESTION FOR ASSIGNMENT

For a first assignment, probably some fairly structured questions are appropriate, based on Dr. O'Malley's interests as given in the case. Various cross-tabulations give first answers to questions about the effect of STAT or CAT, INC and CARD. You may wish to ask students to think about how the information relating to differences among cardiologists might best be summarized as a way of getting them thinking about the context.

Note that the size of the data set will tax some student versions of statistics programs. If you don't want students to wrestle with this, you will want to be quite specific about which questions you do want answered, as students may have to omit certain variables or take samples of the observations. You could also use this as a way of raising the questions of when to use multiple regression versus separate simple regressions.

For second assignments using multiple or logistic regression, you can simply ask the students to advise Dr. O'Malley, and let it go at that, assuming they are familiar with dummy variables.

SUGGESTIONS FOR LEADING THE DISCUSSION

We recommend gearing the discussion to the questions you asked in this case. There's enough complexity to discuss at either the elementary or advanced levels. For advanced students, just ask for a report. For the elementary discussion, ask what techniques would apply to each of your assigned questions and what the results are.

TYPICAL RESULTS AND APPROACHES

Results will vary quite a lot on this one, particularly if students are using different packages to do the calculations. Regression with dummy variables, cross-tabulations, one- and two-way ANOVAs are all possible, though for two-way ANOVAs, the empty cells (and unequal cell sizes in general) will cause difficulty for most packages. The data set is large enough that some student versions of common statistics packages will have trouble holding everything, too.

Starting from the cross-tabulations below (in each case, the entries in the table are the count of cases, the average length of stay, and the proportion of cases with complications), we see that DRGs 483 (Tracheotomy) has a dramatically higher incidence of complications, on average, and a longer length of stay than the others, though it doesn't happen very often. These results are pretty much the same for all cardiologists. A typical length of stay for the other DRGs is perhaps 10 days. Cardiologist 13 has a higher average length of stay than the others, but this is based on only six cases.

Average Length of Stay and Percentage Complications, by DRG and CARD (first part)

ROWS: DRG COLUMNS: CARD

	0	1	2	3	4	5	6	7
104	0	9	1	0	1	0	5	0
	--	10.667	7.000	--	12.000	--	16.200	--
	--	0.44444	0.00000	--	1.00000	--	0.40000	--
105	3	5	0	1	3	1	0	0
	4.667	14.400	--	8.000	9.667	4.000	--	--
	0.66667	0.60000	--	0.00000	0.66667	1.00000	--	--
106	2	33	9	0	44	1	37	0
	8.500	9.636	9.333	--	9.386	7.000	9.216	--
	0.00000	0.30303	0.33333	--	0.38636	0.00000	0.29730	--
107	24	15	1	0	12	0	3	1
	6.583	7.800	7.000	--	8.500	--	10.333	5.000
	0.20833	0.46667	0.00000	--	0.33333	--	0.33333	0.00000
108	1	1	0	0	0	0	0	0
	6.000	18.000	--	--	--	--	--	--
	1.00000	1.00000	--	--	--	--	--	--
483	1	3	1	0	2	0	0	0
	19.000	23.333	16.000	--	44.000	--	--	--
	1.00000	1.00000	1.00000	--	1.00000	--	--	--
ALL	31	66	12	1	62	2	45	1
	6.903	10.470	9.500	8.000	10.387	5.500	10.067	5.000
	0.29032	0.42424	0.33333	0.00000	0.41935	0.50000	0.31111	0.00000

Average Length of Stay and Percentage Complications, by DRG and CARD (second part)

ROWS: DRG COLUMNS: CARD

	8	9	10	11	12	13	14	15	ALL
104	2	1	5	0	0	1	0	4	29
	6.500	19.000	7.800	--	--	8.000	--	11.750	11.103
	0.00000	1.00000	0.40000	--	--	0.00000	--	0.75000	0.44828
105	1	0	5	0	1	0	1	1	22
	5.000	--	11.400	--	8.000	--	15.000	6.000	9.909
	0.00000	--	0.40000	--	0.00000	--	1.00000	0.00000	0.50000
106	1	14	32	2	0	1	6	33	215
	8.000	8.429	8.094	9.500	--	13.000	9.167	8.576	9.000
	0.00000	0.28571	0.37500	0.50000	--	0.00000	0.83333	0.15152	0.31628
107	0	5	20	1	3	3	4	6	98
	--	8.000	5.950	6.000	7.667	12.333	5.250	7.000	7.224
	--	0.60000	0.25000	0.00000	1.00000	0.66667	0.50000	0.33333	.34694
108	0	0	1	0	0	0	1	0	4
	--	-	7.000	--	--	--	11.000	--	10.500
	--	--	0.00000	--	--	--	1.00000	--	.75000
483	0	1	0	0	0	1	1	0	10
	--	32.000	--	--	--	16.000	23.000	--	26.400
	--	1.00000	--	--	--	1.00000	1.00000	--	1.00000
ALL	4	21	63	3	4	6	13	44	378
	6.500	9.952	7.635	8.333	7.750	12.333	9.615	8.591	9.230
	0.00000	0.42857	0.33333	0.33333	0.75000	0.50000	0.76923	0.22727	.36772

No great differences between Status groups emerge for length of stay (most cases are Status 1, anyhow); both cases in status 4 did have complications. Again, tracheotomies mean trouble, across all status groups. Category 2 (immediate surgery required) seems high on both length of stay and complication rates. An informal review of the others suggests nothing new. Increased incidence seems to require a longer stay and a higher complication rate, though (mercifully?) incidences beyond two are rare.

Average Length of Stay and Percentage Complications, by DRG and STAT

DRG	STAT=1	STAT=2	STAT=3	STAT=4	ALL
104	23	4	2	0	29
	9.957	18.500	9.500	--	11.103
	0.43478	0.75000	0.00000	--	0.44828
105	20	0	2	0	22
	10.050	--	8.500	--	9.909
	0.45000	--	1.00000	--	0.50000
106	156	34	23	2	215
	9.000	8.853	9.174	9.500	9.000
	0.28205	0.41176	0.34783	1.00000	0.31628
107	92	2	4	0	98
	7.043	10.000	10.000	--	7.224
	0.31522	1.00000	0.75000	--	0.34694
108	4	0	0	0	4
	10.500	--	--	--	10.500
	0.75000	--	--	--	0.75000
483	6	2	2	0	10
	29.833	25.500	17.000	--	26.400
	1.00000	1.00000	1.00000	--	1.00000
ALL	301	42	33	2	378
	8.980	10.619	9.727	9.500	9.230
	0.33555	0.50000	0.45455	1.00000	0.36772

Average Length of Stay and Percentage Complications, by DRG and CAT

DRG	CAT=1	CAT=2	CAT=3	CAT=4	CAT=5	CAT=6	CAT=7	ALL
104	3	0	10	2	0	12	2	29
	11.333	--	14.600	11.000	--	8.250	10.500	11.103
	0.66667	--	0.50000	1.00000	--	0.25000	0.50000	0.44828
105	0	1	8	0	0	12	0	21
	--	4.000	10.750	--	--	10.167	--	10.095
	--	1.00000	0.50000	--	--	0.50000	--	0.52381
106	210	3	1	0	0	0	0	214
	8.929	13.667	6.000	--	--	--	--	8.981
	0.31905	0.33333	0.00000	--	--	--	--	0.31776
107	91	5	0	0	1	0	0	97
	6.813	15.000	--	--	7.000	--	--	7.237
	0.30769	1.00000	--	--	1.00000	--	--	0.35052
108	1	3	0	0	0	0	0	4
	11.000	10.333	--	--	--	--	--	10.500
	1.00000	0.66667	--	--	--	--	--	0.75000
483	3	3	2	0	1	1	0	10
	21.333	31.000	32.000	--	11.000	32.000	--	26.400
	1.00000	1.00000	1.00000	--	1.00000	1.00000	--	1.00000
ALL	308	15	21	2	2	25	2	375
	8.455	16.267	14.381	11.000	9.000	10.120	10.500	9.237
	0.32792	0.80000	0.52381	1.00000	1.00000	0.40000	0.50000	0.37067

Average Length of Stay and Percentage Complications, by DRG and INC

ROWS: DRG COLUMNS: INC

	1	2	3	4	ALL
104	27	1	1	0	29
	10.852	10.000	19.000	--	11.103
	0.44444	0.00000	1.00000	--	0.44828
105	21	1	0	0	22
	9.667	15.000	--	--	9.909
	0.47619	1.00000	--	--	0.50000
106	203	11	1	0	215
	8.650	15.636	7.000	--	9.000
	0.30542	0.54545	0.00000	--	0.31628
107	89	9	0	0	98
	7.371	5.778	--	--	7.224
	0.35955	0.22222	--	--	0.34694
108	4	0	0	0	4
	10.500	--	--	--	10.500
	0.75000	--	--	--	0.75000
483	6	3	0	1	10
	28.833	25.000	--	16.000	26.400
	1.00000	1.00000	--	1.00000	1.00000
ALL	350	25	2	1	378
	8.923	12.960	13.000	16.000	9.230
	0.35714	0.48000	0.50000	1.00000	0.36772

This may be as far as you wish to go with this. The next step would be to incorporate all the factors using some sort of regression. If you include interaction terms, this gets pretty involved, but here is an approach that incorporates the information from the cross-tabulations above, and fits in the 3500-cell student version of MINITAB. DRG was recorded into a dummy variable for DRG 483 and INC was recorded into a dummy variable called REPEAT (for INC > 1). REPEAT turned out to be insignificant in the first regression of LS on some predictors, the printout for which follows.

Regression of Length of Stay on DRG483 and RS

```
The regression equation is
LS = 6.66 + 16.2 DRG483 + 0.157 RS

Predictor        Coef        Stdev     t-ratio        p
Constant       6.6640       0.4846       13.75    0.000
DRG483         16.190        1.622        9.98    0.000
RS            0.15692      0.03057        5.13    0.000

s = 4.984       R-sq = 28.3%     R-sq(adj) = 28.0%

Unusual Observations
Obs.    DRG483        LS        Fit   Stdev.Fit    Residual     St.Resid
  1      0.00     20.000      9.332      0.282      10.668        2.14R
  3      0.00     32.000      9.645      0.311      22.355        4.49R
 11      0.00     20.000      7.449      0.365      12.551        2.53R
 29      0.00      2.000     12.470      0.767     -10.470       -2.13R
 30      1.00     23.000     25.678      1.582      -2.678       -0.57 X
 55      1.00     58.000     25.364      1.589      32.636        6.91RX
 82      0.00     21.000      9.488      0.296      11.512        2.31R
 94      1.00     34.000     31.484      1.861       2.516        0.54 X
 97      1.00     11.000     25.364      1.589     -14.364       -3.04RX
116      1.00     25.000     24.894      1.603       0.106        0.02 X
135      0.00     38.000     11.528      0.598      26.472        5.35R
167      1.00     19.000     26.463      1.576      -7.463       -1.58 X
179      0.00     19.000     14.981      1.239       4.019        0.83 X
186      0.00     19.000      9.018      0.264       9.982        2.01R
210      0.00     25.000      7.449      0.365      17.551        3.53R
217      0.00     21.000      7.919      0.307      13.081        2.63R
223      1.00     16.000     26.149      1.577     -10.149       -2.15RX
227      0.00      8.000     13.098      0.883      -5.098       -1.04 X
245      0.00      8.000     12.941      0.854      -4.941       -1.01 X
246      0.00     38.000     11.215      0.544      26.785        5.41R
248      0.00     20.000      9.332      0.282      10.668        2.14R
258      1.00     30.000     26.934      1.579       3.066        0.65 X
278      0.00     21.000      8.704      0.260      12.296        2.47R
290      0.00     35.000      7.762      0.325      27.238        5.48R
309      1.00     32.000     27.247      1.585       4.753        1.01 X
316      0.00     19.000      8.233      0.280      10.767        2.16R
324      1.00     16.000     24.423      1.622      -8.423       -1.79 X

R denotes an obs. with a large st. resid.
X denotes an obs. whose X value gives it large influence.
```

Standardized Residuals versus CARD

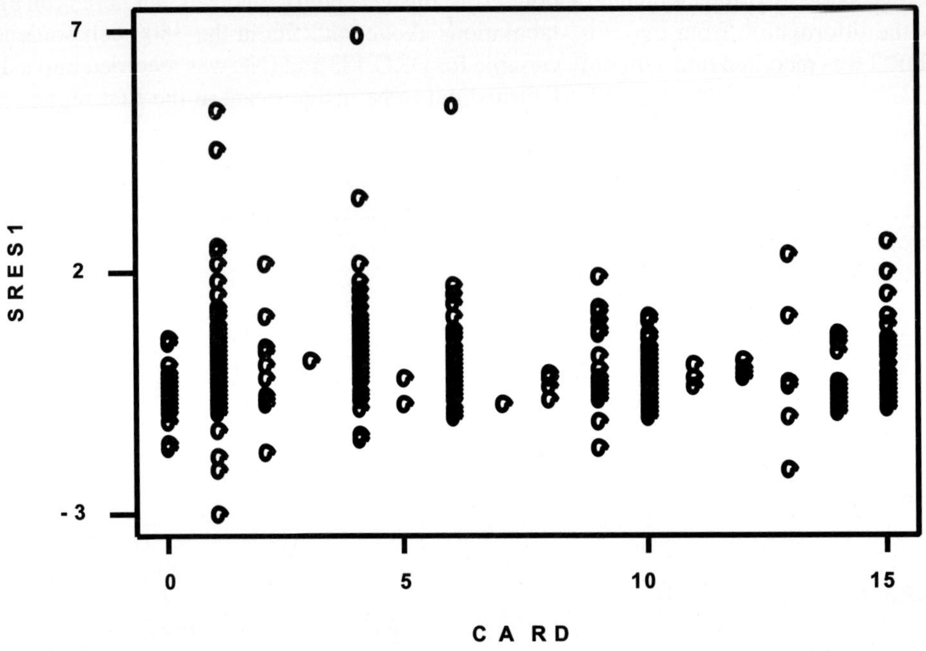

By plotting the standardized residuals against CARD, STAT, and CAT, we see little or no relationship. (Students will have to be able to interpret the residuals correctly to deduce this, of course.) The big outliers are observations 3, 55, 135, 246, and 290, which are spread among four DRGs, three cardiologists, two STATs, three CATs, and two INCs, so there's no apparent evidence in the data to explain these rather large deviations.

Standardized Residuals versus STAT

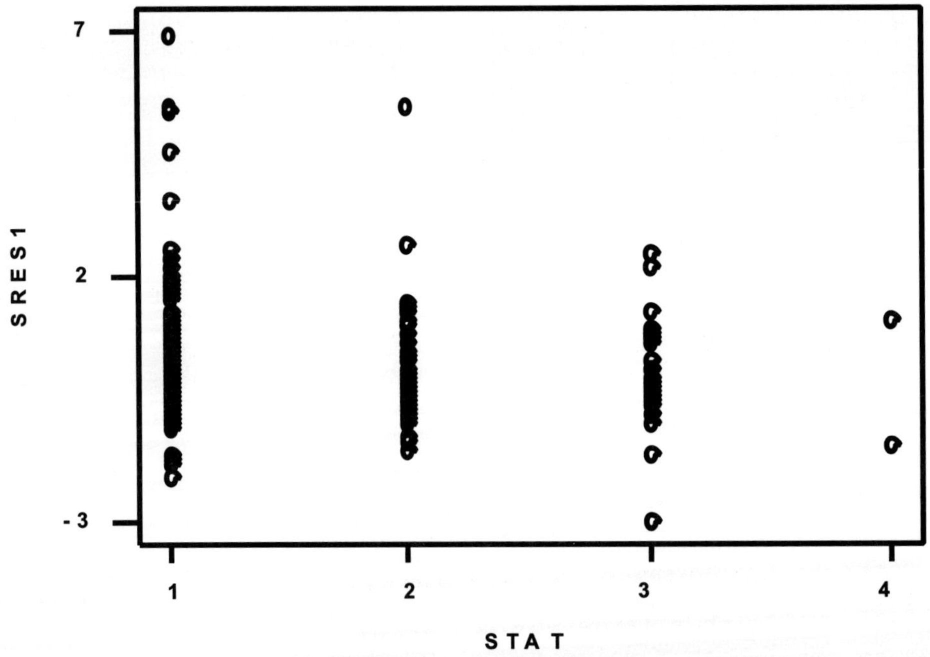

Standardized Residuals versus CAT

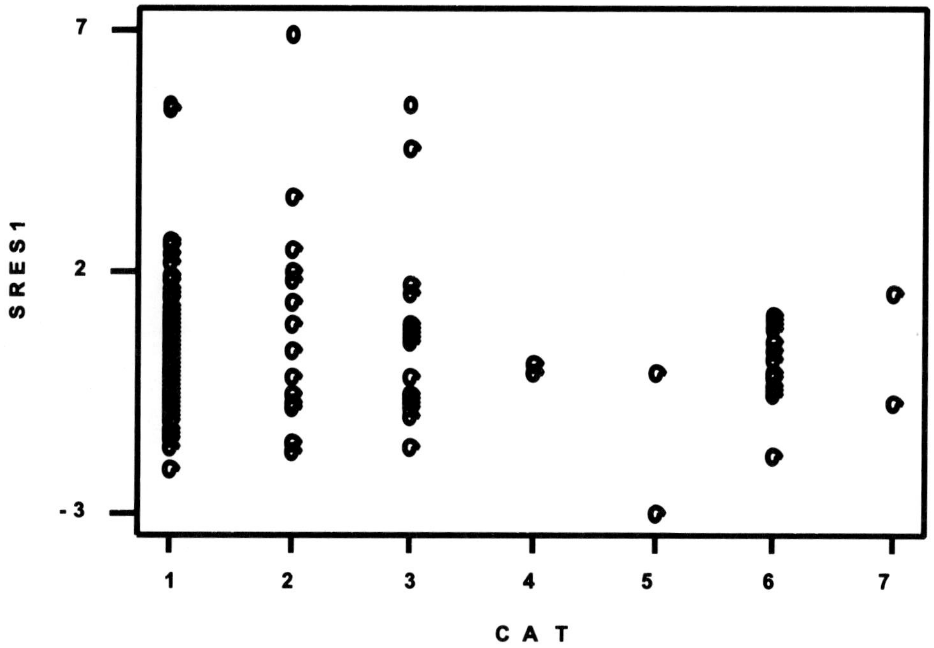

Histogram of Standardized Residuals

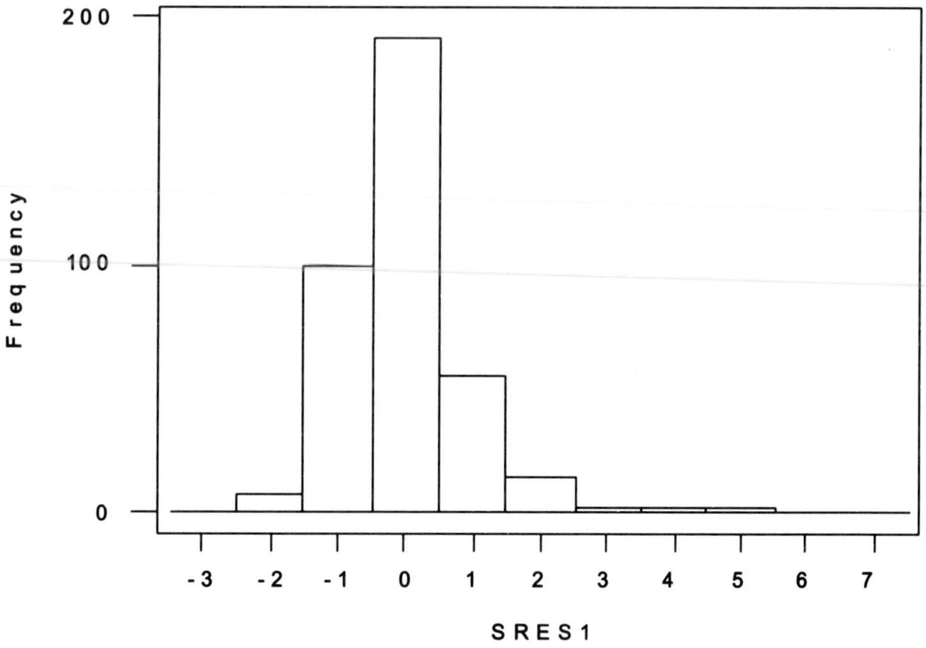

A tentative conclusion might be that DRG 483 increases stay length by about 16 days and that every 10 points of risk stratification score increase length of stay by perhaps 1.5 days, while the evidence linking other factors to length of stay is noticeably weaker.

CASE **71**

THE BAY AREA ATHLETIC CLUB, INSTRUCTOR'S NOTE

Abstract: This case uses descriptive statistics and/or inferential methods to provide a profile of a business customer. Much of the struggle will involve defining who is what type of customer.

Level:	Beginning, intermediate.
Techniques:	Descriptive statistics, hypothesis tests, cross-tabulations, chi-squared analysis.
Typical Class Discussion Time:	30 minutes.
Data Set Size:	94 cross-sectional observations, 9 variables.

GENERAL COMMENTS

The case inquires about the profile of a nonpeak user of the Bay Area Athletic Club. Defining who is a peak user and who is not is part of the challenge. The data set contains a variable giving the percentage of times a member used the club during off-peak hours. Should a nonpeak member be defined as one who used the club exclusively during off-peak hours? Or should it be someone who used the club at least 50% of the time during off-peak hours? Different judgments generate differ answers to the case.

In our experience, most students will define a nonpeak user as one who used the club at least some of the time during nonpeak hours. To obtain a profile of the off-peak user so defined, the student must create a binary variable which is 0 if the variable NonPeak is 0 and 1 otherwise. The case then becomes one with a binary response variable. Cross-tabulations are appropriate to examine the relationship between Peak/NonPeak usage and the variables Age, Marital, Gender, Kids, Location, and Type. Two-population means tests should be applied to frequency of use (Freq) and length of membership (Length).

This case is a beginning-level case if only descriptive statistics are to be used. It is an intermediate case if hypothesis-testing methods are available to the students.

QUESTIONS FOR ASSIGNMENT

1 Who should be designated a peak user and who should be a nonpeak user in this data set? Why?

2 What statistical test should be used to determine whether location and usage type (peak versus nonpeak) are related? What assumptions underlie this test? Which other variables in the data set should be subjected to the same statistical method? Why?

3 Which of the variables in the data set are nominal? Ordinal? Metric?

4 Do you see any problem with the way the data were collected? If so, how might this problem influence your chosen statistical methodology or recommendations?

SUGGESTIONS FOR LEADING THE DISCUSSION

Begin the class discussion by asking for a profile of the off-peak user. If students have different answers, this is likely due to varying definitions of an off-peak member. It's important to understand any differences before going further in the discussion. If it happens that different students defined nonpeak use differently, it will be useful to group them together in the classroom. That way, subsequent differences in recommendations can be attributed in part to the underlying definitions used.

Before leaving the discussion, ensure that the class enumerates a set of managerial actions to be followed. It will be helpful to list these on the board. For instance, some may recommend management provide child care. Write this suggestion on the board, then ask the student what part of their statistical analysis suggests this outcome. You will likely find many students suggesting many things, some of which are not data-driven. Single out the data-driven recommendations from the commonsense recommendations.

Expect one or more students to point out that a member might check in one minute before peak hours begin, thereby using the club mostly during peak hours but being recorded as using the club during off-peak hours. A more accurate measure of usage would require scanning membership cards as members both enter and leave the club. Also expect students to question whether one month of data is representative.

TYPICAL RESULTS AND APPROACHES

A typical nonpeak user (defined as someone who used the club at least part of the time during nonpeak hours) used the club during off-peak hours 45% of the time. Only 8 of the 94 members used the club exclusively during off-peak hours. A typical nonpeak member is an unmarried male in his 20s with no dependents at home. He holds a Silver membership, has been a member for 2.6 years, and comes to the gym 7.5 times per month. He lives in the location 2, the median distance from the club:

Descriptive Statistics of All Variables by Peak/Off-Peak

		N	MEAN	MEDIAN	STDEV	MIN	MAX
Age	Peak	34	3.176	3.0	0.936	2	6
	Off-Peak	60	2.950	2.0	1.383	1	6
Marital	Peak	34	0.471	0.0	0.507	0	1
	Off-Peak	60	0.317	0.0	0.469	0	1
Gender	Peak	34	0.382	0.0	0.493	0	1
	Off-Peak	60	0.483	0.0	0.504	0	1
Kids	Peak	34	0.559	1.0	0.504	0	1
	Off-Peak	60	0.333	0.0	0.476	0	1
Location	Peak	34	1.735	1.5	1.620	0	4
	Off-Peak	60	1.767	2.0	1.442	0	4
Freq	Peak	34	4.912	4.0	3.562	1	15
	Off-Peak	60	7.517	6.0	3.860	1	16
Type	Peak	34	0.588	1.0	0.500	0	1
	Off-Peak	60	0.617	1.0	0.490	0	1
Length	Peak	34	2.647	2.0	1.495	1	6
	Off-Peak	60	2.600	2.0	1.870	1	9

Twenty-eight % of the people in the sample were in their 20s and used the club during off-peak hours. Among the peak-time users, 44% were in their 30s. Although the Pearson chi-square rejects independence of age and usage pattern ($\chi^2 = 16.1$, p-value = 0.7%), small expected cell frequencies limit the reliability of the test:

Cross-tabulation of Peak/Nonpeak and Age

	Age						
Count Total % Row % Col %	1	2	3	4	5	6	
Non-Peak	5 5.32 8.33 100.00	26 27.66 43.33 76.47	10 10.64 16.67 40.00	8 8.51 13.33 47.06	8 8.51 13.33 88.89	3 3.19 5.00 75.00	60 63.83
Peak	0 0.00 0.00 0.00	8 8.51 23.53 23.53	15 15.96 44.12 60.00	9 9.57 26.47 52.94	1 1.06 2.94 11.11	1 1.06 2.94 25.00	34 36.17
	5 5.32	34 36.17	25 26.60	17 18.09	9 9.57	4 4.26	94

Marital status ($\chi^2 = 2.2$, p-value = 13.8%), Gender ($\chi^2 = 0.9$, p-value = 34.4%), Location ($\chi^2 = 2.3$, p-value = 68.7%), and Type ($\chi^2 = 0.1$, p-value = 78.6%) are all independent of usage pattern according to the Pearson chi-squared statistic. Having children in the home is related to usage pattern ($\chi^2 = 4.5$, p-value = 3.3%). Forty-three % of all users have no children and use the club off-peak:

Cross-tabulation of Peak/Nonpeak and Kids

	Kids		
Count Total % Row % Col %	0	1	
Non-Peak	40 42.55 66.67 72.73	20 21.28 33.33 51.28	60 63.83
Peak	15 15.96 44.12 27.27	19 20.21 55.88 48.72	34 36.17
	55 58.51	39 41.49	94

The average number of years with the club (Length) is virtually identical for peak and off-peak users: 2.6 years for each group. The t test of equal population means (assuming equal variances) is 0.13.

Frequency of use per month is statistically higher on average for off-peak than peak users. Off-peak customers use the club an average 7.5 times per month; peak users only 4.9. The t ratio for the test of equal population means (assuming equal variances) is -3.23 (p-value $= 0.2\%$):

Scatterplot of Freq and Peak/Non-Peak

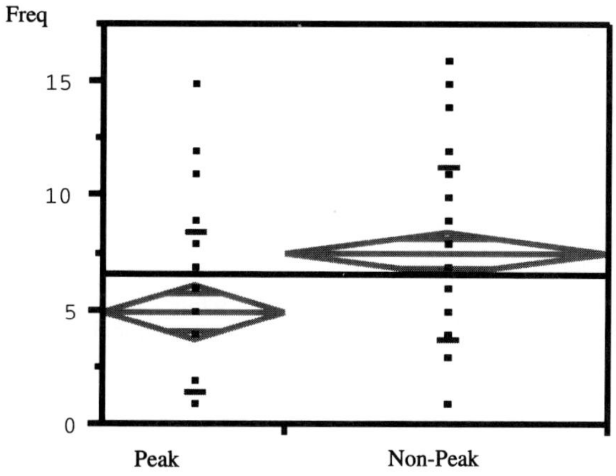

CASE **72**

BLOOD PRESSURE, INSTRUCTOR'S NOTE

Abstract: A very straightforward case, which can be handled with simple plots, descriptive statistics and/or a *t* test. Requires no specific business knowledge, and has few surprises.

Level:	Beginning.
Techniques:	Descriptive statistics, tests of hypotheses.
Typical Class Discussion Time:	15–30 minutes.
Data Set Size:	92 cross-sectional observations, 3 variables.

GENERAL COMMENTS

Simple descriptive measures seem to be all that are necessary here. Histograms, means, and standard deviations probably tell the story as well as anything.

QUESTIONS FOR ASSIGNMENT

You can either let the students decide for themselves what kind of patterns might be of interest to the company, or you can focus the exercise a bit more by assigning questions like:

1 Summarize the blood pressures of those taking part in the health fair by using appropriate descriptive methods.

2 Assess the number of individuals who appear to have blood pressures above the normal (satisfactory) ranges.

3 Does blood pressure tend to be different for women versus men in these data? Significantly so?

4 Comment on any possible biases to be aware of in the data.

SUGGESTIONS FOR LEADING THE DISCUSSION

The questions above work pretty well as discussion questions here, too. It may be worth pointing out that the differences observed are meaningful in the aggregate, but probably not a very useful indicator for any individual involved. If students have a background in the area, they may offer their thoughts on how reliable individual blood pressure measurements might be (not very).

TYPICAL RESULTS AND APPROACHES

The histograms are not perfectly symmetric, but they are generally mound-shaped, for both systolic and diastolic pressure. The means are different for the males and females, in both cases:

GENDER	MEAN SYST	MEAN DIAST
Male	128.54	84.17
Female	114.77	78.68
T	4.6	2.6

The *t* statistics given for testing differences between these means would normally be considered significant. The histograms of SYST and DIAST for males and females separately are somewhat more normal than for the combined data set, and there are three severe cases (SYST > 170) among the males, which could be pointed out.

Students can also approach this problem by counting the number (of each gender, perhaps) who fall in a danger zone. The corresponding table of danger cases by sex could be tested for independence by using a chi squared test.

In general, this case illustrates standard techniques in standard ways. The main problem with these data is in their selection: only those employees who chose to come to the health fair were included. What bias this may have introduced isn't clear, but it ought to raise students' suspicions.

CASE **73**

MATRIMONIAL UNREST, INSTRUCTOR'S NOTE

Abstract: This case has the flavor of a quality improvement exercise, but deals with potentially embarrassing issues for some students. Mature students will find this a light-hearted look at data analysis which, in the end, delivers equal verdicts against both genders.

Level:	Intermediate.
Techniques:	Descriptive statistics, two-population means tests, cross-tabulations.
Typical Class Discussion Time:	30 minutes.
Data Set Size:	75 daily observations, 13 variables.

GENERAL COMMENTS

This case demonstrates how data can be used in quality improvement exercises. In this instance, a married couple having ongoing arguments collects behavioral, biological, and financial data to identify the source of the disagreements. Only two variables, a bad day at work for the husband or premenstrual symptoms of the wife, seem to be related to arguments.

The student who wrote the paper says of this exercise, "My conclusion is that the subject matter is too complex to be easily explained through statistical analysis. However, depersonalizing the situation through this type of analysis has enabled the two of us to accept our own responsibility for our quality of life . . . . Finger-pointing was fun and it allowed my wife and I to laugh at ourselves." Perhaps that says it all.

Don't assign this case if you feel your students will be uncomfortable with discussing or writing about premenstrual symptoms of women. Graduate students, even those composed of international students, have handled the issue with no apparent difficulty. Consultation with our local international student adviser revealed that, for some cultures, the issue of PMS is far less embarrassing than the public admission and discussion of marital arguments.

QUESTIONS FOR ASSIGNMENT

1 Which variables in the data set are nominal? Ordinal? Continuous?

2 What method should be used to analyze the relationships between the variable Fight and the other nominal variables? What assumptions underlie these statistical tests? Do these assumptions appear to be valid for this data set?

3 What method should be used to analyze the relationships between Fight and the continuous variables in the data set? What assumptions underlie these statistical tests? Do you feel these assumptions are satisfied in this data set?

4 Who is most at fault for these fights, Ted or Maria? What led you to this conclusion?

SUGGESTIONS FOR LEADING THE DISCUSSION

This depends on your playfulness and the maturity level of your students. We have tried taking on the role of either Ted or Maria during the discussion. You can open the discussion by saying something like, "Ted [Maria] and I had another awful fight last night. Can you help us understand what's causing these things?" If you are brave, become increasingly annoyed as the evidence suggests some role on the part of the character you are playing.

You may also wish to conduct the discussion as if it were a clinical analysis. Lead the discussion toward statistical methods and outcomes. Regardless of which approach you take, you will help alleviate students' discomfort if you utter "premenstrual symptom" or "PMS" before they do.

TYPICAL RESULTS AND APPROACHES

The variable Fights is consistent with randomness according to the runs test:

Runs Test Results for Fights

```
Average = 0.2667

The observed no. of runs = 28
The expected no. of runs = 30.3333
20 Observations above the average, 55 below
     The test is significant at 0.4864
     Cannot reject at alpha = 0.05
```

None of the metric variables differ on average between days with and without arguments. The t tests of equality of population means (assuming unequal variances) are: 0.15 for Calories, 0.47 for Sleep, −1.47 for Coffee, −1.05 for Bank, and −0.45 for Checkbk. All have P-values exceeding 15%.

Descriptive Statistics for Metric Variables Grouped by Fight

	Fight	N	MEAN	MEDIAN	STDEV	MIN	MAX
Calories	0	55	3000.8	2950	730.6	1661	4491
	1	20	2971	2993	740	1160	3986
Sleep	0	55	6.891	6.5	1.593	4	11
	1	20	6.738	6.5	1.090	5	8.5
Coffee	0	55	0.891	1	0.832	0	3
	1	20	1.250	1	0.967	0	3
Bank	0	55	804.8	901	460.0	19	1737
	1	20	940	992	506	4	1737
Checkbk	0	55	335.6	320	529.5	-595	1560
	1	20	420.0	156	618.0	-595	1560

Ted's bad days at work and Maria's premenstrual symptoms are related to Fight:

Chi-Square Analysis of Fight and Work

```
ROWS: Fight          COLUMNS: Work

            0          1         ALL

 0         48          7          55
        87.27      12.73      100.00        CELL CONTENTS --
        78.69      50.00       73.33            COUNT
        64.00       9.33       73.33            % of Row
                                                % of Column
                                                % of Table
 1         13          7          20
        65.00      35.00      100.00
        21.31      50.00       26.67
        17.33       9.33       26.67

ALL        61         14          75
        81.33      18.67      100.00
       100.00     100.00      100.00
        81.33      18.67      100.00

CHI-SQUARE =      4.792     WITH D.F. =      1
   p-value =      2.86%
```

Chi-Square Analysis of Fight and PMS

```
ROWS: Fight          COLUMNS: PMS

              0           1          ALL

 0           51           4           55
          92.73        7.27       100.00        CELL CONTENTS --
          78.46       40.00        73.33           Count
          68.00        5.33        73.33           % of Row
                                                   % of Column
 1           14           6           20           % of Table
          70.00       30.00       100.00
          21.54       60.00        26.67
          18.67        8.00        26.67

ALL          65          10           75
          86.67       13.33       100.00
         100.00      100.00       100.00
          86.67       13.33       100.00

CHI-SQUARE =      6.556      WITH D.F. =     1
   p-value =      1.05%
```

School ($\chi^2 = 0.16$, p-value $= 69.4\%$), Events ($\chi^2 = 1.14$, p-value $= 28.6\%$), and Bills ($\chi^2 = 0.15$, p-value $= 70\%$) are independent of Fight according to the Pearson chi-squared statistic. Small expected cell frequencies are problematic, though. Day of the week is also unrelated to Fight. The Pearson statistics are $\chi^2 = 1.95$ (p-value $= 16\%$) for Weekend and $\chi^2 = 3.08$, p-value $= 80\%$) for Weekday. This latter test is also suspect due to low expected cell counts.

CASE **74**

PRODUCTIVITY IN THE WORD PROCESSING CENTER, INSTRUCTOR'S NOTE

Abstract: This case is good for sending two messages to introductory statistics students: graphs often tell more than summary statistics, and interpretation of statistical results depends on the eye of the beholder. In this case, a lengthy description of the scenario will lead students to conclusions that seem to coincide with the data. Upon further reflection, alternative explanations are equally plausible. This ambiguity leads to wonderfully lively classroom discussions.

Level:	Beginning.
Techniques:	Descriptive statistics, time series plots.
Typical Class Discussion Time:	30 minutes.
Data Set Size:	40 time series observations, 5 variables.

GENERAL COMMENTS

Assign this case early in the semester. It will inculcate a healthy dose of skepticism that will serve you well if you want students to talk back to you and data.

The case involves monitoring four employees' work productivity. The experimental design is horrid. Ethical problems associated with data collection lurk in the background. All in all, this case dispels any preconceptions that statistics is just an applied mathematics course.

QUESTIONS FOR ASSIGNMENT

1 Examine a time series plot for each of the four monitored employees. What do these plots tell you about the problem described in the case?

2 Calculate the mean, median, and standard deviation for all four employees. What do these numbers tell you about the problem presented in the case?

3 Should one or more of the employees in the word processing center be fired? If so, who, and why?

SUGGESTIONS FOR LEADING THE DISCUSSION

The class discussion of this case will quickly take on a life of its own. You won't need to do much more than ask the question posed in the case: has the software changed the behavior of the employees in the center, and if so, how? Play the devil's advocate as often as possible. If you feel brave, go into class announcing that you, the boss, intend to fire one of the four employees being monitored, and ask the class to tell you which one it should be.

Wrap up the case discussion with an overview of the assumptions behind different recommendations. Assure students that ambiguity arising from statistical analyses is not only commonplace, but perhaps even healthy.

TYPICAL RESULTS AND APPROACHES

Four employees are monitored for their word processing efficiency in words per minute. They perform about the same on average. Mean words typed per minute is about 71 or 72 for all.

Descriptive Statistics

	N	MEAN	MEDIAN	STDEV	MIN	MAX
Sue	40	71.87	74	18.44	0	99
Bob	40	71.22	70	8.73	58	86
Sandy	40	72.43	72	6.71	62	85
Ted	40	71.58	73	5.17	59	79

However, the sequence plots reveals distinct patterns. The supervisor, Sue, might be presumed to have an attitude problem revealed by declining productivity following the announcement of the monitoring program. It might also be that she simply had other duties to perform. We don't really know which is which from the data. Two observations of zero will influence students' views of her productivity:

Sequence Plot for Sue

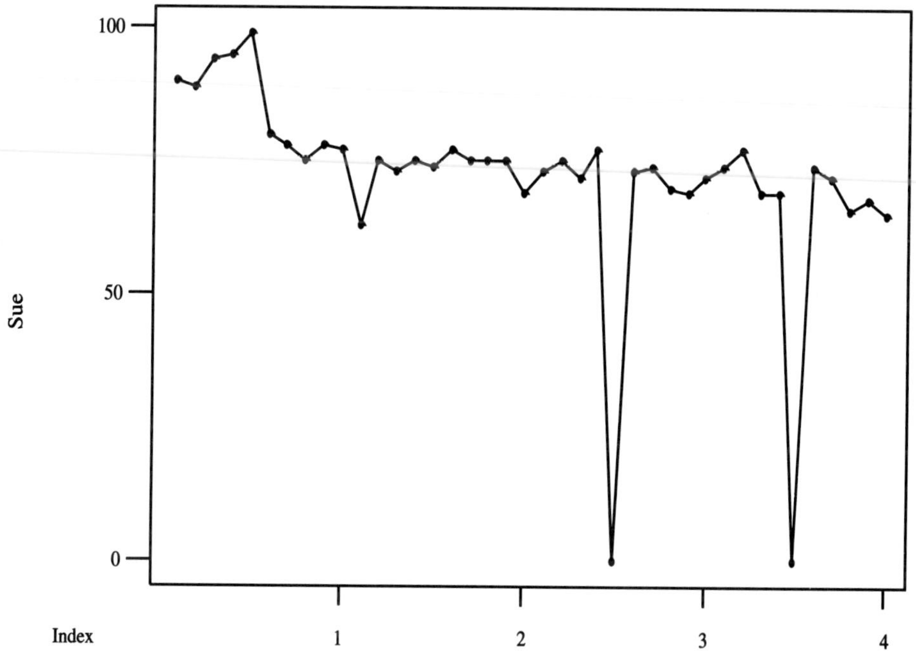

Bob has a high standard deviation relative to his peers. Is this because he is best at doing other duties (e.g., answering the phones), or because he is unfocused? Sandy, as a new hire, seems to be in an adjustment period. And Ted, described in the case as the company sycophant, immediately increases his productivity when told that he is being monitored. Is this a real increase in productivity, or is there another explanation? The answer would rest with Sue, who presumably has been alienated by the entire affair. As we said, expect a lively classroom discussion.

Sequence Plot for Ted

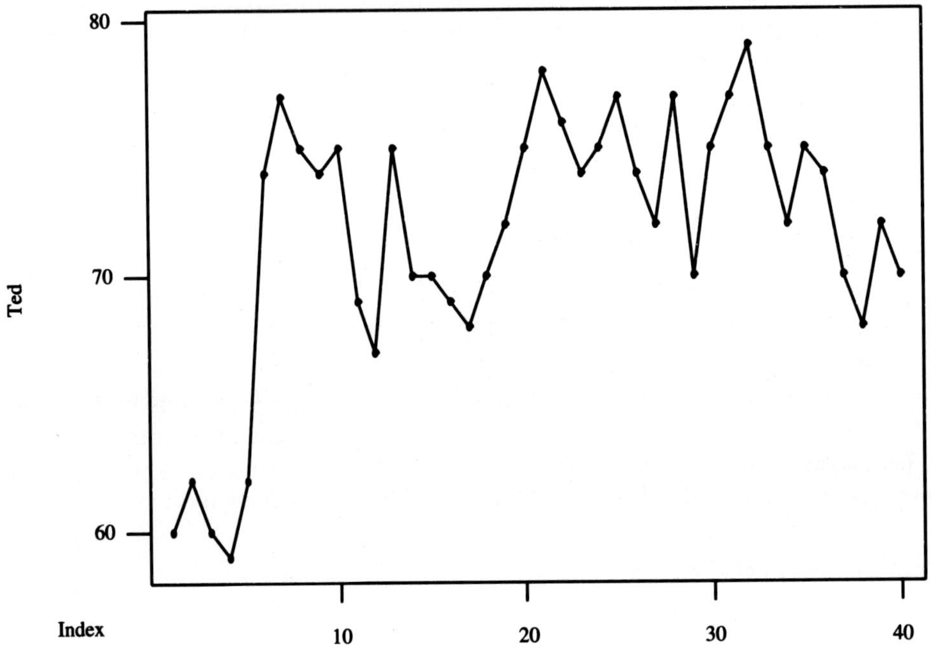

Sequence Plot for Bob

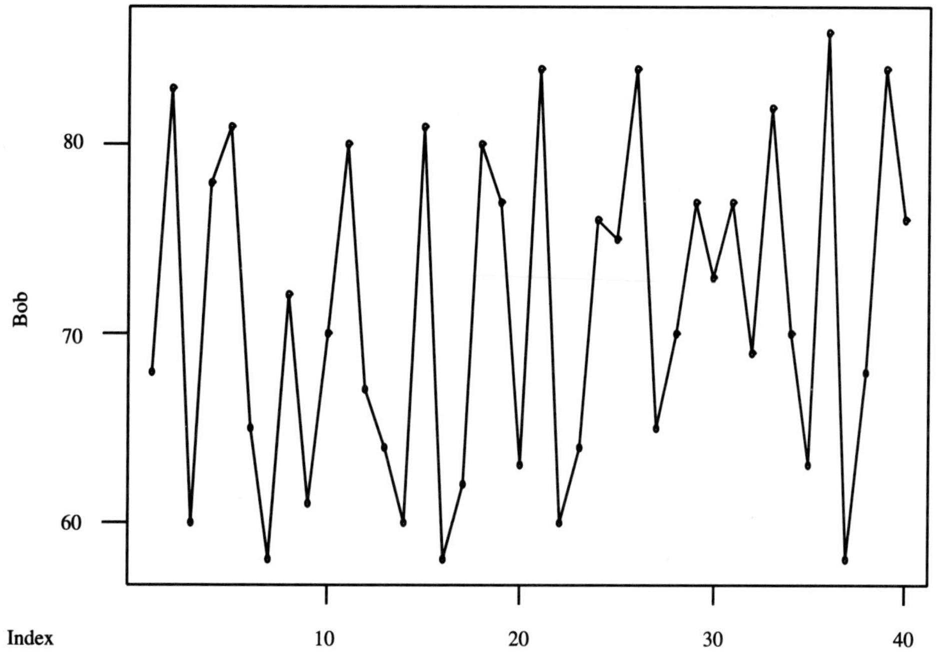

Sequence Plot for Sandy

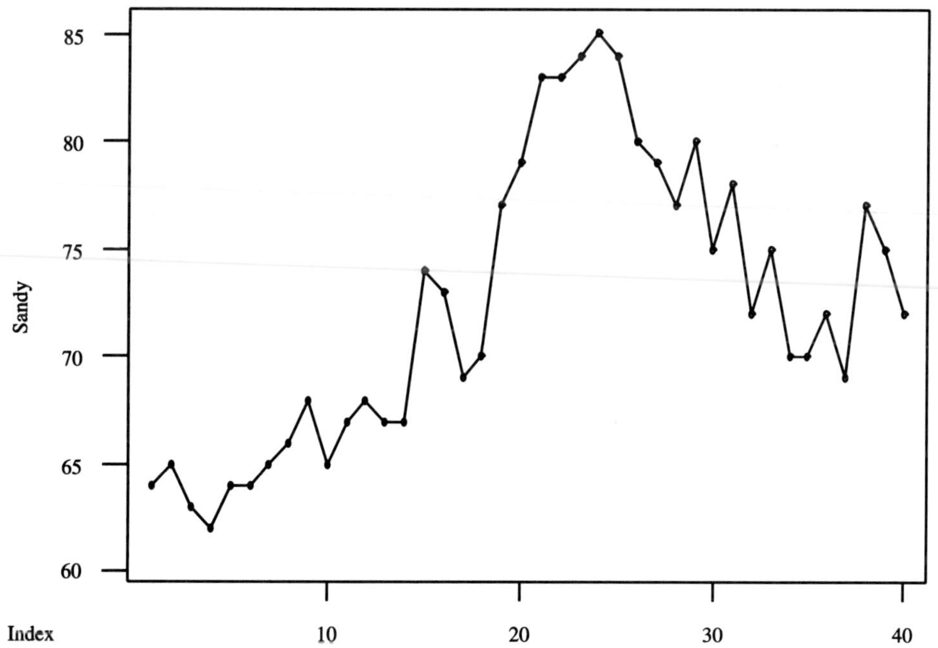

Has the software induced a change in behavior? Not much is observed in the aggregate. Sue and Bob's productivity has declined while Ted and Sandy's has increased. Overall, the average productivity dropped from 291 words per minute before employees knew they were being monitored to 286 afterwards.

CASE **75**

CHURCH ATTENDANCE, INSTRUCTOR'S NOTE

Abstract: A straightforward case suitable for introductory discussions of time series plots and/or various forms of simple or multiple regression. Also provides a vehicle for discussion of assumptions and/or residual analysis.

Level:	Beginning, intermediate.
Techniques:	Time series plots, regression.
Typical Class Discussion Time:	30 minutes.
Data Set Size:	28 observations, 4 variables.

GENERAL COMMENTS

This simple case can be used in many different ways. You can illustrate simple visual analysis of a time series plot, simple regression, multiple regression with dummy variables, or various forecasting techniques. Finally, it's a good case for discussion of the underlying assumptions for regression or for analysis of residuals in a fairly straightforward setting. We rank it as beginning or intermediate, depending on how much you ask students to do with it.

QUESTIONS FOR ASSIGNMENT

For mature audiences, something general works well here:

1 Provide an appropriate summary for Al to use.
2 Forecast church attendance and defend the forecast.

For beginning students you may wish to provide a more structured approach:

3 Plot church attendance over time and interpret the results.
4 Would a regression line be an appropriate summary of the trends in this case? Why or why not?
5 Compute such a regression line and interpret the results.
6 Analyze the residuals of this regression. What do they suggest?
7 Compute a multiple regression of church attendance on the factors given in the case, and interpret the result. Analyze the residuals of this regression. Compare the results with the results of the simple regression you computed in the previous steps.

SUGGESTIONS FOR LEADING THE DISCUSSION

This is a straightforward sort of case, so don't do too much for the students. The assignment questions above should serve pretty well as the basis for a discussion. You may need to introduce the topic of regression assumptions a bit more carefully. The residuals here (see below) clearly are not exactly normally distributed, though those from the multiple regression probably aren't seriously alarming, but the individual observations are averages of differing numbers of observations, so in principle we ought to be a bit suspicious. We see no unambiguous choice here: there's enough noise that averages of 4 and 5 observations may not be dramatically different, but it's a good way to introduce the topic of assumptions and residual analysis. It also provides a partial answer to the case question: other things being equal, we would probably have preferred individual weekly observations (and a corresponding time measure in weeks instead of months) instead of the monthly averages.

TYPICAL RESULTS AND APPROACHES

Time series plots show that attendance has been increasing more or less steadily:

Time Series Plot of Average Attendance

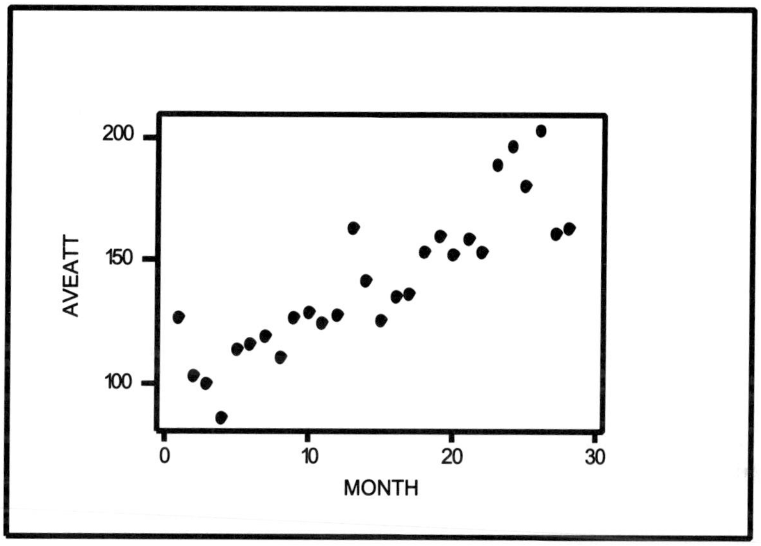

Time Series Plot of Average Attendance (alternate form)

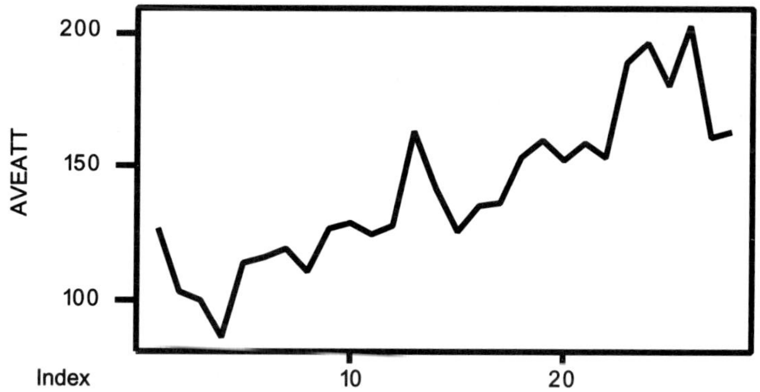

For statistical purposes the first plot may well be more helpful, as it suggests outliers that correspond to things like months with Easter a bit better than the second plot, though not everyone need agree.

A simple regression of AVEATT on MONTH (however questionable its assumptions may be) yields

Regression of AVEATT on Month

```
The regression equation is
AVEATT = 96.1 + 3.13 MONTH

Predictor        Coef        Stdev     t-ratio        p
Constant       96.071        5.654       16.99    0.000
MONTH          3.1256       0.3406        9.18    0.000

s = 14.56         R-sq = 76.4%      R-sq(adj) = 75.5%

Unusual Observations
Obs.     MONTH      AVEATT         Fit  Stdev.Fit     Residual      St.Resid
  1        1.0      127.00       99.20       5.36        27.80         2.05R

R denotes an obs. with a large st. resid.
```

The corresponding residual plots (see below) suggest some problems. You can use these results or their equivalents to suggest the need for multiple regression. The multiple regression they hint at, including the dummy variables for EASTER, THANKS, and SUMMER, yields the printouts on the next page. The multiple regression is considerably better-fitting, and the residuals better-behaved, though by no means completely satisfactory.

Average attendance is going up about 3 per week, months containing Easter are about 25 higher than others, and so on.

Residuals from Simple Regression of AVEATT on MONTH

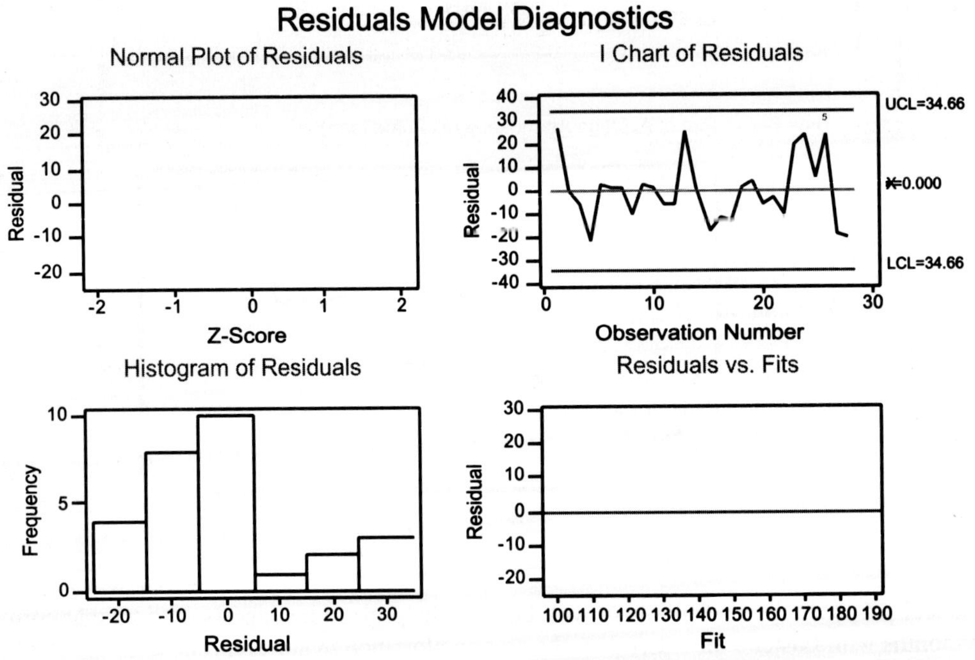

Regression of AVEATT on MONTH, THANKS, EASTER, and SUMMER

```
The regression equation is
AVEATT = 96.5 + 3.23 MONTH + 24.8 EASTER - 18.0 SUMMER - 9.67 THANKS

Predictor        Coef       Stdev    t-ratio        p
Constant       96.493       3.525      27.38    0.000
MONTH          3.2270      0.1987      16.24    0.000
EASTER         24.813       5.309       4.67    0.000
SUMMER        -18.012       4.028      -4.47    0.000
THANKS         -9.671       6.332      -1.53    0.140

s = 8.470        R-sq = 92.9%      R-sq(adj) = 91.7%

Unusual Observations
Obs.    MONTH      AVEATT        Fit  Stdev.Fit   Residual    St.Resid
 23      23.0      189.00     170.71       2.67      18.29       2.27R
 24      24.0      196.00     173.94       2.80      22.06       2.76R

R denotes an obs. with a large st. resid.
```

Residuals of Regression of AVEATT on MONTH, THANKS, EASTER, and SUMMER

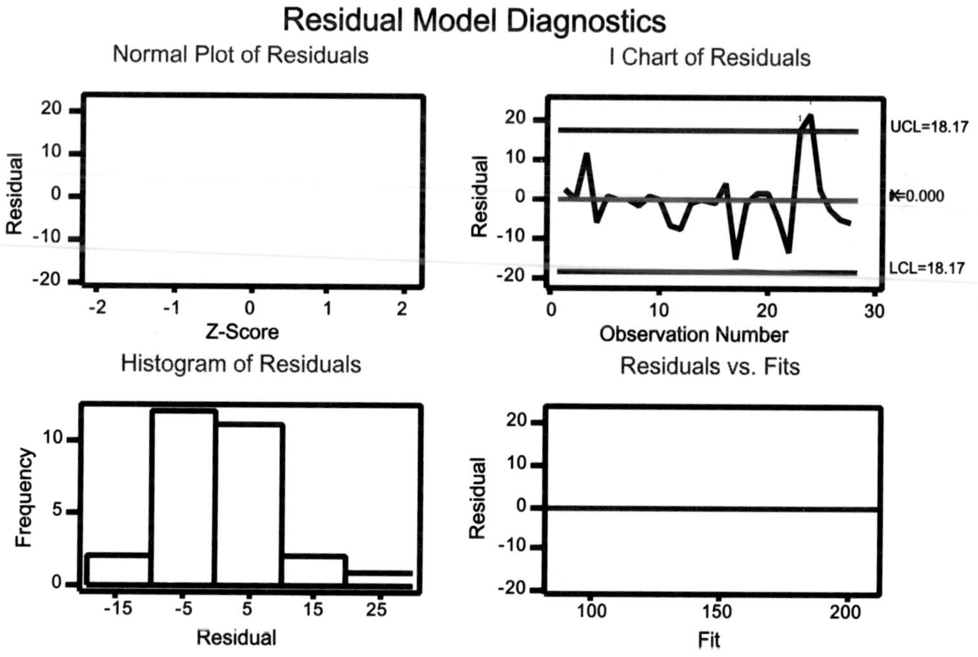

APPENDIX **A**

SAMPLE EXAMINATIONS

Two sample examinations follow. The first is a take-home exam with an in-class follow-up. The second is an in-class exam. Both are based on cases in *PDA*. You may wish to use a case not in the *PDA* volume currently assigned to your students. The texts of all 75 cases are included in this *Instructor's Edition,* so there should be enough to choose from.

We have used both of the exams below in class. We've edited them a bit to fit the format here, but you should be able to get the general idea.

SAMPLE TAKE-HOME EXAMINATION, WITH IN-CLASS FOLLOW-UP

This exam was given to graduate students in the required MBA statistics class. It's based on *Case 39, "Software Service Calls,"* from Volume II of *PDA*. We've given a similar exam using *Case 64, "The Advertising Experiment."*

The handout was distributed one week before the exam, and the exam was distributed at the examination at 6 P.M. on July 26. When the students left the exam, a summary sheet (similar to the instructor's note "Typical Results and Approaches" section) was distributed.

Sample Take-Home Exam, Handout

FINAL EXAMINATION INSTRUCTIONS

Instructions

1. Imagine that you have been told you are to study some data in preparation for a meeting with your supervisor at (date, time).

2. Be prepared to answer in writing questions about the data set, questions that will be given to you at the exam on (date). This will simulate your response to your supervisor's questions about the data set. The questions will be directed only to the data set described below.

3. This is (at least in part) a take-home exam, and is for individual effort only. You are expected to do all your own work (including all computer work), and consult with no one else in preparing for the exam. Any materials you bring to the exam should be yours and only yours. You may consult textbooks and the like, but you may not consult with anyone else in any direct or indirect way.

Please take this restriction seriously. Any apparent violations have to be dealt with under the college rules for academic (dis)honesty.

4. If you have any questions, please ask them early on, so that any confusion or other problems may be cleared up as soon as possible.

5. The situation you are to examine is described in Case 39, "Software Service Calls." The data file is in file SERVCALL and is on your data disk. Please check your disk as soon as possible to verify that you can access the file. If there is any problem with this, let me know as soon as possible.

Sample Take-Home Exam, In-Class Section

At the meeting, Sam's supervisor asked, "Well, Sam, based on the records, what have you come up with for predicting the service calls better than we do now? We really have to improve our handling of those calls."

(a) As Sam, answer the supervisor's question and defend the answer.

(b) Are there any other features of the data that you think Sam or the supervisor should know about? If so, what are they?

Limit: 3 written pages total (both sides, if you like). This is an upper limit, not a lower limit.

SAMPLE IN-CLASS EXAMINATION

This exam presents the students with a case and a series of computer printouts based on the case. It asks them to interpret the output and write a report. It would be appropriate for an in-class final exam for a case-based course. If the students could have access to a computer during the final exam period, the printouts wouldn't be needed, of course, and the exam would be even better, but that's not always practical.

This exam was given to a class of about 80 MBA students in a 75-minute exam period. It's based on Case 26, "Dave's Truck Washing Service," from Volume II of *PDA*.

The printouts were provided by the instructor. For this kind of exam, you should do some background computer work. Include a few printouts leading to ambiguous or odd outcomes, so that you force the student to think carefully about differences in models and assumptions, and explain those thoughts clearly.

In this exam, we asked the students to write a nontechnical executive summary and then forecast sales. If your class needs more direction, you could ask them more specific questions. "Which of the printouts on the following pages describes the best model for the data? Why do you say so?" would be one way to start. You could also ask students to interpret specific numbers on specific pages. "Describe in plain English what the statement *s=12.403* on page 14 means in the context of the problem" would be one level more precise, for example. The idea is to get the students to demonstrate the ability to understand, define, interpret, and articulate. This exam was very successful in that respect. Student responses varied a lot, and it became clear who did or did not digest the material from the class. Further, the students were familiar with this format of work (they'd been doing it all semester), so test anxiety didn't appear to be much of a factor.

Sample In-Class Exam Dave's Truck Washing Service

(here followed the text of *Case 26, "Dave's Truck Washing Service"*)

First, write an executive summary describing what you believe Dave needs to know about revenues of his firm. You may use any of the MINITAB printouts that follow to write this summary. Because Dave knows nothing about statistics, you are not to use technical terms in the executive summary.

Second, help Dave use your results to make short-term forecasts. Forecast sales for this March, and explain to Dave how he could make similar calculations on his own for the future.

Part One: EXECUTIVE SUMMARY

(blank page here for answer)

Part Two: FORECASTED SALES

(on this page, give your forecast and explanation)

(Printouts were attached giving basic descriptive summaries of all the variables in the data set, a scatterplot of SALES versus TEMP, and regression printouts for seven different regression models for SALES.)

APPENDIX **B**

SAMPLE CASE AND CASE WRITE-UP, WITH CRITIQUE

Guiding students toward an appropriate case write-up can be difficult. We have found that graduate students sometimes need as much guidance on writing reports as undergraduates. Graduate students who have had an undergraduate business statistics course in the past may think that statistics homework involves pencil-and-paper, mathematical, and formula-based responses. When we now ask them to produce a well-written and grammatically correct report in a statistics course, it can come as a surprise.

As the instructor, you can assist your students in several different ways:

1 Encourage them to read the section in the casebook entitled "Preparing Written Cases and Business Reports" before the very first assignment. That material includes tips about writing business reports, which may be sufficient for most graduate students. One disadvantage of giving prototype write-ups is that students tend to follow them too closely. Sometimes a model case write-up is taken as the one and only method for writing cases. That stifles creativity, thoughtfulness, and originality. Nonetheless, those examples can be used for guiding students within some boundaries, which you as the instructor may extend, limit, or renounce completely, as appropriate to your pedagogical style.

2 Another way to help your students in their case write-ups might be to hand out copies of the following sample case, write-up, and critique. The next few pages include a case not found in the student case books. The case is followed by a mock student write-up, a critique of that write-up, and an alternative "corrected" version. This may help your students, particularly undergraduates, produce higher quality work at the very beginning of the semester.

3 You may wish to ask your students to prepare their first case without the guidance of the sample case or critique. Buckling down on their own may force them to decide for themselves how to present the material. You might distribute the example after their first assignment has been completed. Regardless of when and how you distribute this material, it is important that you make lots of written comments on their first case to assist them in recovering quickly from any crucial flaws. Sometimes producing an appropriate report is *the* major frustration for students, even beyond the statistical or computational requirements of the course. We hope the following pages are helpful.

The sample case is purposely not complex. We've assumed that you will hand this out early in the semester, and that your students won't know much about advanced topics at the beginning of the course. The case requires nothing more than interpreting data and simple graphics, but we hope the critique will get them moving in the right direction.

CASE **A**

CIGARETTE SMOKING IN THE UNITED STATES

Cigarette smoking has become a contentious topic in the United States. Earlier in this century, smoking was a perfectly acceptable habit for men even though it was considered scandalous for a woman to be caught with a cigarette. Smoking later became fashionable for both sexes, and the incidence of smoking in the overall population increased over time and across all economic, social, and age groups.

Times have changed, though, as medical evidence has linked cigarette smoking to a number of major health problems. In spite of clear evidence about the health risks of smoking, many people continue to smoke in the United States. One reason may be that nicotine is addictive; a medical report in the 1980s equated cessation of smoking with the withdrawal symptoms of hardcore street drugs. It's *hard* to quit smoking.

Given the addictive nature of nicotine, it has been argued that cigarette advertising directed at teenagers should be minimized. Cigarette companies are no longer allowed to advertise on television in the United States. Other social and legislative changes have occurred in the past few years regarding smoking: no-smoking sections in restaurants and nonsmoking hotel rooms are commonplace, some public buildings prohibit smoking indoors, and airlines have banned smoking on most domestic flights. Meanwhile, smoker's rights groups are being organized in small numbers across the United States.

The tables that follow present some statistics on the incidence of smoking in the United States over time and across socioeconomic groups. The sample covers persons 20 years and older. The data also provide evidence of attempts to quit smoking, as shown in the table labeled Quit Ratio. This variable is the percentage of people who ever smoked who no longer smoke. These data come from the *1992 Statistical Abstract of the United States,* U.S. Department of Commerce, Bernan Press, Lanham, MD, page 129.

Use the data in these tables to prepare a report of cigarette smoking habits in the U.S. Numbers in the Present Smokers category are in terms of percentages of the total population. Quit Ratio is the percentage of persons who have smoked but no longer do. NA is not available. A simple, one-page summary of this information has been requested. In your summary, highlight any interesting aspects of the data that you discover.

Present Smokers

	1965	1970	1978	1979	1980	1987	1988
Age:							
20 to 24 years	47.8	41.5	35.8	35.9	36.1	29.5	27.6
25 to 44 years	49.5	44.5	39.5	39.3	38.0	33.2	32.9
45 to 64 years	39.9	38.5	37.1	35.3	36.3	30.9	29.4
Over 65 years	17.1	16.2	16.4	16.5	17.3	15.2	14.9
Sex:							
Male	50.2	44.3	39.0	38.4	38.5	31.5	30.9
Female	31.9	30.8	29.6	29.2	29.0	26.2	25.3
Race:							
White	40.0	36.5	33.6	33.2	32.9	28.3	27.5
Black	43.0	41.4	38.2	36.8	37.2	33.5	32.3
Educational Attainment:							
Less than high school	NA	34.8	35.3	34.9	35.5	33.9	32.8
High school graduate	NA	38.3	36.5	35.4	35.7	32.9	32.6
Some college	NA	36.7	32.7	33.3	31.2	25.8	25.4
College grad.	NA	28.1	23.8	23.4	24.6	16.1	15.6

Quit Ratio

	1965	1970	1978	1979	1980	1987	1988
Age:							
20 to 24 years	17.8	20.8	22.8	22.6	22.2	23.8	27.5
25 to 44 years	23.6	29.8	31.9	31.8	33.0	37.2	36.8
45 to 64 years	30.9	36.1	40.1	42.4	40.9	49.2	52.8
Over 65 years	48.7	56.9	62.4	61.7	61.0	69.2	69.8
Sex:							
Male	31.4	37.9	41.3	41.5	41.5	46.4	47.3
Female	24.6	29.2	33.8	34.0	34.0	41.4	43.4
Race:							
White	30.5	36.7	39.9	40.3	40.4	46.6	48.0
Black	22.8	23.2	27.5	28.0	27.7	32.0	32.8
Educational Attainment:							
Less than high school	NA	38.1	38.7	40.8	39.4	44.3	46.3
High school graduate	NA	33.6	36.3	36.7	36.5	41.1	41.6
Some college	NA	34.9	41.0	37.5	40.6	45.5	46.4
College grad.	NA	48.2	49.7	50.6	48.7	59.1	60.6

A Case Report in Need of Improvement

A REPORT ON CIGARETTE SMOKING IN THE U.S.

The following report gives an analyses of smoking habits in the U.S. from 1965 to 1988. The study covers adult smokers (those at least 20 years old), which was initially reported in *The Statistical Abstract of the United States.* Two different aspects of cigarette smoking were recorded: the incidence of smoking among the general U.S. population and the percentage of former smokers who eventually quit smoking.

The graph shown below is typical of the trend in the incidence of smoking over time for most groups in the U.S. This graph shows the drop in smokers from 1965 to 1988 for the 20- to 24-year-old age group. The decline shows a clear downward trend over the 23-year period. For this age group, the decline was 20%. Other age groups also declined, but some, such as the 65-year-and-older group, didn't register the same percentage drop.

In each of the years in which data was available, a higher percentage of men than women smoked, although the gap is closing: 18.3% in 1965 and 5.6% in 1988. More whites than blacks were smoking in 1965 (in percentage terms), but that ranking had reversed by 1988. In 1970, high school graduates had the highest incidence of smoking over all other educational attainment groups, but by 1988 the group who had not graduated from high school was highest.

More smokers are quitting these days than ever before. In 1965, for example, 47.8% of the 20- to 24-year-olds were smoking, but 17.8% quit. However, by 1988, only 27.6% of that same age group smoked and 27.5% quit, so that essentially none of the people in the 20-to-24 age bracket were smoking by 1988. This clearly indicates that smoking will soon become nonexistent in the U.S.

I'd be happy to provide any further information.

```
                              GRAPH #1
         -     1
         -
     45.0+
         -
 smoking -          2
         -
         -
     37.5+
         -              3    4    5
         -
         -
         -
     30.0+                             6
         -
         -                                  7
         -
         -
         +---------+---------+---------+---------+
         0         2         4         6         8
```

A CRITIQUE

There are a number of problems with the preceding case write-up. We can classify typical mistakes into several general categories.

- *Grammatical errors.* When your report has style and grammatical errors, it sends the message that your work can be expected to be sloppy or unimportant. (In the report on the previous page, "analyses" should be "analysis," and the word "data" is plural.)
- *Inappropriate or misused graphs and tables.* Make sure that your graphs are as informative as possible. Label all axes on your graphs, and provide headings and other useful information. (In the graph on the previous page, the 17 missing years of data have been ignored, so the time series graph is deceptive. The reader of your report would probably want to know why there are only seven data points shown on the graph. The graph also suggests that the incidence of smoking leveled off during the middle third of the time span. In fact, those are simply the three years 1978, 1979, and 1980, with big gaps in the data from 1970 to 1978 on the left side, and 1980 to 1987 on the right side, of the "flat" portion of the curve.)
- *Numerical errors or misrepresentations.* It is easy to make silly little mistakes while interpreting your statistical results. (There are 24, not 23, years from 1965 to 1988. The 20-to-24-year-old group declines 20 percentage points, not percent, from 1965 to 1988.)
- *Unclear or ambiguous sentences.* Avoid ambiguous statements. (What does "18.3% in 1965 and 5.6% in 1988" refer to in paragraph three?)
- *Reaching incorrect conclusions.* Often simply not understanding the problem, or not understanding the data, will lead you to draw wrong conclusions. (Paragraph four compares a 27.6% smoking rate with a 27.5% quit rate, and concludes that only 0.1% are left smoking. This is wrong because the 27.5% quit rate does not refer to the whole population, but only that subset of the population who were smokers.)
- *Jumping to conclusions.* Avoid the tendency to read more into the data and statistical results than is really there. Provide a statement of the limitations of the data and shortcomings of the statistical analysis when appropriate. (We would not be willing to conclude from this data, even if we had not made the earlier error, that "smoking will soon become nonexistent in the U.S.")
- *Making a big thing out of nothing.* There is a tendency to look for things in data that aren't really there, and infer that some trend or difference is somehow important or revealing. (The difference in the two high school groups in 1988 is only 0.2 percentage points.)

The following revision accommodates some of these concerns.

An Improved Case Report

A (REVISED) REPORT ON CIGARETTE SMOKING IN THE U.S.

This report analyzes cigarette smoking in the U.S. from 1965 to 1988. The subjects were adult smokers (those at least 20 years old). Two aspects of cigarette smoking behavior were recorded: the incidence of smoking among the general U.S. population and the percentage of smokers who eventually quit smoking.

The graph below typifies the downward trend in smoking over time for most age groups in the U.S. From 1965 to 1988 for the 20-to-24-year-old age group, the rate of smoking declined from 47.8% to 27.6%, over 20 percentage points. (Our original data set did not include data for 17 of the 24 years, so the graph is incomplete.) Other age groups followed the same decline, but different age groups demonstrated less dramatic changes. For instance, the 65 years and older age group went from 17.1% in 1965 to 14.9% in 1988.

In the years for which data were available, a higher percentage of men than women smoked, although the gap is closing: the difference between men's and women's smoking behavior was 18.3% in 1965, but only 5.6% by 1988. Whites are less likely than African Americans to be smokers, and more likely to quit. College graduates tend to smoke less than those groups of lower educational attainments, and this difference appears to continue even as late as 1988.

Recently, more smokers are quitting. This trend is clear for all age groups and educational attainment groups and for both races and genders. In 1965, for example, only 17.8% of all smokers in the 20-to-24 age group had quit smoking by 1965, whereas this quit rate was 27.5% in 1988. Twenty-four point six percent of all women smokers had quit smoking according to the 1965 survey, but 43.5% of women smokers had quit according to the 1988 data.

The data set has missing years of data; with only 7 years of data for each category, it is difficult to do any further statistical analysis. The case specifically mentions the problem of teen smokers, but the data set does not include teenagers, so we can't really infer anything about this issue. Nonetheless, an examination of both the smoking rates and quit rates suggests that Americans have made headway in quitting smoking since 1965.

```
                Percentage of 20-to-24-year-olds who smoke: 1965-1988
       49.0+                (17 years of data are unavailable)
           -      *
 SMOKING   -
   (%)     -
           -
       42.0+               *
           -
           -
           -
           -                              *  *  *
       35.0+
           -
           -
           -
           -                                                    *
       28.0+                                                        *
           -
        ------+---------+---------+---------+---------+---------+YEAR
            1965      1970      1975      1980      1985      1990
```

APPENDIX **C**

USING THE DATA DISK

GENERAL INSTRUCTIONS

This *Instructor's Edition* comes with a data disk containing files for most of the cases in *PDA*. Installation procedures vary among software packages and among installations, and each instructor should distribute copies of the data sets in a form appropriate to the particular installation and software package used in the course. This avoids our giving out instructions that might be wrong for a particular installation. It also keeps down the costs to the student. Each volume of *PDA* contains a brief section entitled *"The Data Disk,"* advising the student to expect data sets and instructions from the instructor.

The data disk is formatted for use in IBM-compatible (MS-DOS) systems. It contains the data files for all three volumes. *Be sure to consult the README file on the disk for technical details about extracting the files described below.* Each file is provided in five forms:

- *ASCII files.* The ASCII files are in unformatted form, consisting of one line for each row of data. Most statistics packages can read such files. Variable names are not included in these files, which makes them readable by a variety of programs. The ASCII files are in directory \ASCII on the disk. Each file has file extension ASC. For example *Case 64, The Advertising Experiment,* says that data are available in file ADVERT. For the ASCII files, this is the file named ADVERT.ASC. The alphanumeric fields for state names in *Case 9, Healthy Lifestyles* do not contain quotes around the names.
- *LOTUS files.* The LOTUS files are in worksheet (WK1) form. They contain one row of data for each observation. Variable names are not included in the worksheets. Most spreadsheets and many statistics packages can read such files. The LOTUS files are in directory \LOTUS on the disk. Each file has file extension WK1. For example *Case 64, The Advertising Experiment,* says that data are available in file ADVERT. For the LOTUS files, this is the file named ADVERT.WK1. *Note that the regressions built into many spreadsheets will not operate on matrices with missing data.* In such cases either you or the students will need to edit the data set appropriately, unless you are using a macro package which does this automatically.
- *Excel files.* The Excel files are in worksheet (XLS) form. They contain onc row of data for each observation. Variable names are not included in the worksheets. Most spreadsheets and many statistics packages can read such files. The Excel files are in directory \XLS on the disk. Each file has file extension XLS. For example *Case 64, The Advertising Experiment,* says that data are available in file ADVERT. For the Excel files, this is the file named ADVERT.XLS. *Note that the regressions built into many spreadsheets will not operate on matrices with missing data.* In such cases either you or the students will need to edit the data set appropriately, unless you are using a macro package which does this automatically.

• *MINITAB workspace files.* The MINITAB files are in MINITAB workspace (MTW) form. Variable names are included in the workspace. The MINITAB workspace files are in directory \MTW on the disk. Each file has file extension MTW. For example *Case 64, The Advertising Experiment,* says that data are available in file ADVERT. For the MINITAB workspace files, this is the file named ADVERT.MTW.

• *MINITAB portable files.* The MINITAB portable files are in ASCII. Variable names are included in the workspace. These files are useful when converting MINITAB workspaces to mainframes or other computers, as the MTW files are not directly usable in those situations. The MINITAB portable files are in directory \MTP on the disk. Each file has file extension MTP. For example *Case 64, The Advertising Experiment,* says that data are available in file ADVERT. For the MTP files, this is the file named ADVERT.MTP.

NOTES

Case 9, Healthy Lifestyles, Case 23, Money Supply and Interest Rates, Case 25, Emergency Admissions, Case 29, Books, Case 30, Elisa Plate Uniformity, Case 34, Leading Indicators, Case 36, Violence on TV–II, Case 47, Westar Beverage Sales, Case 51, Abandoned Calls, Case 55, Predicting Snowfall, Case 57, Utility Bills, Case 61, Deer-Train Collisions in Eastern Hokkaido, Case 62, Public Debt in Germany, and *Case 82, Heart Surgery,* contain missing values. These appear as asterisk (*) in the ASCII files and MINITAB worksheets, as empty cells in the WK1 and XLS files, and as question mark (?) in the JMP files.

USING THE DATA DISK WITH STATISTICS PROGRAMS

The four forms of data set provided with *PDA* will get most classes off and running. You may wish to provide your students a specific sheet of instructions for your installation. You may also wish to make copies of the data disk specifically for your installation. For example, if you're using MINITAB, you might wish to copy the MTW files and provide them to your students. That way, the students can concentrate on analyzing data, not the clerical details. On the other hand, you may want them to get used to handling those details.

We've tried different packages. Here are a few, with brief notes on the conversion.

• *MINITAB, Versions 7 and 8.* The RETRIEVE command in MINITAB will read the MTW files into a workspace.

• *STATISTIX, Version 4.* STATISTIX will read either the ASCII files or the WK1 files. You will need to assign variable names by typing them in. The appropriate menu items to invoke this are File Management / Import. *Be sure to specify all variable names when importing WK1 files.* Otherwise, STATISTIX will attempt to determine the kind of variable (integer or real) automatically from the first few rows of the worksheet. For a few PDA data sets, we have found this to be unreliable. It is perhaps preferable to import the ASCII files directly, specifying names and types of variables explicitly.

• *SPSS.* SPSS will read the ASCII files. DATA LIST FREE format is probably the most useful way. You will need to provide the DATA LIST statement indicating that the data is to be read from the *PDA* ASCII files. The SAVE FILE= command will save the resulting data set in SPSS system file format.

• *Quattro.* The various Quattro's we've tried will read the WK1 files directly, and that's probably the simplest method.

• *SAS.* The ASCII files may be used wherever SAS permits data in unformatted form.

• *JMP files.* The JMP files are in JMP worksheet form. Variable names are included in the data sets. All variables, except alphanumeric, read into JMP files are continuous. We have not changed the JMP modeling types–this is part of the learning process for those using JMP. The JMP files are in the directory \JMP on the disk. Each file has the extension JMP. For example *Case 64, The Advertising Experiment,* says that data are available in file ADVERT; for the JMP files, this file is named ADVERT.JMP.

Instructors wishing to use the data sets in Macintosh rather than IBM formats should write or call the publisher.

CASE 1

WASTE PAPER

Waste accompanies any production process. In this case, the production process is a printing press, which produces small books for a variety of customers. Data from 127 runs of this press, gathered over about a month, are listed on the next pages and are available in file WASTEPAP.

In most press runs, some books are wasted, and the number of wasted books (WASTE) is influenced by such factors as NET (the number of books requested by the customer) and GROSS (the total number of books actually produced to satisfy the customer's request. GROSS = WASTE + NET). Moreover, some books are more complicated to produce than others. Books of 8 or 16 pages (PAGES) can be produced using a single "web" of paper, while books of 24 or 32 pages require two webs, and this increases the chance of some sort of error in the final book. Similarly, books with larger numbers of colors (COLORS) may be more prone to errors, as the separate colors must be properly aligned with each other to produce a satisfactory product. The company's usual book is four-color, but occasionally other variations occur. The basis weight of the paper used (PAPER, in pounds) may also influence waste.

In pricing its four-color services, the company assumes a common "waste factor" for all of the different sizes of book (8-paged, 16-paged, etc.). Price quotes for other than four-color books are handled separately, but the analyst wants to know whether the "waste factor" really should be the same for all sizes of four-color books. Can you advise her?

The case is based on a real situation, described to us by Don W. Hansen, V.P. Operations, American Web, Inc., who also provided the data.

DATA SET

OBS	NET	GROSS	WASTE	PAGES	PAPER	COLORS
1	52700	64700	12000	24	70	1
2	4800	7700	2900	8	50	2
3	13100	16700	3600	16	60	4
4	72000	77400	5400	16	50	4
5	72000	77200	5200	16	50	4
6	72000	80100	8100	16	50	4
7	26700	33100	6400	8	80	4
8	26700	31900	5200	32	60	4
9	29100	37100	8000	24	70	1
10	10300	14700	4400	32	50	1
11	10300	14000	3700	16	50	4
12	10300	15600	5300	16	70	4
13	7900	11500	3600	32	70	4
14	7900	10600	2700	16	60	4
15	20800	23700	2900	16	40	1
16	16300	19900	3600	8	40	4
17	17900	20200	2300	8	40	4
18	53100	58600	5500	16	47	4
19	53100	60000	6900	16	47	4
20	26200	37300	11100	16	60	4
21	26200	34400	8200	32	60	4
22	37300	45800	8500	32	60	4
23	52500	57500	5000	32	60	1
24	23500	28300	4800	32	60	1
25	23690	31000	7310	8	40	4
26	17900	24000	6100	16	40	4
27	28325	34400	6075	32	47	4
28	28325	33500	5175	32	47	4
29	28325	33000	4675	16	47	4
30	50000	59400	9400	16	47	4
31	51200	57900	6700	16	40	5
32	51200	57900	6700	16	40	4
33	11000	14600	3600	16	40	4
34	7800	10800	3000	16	60	4
35	3200	6300	3100	16	60	4
36	20000	25300	5300	16	70	4
37	26550	32200	5650	16	60	4
38	26550	32200	5650	16	40	5
39	26550	36300	9750	24	40	4
40	19570	24700	5130	32	70	4
41	16000	18100	2100	16	70	2
42	24383	34000	9617	32	70	4
43	24383	30600	6217	32	50	4
44	36300	41800	5500	32	50	4
45	36380	42900	6520	32	50	4
46	14200	18200	4000	16	50	4

(continues on the following page)

OBS	NET	GROSS	WASTE	PAGES	PAPER	COLORS
47	14200	17200	3000	8	50	4
48	10800	13900	3100	32	50	1
49	11900	14800	2900	16	50	4
50	10800	15400	4600	32	70	4
51	31300	38000	6700	24	40	4
52	52700	62300	9600	16	70	6
53	13100	16300	3200	16	60	4
54	4350	9200	4850	32	50	4
55	66560	82000	15440	16	60	5
56	66560	82100	15540	24	60	2
57	66560	83300	16740	32	60	2
58	36000	46600	10600	8	100	4
59	72000	83700	11700	16	50	4
60	43000	48400	5400	16	50	4
61	72000	78500	6500	16	50	4
62	72000	81400	9400	16	50	5
63	11300	14500	3200	16	80	4
64	5700	9800	4100	8	80	4
65	29000	34500	5500	16	70	4
66	29000	35400	6400	32	47	4
67	29000	35700	6700	32	47	1
68	10300	17000	6700	24	60	1
69	10300	19700	9400	32	60	4
70	17300	27900	10600	32	50	1
71	17300	22500	5200	32	70	4
72	14000	21900	7900	24	60	4
73	14000	20800	6800	16	70	4
74	15600	19100	3500	8	60	1
75	20900	28900	8000	24	60	4
76	15600	20400	4800	24	60	4
77	15600	24400	8800	32	40	4
78	11300	16500	5200	8	80	4
79	5150	11200	6050	8	100	4
80	53100	61900	8800	16	47	4
81	53100	58600	5500	16	47	4
82	53100	57300	4200	16	47	4
83	26200	34700	8500	32	60	1
84	8659	14900	6241	16	50	4
85	15028	18700	3672	8	47	4
86	52500	55100	2600	8	60	1
87	52500	57400	4900	16	60	4
88	52500	56800	4300	16	60	4
89	52500	58900	6400	16	60	4
90	23690	26000	2310	16	40	4
91	41000	45200	4200	8	40	1
92	41000	54100	13100	32	47	4
93	17100	22300	5200	32	60	4
94	12650	16600	3950	16	70	4

(concludes on the following page)

OBS	NET	GROSS	WASTE	PAGES	PAPER	COLORS
95	12650	16600	3950	24	47	4
96	14162	19500	5338	8	100	4
97	14400	18500	4100	16	45	4
98	14400	19200	4800	16	45	4
99	14400	20700	6300	16	60	5
100	20850	26200	5350	32	70	4
101	28325	35200	6875	16	47	4
102	11400	14200	2800	16	40	4
103	5625	10000	4375	16	50	4
104	5625	8200	2575	16	50	4
105	5625	11300	5675	16	50	4
106	50000	56800	6800	16	47	4
107	50000	60300	10300	16	47	4
108	50000	56000	6000	16	47	4
109	50000	56000	6000	16	47	4
110	51200	61000	9800	16	70	4
111	38100	48000	9900	24	50	2
112	38700	51100	12400	32	60	4
113	54075	62400	8325	24	45	4
114	54075	63000	8925	32	45	2
115	27037	31100	4063	32	100	4
116	20000	35000	15000	32	47	4
117	27150	37400	10250	24	40	4
118	20000	32000	12000	32	47	4
119	18050	26000	7950	8	60	4
120	36380	45200	8820	16	50	4
121	7200	11100	3900	8	70	4
122	36380	40500	4120	16	50	4
123	36380	42000	5620	32	50	4
124	36380	41700	5320	32	50	4
125	24000	35500	11500	32	60	4
126	15450	21100	5650	16	60	4
127	15450	23300	7850	16	80	4

CASE **2**

DOME LOADING

Circle Electronics manufactures keyboards and data-entry devices. Most of the company's products contain one or more bubble-like steel parts called domes. The domes are placed on a mylar surface. When a key on the keyboard is pressed, it physically pushes down on a dome and that causes appropriate electrical contacts. The computer receives electronic signals (corresponding to the key pressed) as a result.

A robot assembly called (naturally) a *dome loader* loads the domes onto the mylar surface. The process works as follows. The mylar sheet comes in 20" × 27" sheets. These sheets are fed manually into the dome loader, which takes domes one at a time from a reel and places them on the mylar sheet, at locations specified by X and Y specifications. The X and Y specifications give the horizontal and vertical distances, respectively, which specify the desired location of the center of the circular dome. The diagram on the next page gives two domes. Dome A has X and Y specifications of 1.425 and .800 inches, respectively, while Dome B has X and Y specifications of 2.300 and 1.625 inches, respectively. Once the domes have been placed on the mylar sheet, the sheet is fed manually into a machine that cuts it to its final size by using a steel rule die-cutting process.

If the resulting keyboard or similar product is to work reliably, the process must produce a sheet with accurately placed domes. The industry currently considers dome placement "acceptable" if it is within plus or minus .01 inches of the X and Y specifications, a range that is generally attainable by hand tools.

Circle Electronics studied one product to see if the dome-loading process was producing acceptable sheets and to provide any information that seemed helpful to the manufacturer of the dome loader in correcting any problems. The X and Y specifications for the domes are listed in the data that follow. Five consecutive sheets were measured. Each

This case is based on a real industrial situation, though the name of the company has been disguised. Brad A. Strong, Vice President, Operator Interface Technology, provided the data.

sheet contained nine domes. The data include the following variables (available in file DOMELOAD). SHEET is the sheet number and DOME is the dome number (the dome loader first places Dome 1 on the mylar sheet, then Dome 2, etc.). XMEAS is the actual X distance to the center of the dome; XSPEC is the X specification for this dome. YMEAS is the actual Y distance to the center of the dome; YSPEC is the Y specification for this dome.

The XMEAS and YMEAS variables were measured after the completion of the manufacturing process by special, accurate measurements. The XSPEC and YSPEC variables are the same for corresponding dome numbers and come from the manufacturing specifications for the particular product.

Is the process behaving acceptably? Are there any patterns that suggest how variation in the process could be reduced?

DIAGRAM

```
                  |<------2.300------>|
                  |                   |
                  |<-1.425->|         |
                  |         |         |
                  ********************************
                  *                              *
  --+-----------  *                   B          *
    |             *                              *
  1.625           *                              *
    |             *                              *
    |----+------  *         A                    *
    |    |        *                              *
    |  0.800      *                              *
    |    |        *                              *
----+----+------  ********************************
```

DATA SET

OBS	SHEET	DOME	XMEAS	XSPEC	YMEAS	YSPEC
1	1	1	1.439	1.425	0.804	0.800
2	1	2	2.321	2.300	1.625	1.625
3	1	3	3.853	3.863	2.262	2.250
4	1	4	5.120	5.113	2.253	2.250
5	1	5	2.928	2.925	2.885	2.875
6	1	6	5.439	5.425	2.896	2.875
7	1	7	7.311	7.300	2.888	2.875
8	1	8	7.299	7.300	1.641	1.625
9	1	9	7.859	7.838	0.195	0.175
10	2	1	1.416	1.425	0.805	0.800
11	2	2	2.285	2.300	1.631	1.625
12	2	3	3.860	3.863	2.258	2.250
13	2	4	5.098	5.113	2.255	2.250
14	2	5	2.916	2.925	2.900	2.875

(concludes on the following page)

OBS	SHEET	DOME	XMEAS	XSPEC	YMEAS	YSPEC
15	2	6	5.413	5.425	2.884	2.875
16	2	7	7.288	7.300	2.884	2.875
17	2	8	7.296	7.300	1.655	1.625
18	2	9	7.826	7.838	0.185	0.175
19	3	1	1.430	1.425	0.805	0.800
20	3	2	2.301	2.300	1.612	1.625
21	3	3	3.867	3.863	2.237	2.250
22	3	4	5.116	5.113	2.254	2.250
23	3	5	2.920	2.925	2.883	2.875
24	3	6	5.432	5.425	2.868	2.875
25	3	7	7.296	7.300	2.868	2.875
26	3	8	7.299	7.300	1.632	1.625
27	3	9	7.833	7.838	0.178	0.175
28	4	1	1.412	1.425	0.793	0.800
29	4	2	2.292	2.300	1.623	1.625
30	4	3	3.860	3.863	2.248	2.250
31	4	4	5.102	5.113	2.252	2.250
32	4	5	2.931	2.925	2.892	2.875
33	4	6	5.425	5.425	2.881	2.875
34	4	7	7.307	7.300	2.883	2.875
35	4	8	7.295	7.300	1.631	1.625
36	4	9	7.828	7.838	0.181	0.175
37	5	1	1.412	1.425	0.793	0.800
38	5	2	2.292	2.300	1.611	1.625
39	5	3	3.852	3.863	2.238	2.250
40	5	4	5.102	5.113	2.236	2.250
41	5	5	2.919	2.925	2.873	2.875
42	5	6	5.413	5.425	2.859	2.875
43	5	7	7.294	7.300	2.862	2.875
44	5	8	7.294	7.300	1.619	1.625
45	5	9	7.834	7.838	0.172	0.175

CASE **3**

SPOT PRICES

All the way home from class, Sarah could hardly concentrate on the traffic. She was so excited about the prospect of her future wealth that she almost got into several accidents.

All this enthusiasm was sparked by Sarah's finance class at the University of Oklahoma. The lecture for the day was about spot markets. Her ears really picked up during the lecture and, even before she was out the door, Sarah had discovered a scheme that could allow her to make money on the spot market for natural gas.

The spot market for natural gas is that market in which natural gas is sold under a contract term of no more than one month. Having worked for some time for an Oklahoma pipeline company, Sarah knows a couple of important things about the spot market for natural gas. First, she knows that the spot price for Louisiana natural gas is generally known on or before the first day of trading on the Oklahoma market. Second, she knows that the Oklahoma spot price typically runs about 85% of the Louisiana spot price, even throughout the seasonal variations in these two markets. Therefore, in Sarah's mind, it should be easy to predict the current Oklahoma spot price and buy or sell futures based on the current Louisiana spot price. This would allow her to pay off her student loans and go to Jamaica for spring break.

The data set, in the file SPOTS, contains 46 monthly observations on Louisiana and Oklahoma spot prices for natural gas. In this file, MONTH represents the appropriate month (with one equaling January and so forth) of that YEAR. OSPOT is the Oklahoma spot price for natural gas measured in dollars per million British thermal units ($/MMBtu), and LSPOT is the Louisiana spot price, also in $/MMBtu. What formula will allow Sarah to play the spot market successfully? What is Sarah's forecast for the Oklahoma spot price for November 1997 if the November 1997 Louisiana spot price was just announced to be 1.89 $/MMBtu?

DATA SET

MONTH	YEAR	OSPOT	LSPOT
1	94	1.985	2.215
2	94	2.008	2.138
3	94	1.753	1.968
4	94	1.442	1.643
5	94	1.372	1.533
6	94	1.350	1.528
7	94	1.375	1.583
8	94	1.420	1.693
9	94	1.577	1.863
10	94	1.610	1.838
11	94	1.743	2.058
12	94	1.890	2.357
1	95	1.913	2.323
2	95	1.582	1.890
3	95	1.357	1.608
4	95	1.383	1.665
5	95	1.483	1.792
6	95	1.518	1.837
7	95	1.488	1.793
8	95	1.485	1.725
9	95	1.418	1.663
10	95	1.425	1.705
11	95	1.557	1.915
12	95	1.860	2.252
1	96	2.313	2.570
2	96	1.733	2.132
3	96	1.373	1.637
4	96	1.362	1.618
5	96	1.362	1.600
6	96	1.387	1.610
7	96	1.362	1.568
8	96	1.330	1.490
9	96	1.350	1.503
10	96	1.522	1.718
11	96	1.917	2.202
12	96	2.013	2.387
1	97	1.737	1.980
2	97	1.333	1.490
3	97	1.318	1.468
4	97	1.318	1.470
5	97	1.315	1.448
6	97	1.275	1.408
7	97	1.130	1.267
8	97	1.175	1.287
9	97	1.397	1.518
10	97	1.723	1.882

CASE **4**

JOANNE'S FORECASTING PROBLEM

Joanne had written memos to the computer center director for months to complain about unsatisfactory turnaround time for the jobs her department ran on the computer. Basically, her department submitted regular jobs to process orders, write invoices, resupply parts, and/or change customer records. The same computer program processed all of these transactions, in whatever mix of orders, invoices, resupply requests, or customer change requests the department happened to request on that day. Joanne felt the computer center charges were outrageous, anyway. The combination of (what she perceived as) high rates and poor service frustrated her. It also caused tension in the various meetings that she and the computer center manager attended together.

Now, though, she was in a quandary. The computer center manager had just written to her, explaining that he was now prepared to provide whatever capacity was necessary to meet her turnaround time requirements, if she would simply forecast the capacity she needed, in terms of CPU minutes per month. After some inquiries, Joanne determined that CPU minutes were a measure of the computer time used for a job. The computer center manager's memo was a blessing and a curse: a blessing in that the computer center would respond to her needs, and a curse because while she was prepared to forecast her needs in her own units (she forecasted a monthly average of 8,000 orders, 6,500 invoices, 700 resupply requests, and 30 customer changes per month, based on the latest business plan), she had no real notion of how to convert them into CPU minutes, or any of the other kinds of units that the "computer types" seemed to talk about.

Glancing over her monthly bills from the computer center for the last year, she wondered if she could use that data to convert her forecast to the units the computer center

This case is an amalgam of several real situations and data sets from the authors' experience.

required. The relevant information from her computer center bills from the 12 months of last year is listed below. How should Joanne proceed? The data are also available in file JOANNE.

DATA SET

MONTH	ORDERS	INVOICE	RESUPPLY	CUSTOMER	CPUMIN
1	7000	4026	487	28	2281
2	6046	6073	619	46	2203
3	8031	4096	892	29	2602
4	4903	6023	900	11	1806
5	6208	4016	703	18	2072
6	4902	7773	402	47	2163
7	9003	4096	450	48	2689
8	9604	6032	706	57	3204
9	10618	10068	719	12	3921
10	10497	11063	850	9	4016
11	6033	6001	906	18	2168
12	2014	3004	402	18	902

CASE **5**

T-SHIRT DESIGNS

Julio has one semester left in his MBA program and is looking at paying back some hefty student loans. Even though he has had to borrow money to complete the program, he's decided it was a worthwhile investment. Nonetheless, Julio would like to repay some of his loans before graduating if at all possible.

Julio has recently taken entrepreneurial and small-business classes and has decided to put some of this knowledge to work. He suspects that a number of his fellow students would be willing to shell out a few bucks for school memorabilia. Since summer is right around the corner, Julio is pretty sure that he can make some money by selling T-shirts to students on campus.

Informal conversations with people in the hallways have given him some ideas about popular T-shirt designs. Julio has narrowed down the possibilities to two. He's given sketches to a local T-shirt printing company, and they have made two prototype shirts. At this stage, Julio has no idea about the relative demand for the two T-shirt designs. If he's going to all the trouble and expense of having the shirts made up, it would be useful to know which of the two shirts is more popular.

The two shirts look something like this. The red shirt has a scoop-neck and is made out of 100% cotton. On the back of the shirt is a logo of the school. Under the logo are the words "PARTY TIME at PU UNIVERSITY!" Julio suspects that this shirt will mostly appeal to the party-goers.

The other shirt is more conservative. It is a white, button-down-collar shirt made of 50% cotton and 50% synthetic materials. This shirt has a simple, and small, version of the school logo on the upper-right-hand section of the front of the shirt.

Julio has decided to gauge the relative markets for his two T-shirts by running a quick survey. With permission from the student government, he has set up a table in the

This case was based on a survey designed by M.A.Smith, who also wrote the survey questionnaire.

student union. The two prototype shirts are there for inspection, along with a survey designed to gauge people's reactions to the two shirts. He has a big sign over the table offering a free soft drink to all participants to induce people to stop by the table and participate in the survey. On the corner of his table is a big cardboard box with a slit in the top. Students are instructed to put their completed questionnaires in the box, thereby guaranteeing anonymity of the responses.

The actual survey and the resulting data set are shown below. The raw data reside in the file named T-SHIRTS. Julio has hired you to analyze these data. Prepare a report about the preferences revealed by the data. Comment on whether the other survey variables relate somehow to preferences. As a final note, feel free to comment on the survey itself and give any suggestions that you might have for improving it.

SURVEY

I am trying to measure people's preferences for one of two different T-shirt designs. Would you please help me with my study by answering each of the following questions to the best of your ability. Do not put your name on this page because all responses should be anonymous.

1. My gender is

   1 = female
   0 = male

2. My current or most recent scholastic grade point average on a 4-point scale is ________.

3. Circle the number that most accurately describes your attitude about the red and white T-shirts on the table.

   1 = I strongly prefer the red T-shirt to the white T-shirt.
   2 = I somewhat prefer the red T-shirt to the white T-shirt.
   3 = I like both equally, or am indifferent between the two.
   4 = I somewhat prefer the white T-shirt to the red T-shirt.
   5 = I strongly prefer the white T-shirt to the red T-shirt.

4. I am __________ years old.

5. My gross annual income last year was $_______. (If you were gainfully unemployed, enter 0.)

6. I am (circle one)

   1 = left-handed
   2 = right-handed
   3 = ambidextrous (both left- and right-handed)

DATA SET

GENDER	GPA	PREFER	AGE	INCOME	HANDED
0	4.00	4	34	20.0	2
0	2.90	1	23	12.0	2
1	3.70	4	32	27.0	2
1	2.50	4	23	22.0	2
1	3.90	5	27	20.0	1
0	4.00	5	39	75.0	2
1	3.40	2	24	24.0	2
0	3.50	2	27	30.0	2
0	4.00	4	28	24.5	2
0	4.00	2	43	45.0	2
0	3.40	3	28	30.0	2
1	3.40	2	23	0.0	1
0	3.00	2	26	30.5	1
1	3.70	5	40	24.0	2
0	3.90	4	31	30.0	2
0	3.80	2	30	0.0	2
1	3.10	2	36	38.0	2
0	3.00	2	26	32.5	2
0	3.85	2	30	40.0	2
0	3.20	2	33	23.0	2
0	2.00	4	29	24.5	2
0	3.00	3	25	25.0	2
0	3.70	2	29	30.0	2
0	3.50	4	29	40.0	2
1	3.80	2	26	35.0	2
0	3.85	2	46	27.0	1
1	4.00	4	26	24.0	2
1	3.50	2	24	0.0	2
1	3.80	2	33	49.0	2
0	4.00	5	27	34.0	2
0	3.50	2	27	45.0	2
1	3.30	1	24	31.0	2

GENDER	GPA	PREFER	AGE	INCOME	HANDED
0	3.60	3	27	32.0	2
0	4.00	4	26	25.0	1
1	4.00	4	25	25.5	2
1	3.00	4	37	52.0	2
0	3.78	4	29	28.5	2
1	3.00	5	26	0.0	2
1	3.70	2	24	18.0	2
0	4.00	5	31	60.0	2
0	2.97	2	23	14.0	2
1	3.00	5	29	0.0	2
1	3.61	5	32	50.0	2
0	4.00	2	46	65.0	2
0	3.50	4	31	0.0	2
1	3.40	4	25	25.0	2
0	3.50	4	25	20.0	2
1	3.58	4	24	0.0	2
1	3.20	4	27	20.0	2
0	3.00	4	24	25.0	2
1	3.06	4	25	0.0	2
0	3.20	4	24	18.0	2
0	4.00	2	25	25.0	2
1	3.00	2	24	30.0	2
0	4.00	2	23	16.0	2
0	3.20	3	27	32.0	2
0	3.00	3	30	0.0	2
1	4.00	2	29	0.0	2
1	2.85	2	24	0.0	2
1	3.00	4	40	0.0	2
1	4.00	5	29	40.0	2
1	3.30	2	25	15.0	2
1	4.00	2	47	46.0	2
1	3.50	1	37	60.0	2

CASE **6**

STOCKS A AND B

The data that follow (and in file STOCKAB) give the rates of return over a 50-year period for two stocks, which we'll call stock A and stock B. The rate of return is defined as the increase in value of the portfolio (including any dividends or other distributions) during the year divided by its value at the beginning of the year. That's the fraction by which your wealth would have changed had it been invested in that particular combination of securities. The rate of return may be either positive or negative.

What was the average return for stock A? Stock B? In the theory of finance, the standard deviation of the return is often used as a measure of the risk associated with investing in a given security. What are the standard deviations of the returns for stocks A and B?

Assume that the history of stocks A and B is a useful guide to what may be expected of them in the foreseeable future. (That may seem like a big assumption, but assume it anyway.) Suppose you want to use this history as a guide to making an investment decision for the long run. Based on this history, and assuming that standard deviation of return is the appropriate measure of risk, does it make any difference whether you invest wholly in stock A, wholly in stock B, or half in stock A and half in stock B? Assume that average return is to be maximized and risk is to be minimized, if that is jointly possible.

DATA SET

YEAR	A	B	YEAR	A	B	YEAR	A	B	YEAR	A	B
1	0.215	0.199	14	0.122	0.122	27	0.001	0.112	40	0.215	0.045
2	0.194	0.017	15	0.265	0.103	28	0.122	0.294	41	0.315	0.419
3	0.130	0.112	16	0.237	0.218	29	0.230	0.055	42	0.008	0.199
4	0.337	0.218	17	0.151	0.055	30	-0.013	0.179	43	0.058	0.055
5	0.165	0.246	18	0.201	0.294	31	0.044	0.132	44	-0.035	0.151
6	0.115	0.160	19	0.230	0.304	32	0.258	0.112	45	0.265	0.227
7	0.308	0.218	20	0.194	0.246	33	0.280	0.218	46	0.215	0.151
8	0.130	0.294	21	0.008	0.112	34	0.122	0.246	47	0.211	0.093
9	0.330	0.170	22	0.265	0.313	35	0.072	0.227	48	0.230	0.221
10	0.294	0.103	23	0.222	0.017	36	0.187	0.266	49	0.115	0.218
11	0.244	0.055	24	0.144	0.333	37	0.130	0.294	50	0.106	0.025
12	0.401	0.179	25	0.108	0.103	38	0.101	0.074			
13	0.251	0.428	26	0.194	0.122	39	0.308	0.246			

CASE **7**

TOM'S USED MUSTANGS

Tom has recently opened a used car lot. Although he has very little business experience, his passion has always been for vintage automobiles. Tom has been collecting and repairing old cars since he was a teenager and has lots of technical knowledge about carburetors, shock absorbers, air conditioners, and so forth. Tom has decided to turn his hobby into a money-making venture for himself. Unfortunately, Tom knows almost nothing about advertising, taxes, social security payments for his employees, bookkeeping, or other aspects of the *business* of selling cars. For this reason, Tom is taking a few business courses at the local community college.

The car lot has been open for a few months now. Several of Tom's customers have told him that his asking prices are way out of line with the rest of the market—sometimes too high and sometimes too low, but never close to the going rate. Tom typically relies on a gut feeling for determining asking prices. For instance, all else the same, he thinks that red cars appear sportier and can be sold at a premium. He knows that older cars sell for less, but he doesn't know exactly how much less. Apparently this informal scheme is not working very well.

As an alternative, Tom has secretly visited a competitor's lot, but there are so many physical characteristics of a car that he is having trouble pinpointing a useful pricing scheme. Finding perfect matches for his cars is nearly impossible given all of the possible options on cars.

To more thoroughly understand this issue, Tom has hired you to perform a statistical analysis of asking prices for used Mustangs. Tom has already collected information from the local newspapers and other sources. The data file named MUSTANGS contains observations on 35 used Mustangs and 10 different characteristics.

This data set was assembled by Marla S. McClure and Larry Welner as part of a course project. Data collection involved surveying 35 private owners and dealers who advertised in various media on March 27, 1992.

Prepare a report for Tom on the influence of various options on asking price, and show him how he should use this information to set prices on other cars in the near future.

DEFINITIONS

PRICE	Asking price in dollars
CONVERT	1 = convertible 0 = not convertible
AGE	Age of the car in years
MILES	Approximate odometer reading in miles
TRANS	1 = automatic transmission 0 = manual transmission
AIR	1 = air conditioner 0 = no air conditioner
CYL	Number of cylinders in the engine
COLOR	1 = maroon 2 = silver 3 = gray 4 = red 5 = blue 6 = black 7 = white
GT	1 = GT model 0 = not a GT model
OWNER	0 = private owner 1 = dealer owner

DATA SET

PRICE	CONVERT	AGE	MILES	TRANS	AIR	CYL	COLOR	GT	OWNER
12688	0	1	8200	0	1	8	1	1	1
11500	0	1	27000	1	1	8	2	0	0
11650	0	1	3000	0	1	8	3	1	1
10995	1	2	27000	1	1	8	4	0	1
6200	0	3	55000	1	1	4	4	0	0
9500	0	3	22000	0	0	8	3	0	0
8700	0	3	25000	1	1	8	5	0	0
12500	0	2	23800	0	1	8	4	1	1
10488	1	4	20000	0	1	8	7	1	1
5988	0	4	37000	0	1	4	3	0	1
7288	1	5	84000	1	1	4	5	0	1

(concluded on the next page)

PRICE	CONVERT	AGE	MILES	TRANS	AIR	CYL	COLOR	GT	OWNER
3590	0	5	55000	0	0	4	6	0	0
2800	0	6	62000	0	1	4	5	0	0
6995	1	7	78000	1	1	4	1	0	0
4900	0	7	60000	0	1	8	7	1	0
2488	0	7	67000	1	1	4	1	0	1
4200	1	8	65000	1	1	6	5	0	0
12990	1	1	13300	1	1	8	7	0	1
14500	1	3	31000	1	1	8	7	1	0
10900	1	3	4000	1	1	8	6	0	1
13500	1	3	23000	0	1	8	6	1	0
12000	0	3	89000	0	1	8	7	1	1
8500	0	3	34000	1	1	8	7	0	0
4388	0	4	80000	0	1	4	5	0	1
3000	0	7	95000	0	1	4	5	0	0
3900	0	8	90000	0	1	8	1	1	0
3350	0	8	91000	0	1	8	6	1	0
10500	0	3	17000	0	1	8	1	1	0
12988	0	1	21000	1	1	8	7	1	1
6500	0	3	73000	0	1	8	7	0	0
12900	1	2	19000	0	1	8	6	0	0
9000	0	2	32000	0	1	8	5	0	0
7000	0	6	56000	0	1	8	7	1	0
5900	0	4	110000	1	1	8	5	0	0
2500	0	6	92000	0	1	4	5	0	0

CASE **8**

AMERICAN HISTORY ILLUSTRATED

Imagine you are the staff assistant to the publisher of *American History Illustrated.* Imagine further that you reported yesterday to your boss, the publisher, that the renewal rate for subscriptions increased from 0.512 in January to 0.641 in February. The renewal rate is computed as the number of subscriptions renewed in a given month divided by the total number of subscriptions that expired in that month. It gives the fraction renewed of subscriptions that could have been renewed. Imagine, finally, that your boss was pleased to hear this and wanted to know which kinds of subscriptions were contributing most to the increase. He asked you to prepare a further summary by breaking the subscriptions down into several categories: gift subscriptions, previously renewed subscriptions, direct mail subscriptions, subscriptions from a subscription agency, and those from a catalog agent. This follows a standard industry classification scheme. You have obtained the data, as follows.

TABLE

	Gift subs	Previous renewal	Direct mail	Sub. agent	Catalog agent	Total subs
January						
Total	3,594	18,364	2,986	20,862	149	45,955
Renewed	2,918	14,488	1,783	4,343	13	23,545
February						
Total	884	5,140	2,224	864	45	9,157
Renewed	704	3,907	1,134	122	2	5,869

This case is based on real data, originally reported by C. H. Wagner in *The American Statistician* 36, no. 1, (February 1982), pp. 46–48. Reprinted with permission from *The American Statistician.*

For example, the total renewal rate for January was 23,545/45,955 = .512, and the renewal rate for February for previously renewed subscriptions was 3907/5,140=0.760.

Provide the summary the boss requested, and interpret it for him.

CASE **9**

HEALTHY LIFESTYLES

The Centers for Disease Control and Prevention (CDC) in Atlanta, Georgia, is the government agency responsible for disease-related issues in the United States. The CDC coordinates efforts to counteract outbreaks of diseases and funds a variety of medical and health research studies. The CDC also serves as a central clearing house for health-related data.

The CDC conducts the annual Behavioral Risk Factor Surveillance Survey (BRFSS). The survey measures a whole series of lifestyle characteristics that relate to health and longevity, such as smoking and use of seat belts. The survey compiles data on a state-by-state basis.

Your task is to prepare a summary of the latest BRFSS results. Your report is to be issued to major news organizations, such as the Associated Press, and will appear in major newspapers around the globe. For this reason, it would be inappropriate to use technical jargon in your report.

Your boss has suggested a few general ideas about what is likely to appeal to your target audience. As you study the data, you might find other things worth including.

- Report any interesting (e.g., unexpected, humorous, or odd) differences between states.
- Devise a weighted index of all seven lifestyle variables. The weighted index is to serve as an overall or composite measure of healthy lifestyles. Apply your weight to the states of Minnesota, Florida, and California as an example of what your weighted index shows.
- Discuss any noteworthy limitations of the survey or data set.

The data set in the file named HEALTHY contains six measures taken from the BRFSS. The data are described in more detail below, but feel free to visit the BRFSS website (*www.cdc.gov/nccdphp/brfss/about.htm*) for more information. All data are taken from the 1995 BRFSS except for those regarding exercise, which are taken from the 1994 BFRSS. The exercise data for Rhode Island are not available.

DEFINITIONS

BINGE Percentage of people responding yes to the question, "Considering all types of alcoholic beverages, how many times during the past month did you have five or more drinks on an occasion?"

DWI Percentage of people responding one or more times to the question, "During the past month, how many times have you driven when you've had perhaps too much to drink?"

BMI Each respondent's reported height and weight were translated into a body mass index: BMI = (weight in kg)/(height in meters)^2. A BMI greater than or equal to 27.8 for men, and 27.3 for women, is considered unhealthy. The data represent the percentage of people in the unhealthy BMI range.

SMOKER Percentage of people who have smoked at least 100 cigarettes during their lifetime and are current smokers.

SEATBELT Percentage of people who report they do not always wear a seatbelt when they drive or ride in a car.

EXERCISE Percentage of people responding no to the question, "During the past month, did you participate in any physical activities or exercises such as running, calisthenics, golf, gardening, or walking for exercise?"

DATA SET

STATE	BINGE	DWI	BMI	SMOKER	SEATBELT	EXERCISE
Alabama	13.58	2.62	31.83	24.53	32.26	45.81
Alaska	19.20	1.25	31.35	24.96	33.06	22.84
Arizona	13.46	2.69	24.53	22.86	25.93	23.65
Arkansas	8.75	1.49	30.14	25.20	32.65	35.10
California	15.30	1.90	26.43	15.50	14.78	21.75
Colorado	16.30	3.06	21.88	21.79	35.46	17.16
Connecticut	14.41	2.49	24.68	20.79	30.83	21.90
Delaware	8.61	1.38	29.49	25.45	29.04	36.35
Florida	13.12	2.60	29.80	23.13	24.31	27.92
Georgia	11.99	2.15	28.21	20.46	36.11	32.87
Hawaii	12.37	2.07	21.83	17.76	12.63	20.70
Idaho	12.94	2.02	27.18	19.76	42.53	21.84
Illinois	13.58	1.76	30.09	23.08	30.66	33.46
Indiana	12.75	2.58	34.65	27.19	43.33	29.62
Iowa	17.95	3.30	31.58	23.17	37.25	33.25
Kansas	13.86	3.20	27.81	22.01	42.09	34.43
Kentucky	9.70	0.59	28.79	27.83	34.75	45.86
Louisiana	14.01	2.80	30.74	25.21	32.81	33.41
Maine	11.45	0.93	26.94	24.96	48.74	40.72
Maryland	8.21	1.14	29.08	21.22	25.38	30.15

(concludes on the following page)

STATE	BINGE	DWI	BMI	SMOKER	SEATBELT	EXERCISE
Massachusetts	17.83	3.50	21.93	21.74	41.77	24.01
Michigan	18.26	3.30	31.62	25.74	29.70	23.02
Minnesota	18.01	4.89	28.35	20.48	40.82	21.74
Mississippi	8.67	1.12	31.62	24.03	41.90	38.42
Missouri	14.10	2.13	32.88	24.26	36.10	31.84
Montana	14.31	3.40	24.91	21.13	43.25	21.02
Nebraska	15.83	2.76	29.18	21.90	46.91	24.22
Nevada	18.97	3.73	26.91	26.31	28.34	21.59
New Hampshire	16.59	1.62	25.89	21.45	46.98	25.62
New Jersey	13.95	1.98	24.36	19.17	31.54	30.89
New Mexico	14.06	3.26	23.84	21.17	15.58	19.35
New York	12.39	0.90	27.83	21.47	25.80	37.12
North Carolina	5.76	1.11	28.87	25.84	14.06	42.79
North Dakota	16.97	4.19	30.67	22.69	58.31	31.98
Ohio	9.87	1.64	31.53	25.98	31.80	37.97
Oklahoma	6.69	1.17	24.14	21.66	45.93	30.38
Oregon	13.90	1.77	28.76	21.79	17.14	20.76
Pennsylvania	19.38	3.58	30.08	24.17	38.97	26.45
Rhode Island	18.70	3.65	24.90	24.74	49.87	*
South Carolina	9.17	1.39	28.65	23.74	23.20	31.36
South Dakota	14.36	5.24	28.65	21.79	56.83	30.67
Tennessee	5.22	1.01	30.95	26.46	38.20	39.66
Texas	15.27	3.70	28.60	23.66	22.35	27.80
Utah	9.93	1.22	25.04	13.16	39.66	20.90
Vermont	15.96	2.40	25.36	22.11	28.77	23.24
Virginia	14.46	2.54	29.24	22.03	27.67	22.92
Washington	13.44	2.12	25.43	20.16	22.02	18.16
West Virginia	5.92	0.87	31.94	25.73	30.01	45.28
Wisconsin	22.89	4.54	30.15	21.80	43.83	25.89
Wyoming	15.57	3.22	27.26	22.02	52.12	20.95

CASE **10**

LAKE COUNTY LUNCHES

Lake County, Colorado (county seat: Leadville), is located about 100 miles west of Denver in the Colorado Rockies. The county is home to the highest airport in the continental United States; much of the county lies above 9,000 feet. The economy is largely dominated by mining (with its boom and bust cycles) and tourism.

As in many other rural areas, the school lunch program is an important component of public policy. For many poor children, the lunch they get in school provides most of their daily nutrition. Over a recent 11-year interval, the average daily number of lunches served generally declined. Since the average per capita income in the county also declined, the director of the program was eager to understand the reasons for the decline. If the families were getting poorer and simultaneously fewer lunches were being served, it would raise questions about how well the program was meeting its goals.

Eleven years' data are included in file LAKE. The variables include: YEAR, the calendar year; POP, Lake County population; UNEMPL, percentage unemployment in the state of Colorado; LKUNEMP, percentage unemployment in Lake County; LUNCH, average daily lunches served; INCOME, average per capita income in Lake County; and ENROLL, enrollment in Lake County schools. Based on these data, should the director be worried?

This case describes an actual situation. The data and the situation were brought to our attention by Carmen Lamar Arnold, Partner, Langford Partners, A. G.; Terrance L. Bartell, Assistant Business Manager, El Jebel Temple, Shriners; Jeanne Donohue; and Esther Kettering, Corporate Secretary, Sales and Acquisition Director, Golden Securities, Inc.

DATA SET

YEAR	POP	UNEMPL	LKUNEMP	LUNCH	INCOME	ENROLL
1	8830	5.9	9.1	940	10996	1918
2	8677	5.5	9.0	929	11728	1843
3	8523	7.7	29.2	924	10190	1630
4	7788	6.7	31.3	793	8207	1419
5	7325	5.6	16.7	746	9269	1355
6	6980	5.9	13.2	804	9814	1290
7	6667	7.4	18.2	718	9253	1217
8	6292	7.7	21.3	660	9092	1117
9	6010	6.4	15.9	725	8908	1154
10	6197	5.8	12.0	719	8515	1163
11	5975	5.0	11.0	723	8115	1147

CASE **11**

MACHINE PRODUCTION RECORDS

Helen managed a major manufacturing facility that was devoted largely to producing millions of identical small metallic parts. While the parts produced were (intended to be) identical, the facility produced them by using hundreds of presses, of three different types (which we'll call types 1, 2, and 3). The rated production rates for machine types 1, 2, and 3 were 700, 200, and 155 parts per minute, respectively, but actual production rates varied. Helen felt that factors such as quality of input material, worn or "changed-out" dies, and dirty or poorly maintained presses would probably affect production.

The facility had always kept daily production records, but from what Helen could tell, no one ever consulted them or did anything with them other than to put them in file cabinets at the end of each reporting period. She thought that in principle she should be able to monitor production rates for the types of machines, identify any machines that seemed to be in need of adjustment, and characterize the amount of downtime to be expected. Accordingly, she extracted one day's production records for 116 presses at her facility.

The machines were scheduled for a shift of 7.75 hours each day. The operators recorded the hours of operation manually on clipboards kept near each machine. The actual quantities of parts produced were determined from automatic counters on the machines. The data on the next two pages (and in file PROD) were derived from these logs.

Can you advise Helen?

This case is based on a real production line situation, though the name of the organization and the exact nature of the product have been disguised. The data were provided by Ted Manzanares.

DEFINITIONS

MTYPE	Type of press (1, 2, or 3)
PROD	Production, total parts produced
HRSWRK	Hours of production time
HRSDOWN	Hours the press was down, or inoperative, for any reason

DATA SET

OBS	MTYPE	PROD	HRSWRK	HRSDOWN	OBS	MTYPE	PROD	HRSWRK	HRSDOWN
1	1	320242	7.75	0	39	1	127686	3.75	4
2	3	54311	7.75	0	40	3	60159	7.58	0.17
3	2	60820	7.75	0	41	3	48111	6	1.75
4	1	230679	5.16	2.5	42	3	22461	2.5	1.25
5	1	253855	5.67	2	43	1	308209	7.58	0.17
6	1	142151	3.57	4	44	3	82915	7.58	0.17
7	3	16118	0.58	7.17	45	2	53671	7.42	0.33
8	1	267881	7	0.75	46	2	50006	6.95	0.8
9	2	25295	4.08	3.67	47	1	214085	6	1.7
10	2	59360	7.75	0	48	2	53614	6.83	0.92
11	3	65201	6.83	0.92	49	3	82810	7.75	0
12	2	20521	3	4.75	50	1	109567	3	4.7
13	2	22916	3	4.75	51	1	245819	6.5	1.2
14	1	315124	7.75	0	52	3	74643	7.75	0
15	3	47854	6	1.75	53	3	30111	3.67	4.08
16	2	59868	7.62	0.13	54	3	30561	3.75	0
17	3	75393	6.83	0.92	55	3	47696	4.41	3.34
18	2	51374	7.25	0.5	56	2	40289	5.75	2
19	3	30372	4.34	3.41	57	1	297330	7.75	0
20	3	42784	5.75	2	58	3	31593	3.5	0.25
21	1	202003	6	1.7	59	3	46471	4.75	3
22	2	54597	7.75	0	60	3	31499	3.5	0.25
23	2	59066	7.75	0	61	3	60469	7.33	0.42
24	3	83712	7.75	0	62	3	63703	7.75	0
25	3	20530	2.5	1.25	63	2	19769	2.92	4.83
26	3	18188	2.17	5.58	64	1	212743	7.75	0
27	3	87400	7.75	0	65	2	55767	7.75	0
28	1	311877	7.75	0	66	3	15038	1.75	2
29	2	15783	1.75	6	67	3	31556	3.5	0.25
30	1	259945	7.58	0.17	68	3	30380	3.42	0.33
31	2	31034	4.83	2.92	69	3	68982	6.58	1.17
32	3	25569	4.17	3.58	70	2	59850	7.7	0.05
33	2	15771	2.67	5.08	71	1	300891	7.75	0
34	2	6899	1.5	6.25	72	2	56431	7.75	0
35	3	82837	7.75	0	73	3	83171	7.75	0
36	2	58349	7.75	0	74	3	16650	2.25	5.5
37	3	68899	7.75	0	75	2	52644	7.43	0.32
38	3	41909	5.25	2.5	76	3	8954	1.08	2.67

(concludes on the following page)

OBS	MTYPE	PROD	HRSWRK	HRSDOWN	OBS	MTYPE	PROD	HRSWRK	HRSDOWN
77	1	316080	7.75	0	97	3	61092	7.5	0.25
78	2	20200	2.93	4.82	98	3	87114	7.75	0
79	2	61050	7.65	0.1	99	3	8675	1.08	2.67
80	2	30459	4.47	3.28	100	1	301077	7.5	0.2
81	2	25061	2.83	4.92	101	3	5084	0.58	3.17
82	2	60224	7.75	0	102	2	55753	7.62	0.13
83	3	60935	7.5	0.58	103	1	285831	7.75	0
84	3	31233	3.75	0	104	1	302774	7.75	0
85	3	20237	1.92	5.83	105	2	57083	7.25	0.5
86	1	283785	7.25	0.5	106	1	281949	7.25	0.5
87	2	49499	6.75	1	107	3	40469	6.08	1.67
88	3	59722	6.67	1.08	108	1	300039	7.75	0
89	1	271276	6.58	1.1	109	3	88634	7.67	0.08
90	3	62762	7.75	0	110	1	312501	7.75	0
91	2	61075	7.75	0	111	3	75455	6.83	0.92
92	1	220686	6.83	0.92	112	2	27600	3.63	4.12
93	3	78103	7.75	0	113	1	273486	7	0.75
94	2	51300	7.54	0.21	114	3	52164	6.67	1.08
95	3	40967	3.25	4.5	115	2	29367	5.05	2.7
96	3	31056	3.5	0.25	116	3	8142	1.08	2.67

CASE **12**

KERRVILLE RAINFALL

Everybody talks about the weather, but no one *does* anything about it, according to Mark Twain. The data on the next page give the total annual rainfall (TOTAL, in inches) for Kerrville, Texas, from 1931 to 1995. Summarize any patterns you find in this time series. Can these data be used to predict rainfall for Kerrville for 1996? These data are also available in file KERRVILL.

The data used in the case are from the official NOAA-NWS supervised Cooperative Weather Observation Site for Kerrville, Texas, located at the Knipling-Bushland U.S. Livestock Insects Research Laboratory. They were collected by volunteer weather observers, including Al Siebenaler, in cooperation with NOAA.

DATA SET

YEAR	TOTAL	YEAR	TOTAL
1931	35.07	1964	31.85
1932	41.53	1965	40.86
1933	19.14	1966	27.62
1934	24.22	1967	27.84
1935	49.29	1968	33.33
1936	47.68	1969	32.03
1937	27.10	1970	21.43
1938	20.95	1971	34.48
1939	28.42	1972	28.39
1940	39.14	1973	33.33
1941	39.66	1974	38.87
1942	29.16	1975	29.29
1943	21.38	1976	34.71
1944	36.87	1977	23.57
1945	33.89	1978	44.26
1946	34.48	1979	42.31
1947	27.19	1980	27.45
1948	25.55	1981	41.51
1949	41.82	1982	21.54
1950	22.70	1983	25.14
1951	18.15	1984	22.48
1952	40.91	1985	36.03
1953	26.36	1986	38.19
1954	14.64	1987	42.13
1955	28.90	1988	30.90
1956	14.04	1989	23.16
1957	55.09	1990	33.58
1958	36.71	1991	44.52
1959	29.30	1992	41.40
1960	37.07	1993	23.06
1961	26.63	1994	38.76
1962	17.13	1995	28.26
1963	21.34		

CASE **13**

WORKLOAD RATINGS

Many organizations use opinion surveys as one part of their evaluations of managers or executives. The managers being evaluated often wonder how fair these processes are. In particular, they wonder if they will somehow get lower ratings if they strictly enforce rules or set high standards for work.

In universities the same problem affects instructors, who complain about the student evaluations of the instructors and their courses. In particular, they complain that when instructors make students work hard, the students give them low ratings. Is this really true? (The students, in turn, often wonder if anyone ever looks at the evaluations they fill out, but that's another story.)

All of the classes offered in a recent fall semester by the University of Colorado at Denver, College of Business and Administration, were evaluated by the students in those classes, by using a standard form prescribed by the University's Board of Regents. Students responded to 12 questions, each of which was scored from 0 to 4 (0 = "F," 4 = "A"), and the class averages for the responses to each question were tabulated, provided to the instructor, and given to deans, department chairs, and to the student government, which published them for general review.

For each of the 161 sections, the class averages for three of the questions are listed on the next two pages (and are available in file WORKLOAD). The variables on that file are as follows. CRSRTG is the course rating: "Rate this course compared with your other university courses." WORKLOAD is the workload: "Rate the work required for this course" (F = too little, C = about right, A = too much). Finally, INSTRTG is the instructor rating: "Rate this instructor compared to your other university instructors."

Note that for WORKLOAD, a rating of 2.0 would be "about right," as perceived by the students. For CRSRTG and INSTRTG, 0 means a failing rating, 4 means a perfect one. All of the numbers given are class averages for the particular sections offered.

Do WORKLOAD and INSTRTG appear to be related, based on these data? Do they appear related if we adjust first for the effects of the course rating? Do these data support the rumors about instructors' being penalized for making students work hard?

WORKLOAD	CRSRTG	INSTRTG	WORKLOAD	CRSRTG	INSTRTG
2.33	2.48	2.19	3.24	2.43	2.72
2.33	3.17	3.42	3.00	2.41	2.56
3.07	3.60	3.53	3.17	2.78	3.00
2.19	3.15	3.24	2.58	3.58	3.50
2.04	3.53	3.68	1.50	1.69	1.94
2.34	3.29	3.36	2.29	3.23	3.40
2.11	2.28	2.11	2.46	3.32	3.68
2.11	2.11	1.78	2.03	2.97	3.57
1.67	2.00	1.33	2.14	3.13	3.46
2.33	3.83	3.75	2.12	2.87	2.87
1.95	2.91	3.23	2.08	3.23	3.62
2.19	3.00	3.12	3.00	3.55	3.75
1.73	2.62	2.75	2.00	3.13	3.47
2.95	3.05	2.84	2.43	2.76	3.35
2.78	2.51	2.62	2.05	2.55	2.72
2.65	2.69	2.62	2.54	3.69	3.69
2.50	3.25	3.33	2.00	2.80	3.15
2.29	3.75	3.75	2.08	3.08	3.37
1.88	3.03	3.18	2.37	3.62	3.69
1.93	2.79	3.00	3.11	3.67	.67
2.15	3.62	3.87	2.39	3.00	3.24
2.04	3.22	3.52	2.14	2.81	2.86
2.78	2.78	3.35	2.17	3.17	3.34
2.73	2.77	3.00	2.40	3.40	3.56
2.29	2.71	2.43	2.04	2.81	2.96
2.93	3.27	3.19	2.22	3.03	3.12
3.56	2.29	1.96	1.95	2.70	3.05
2.65	3.50	3.57	2.25	2.50	2.25
3.11	3.52	3.59	2.73	3.50	3.82
1.83	2.00	2.00	2.34	1.90	2.03
2.90	3.75	3.65	2.50	2.00	2.24
2.12	3.62	3.69	2.32	2.31	2.31
2.42	3.23	3.19	2.04	2.04	2.00
2.21	3.43	3.43	2.09	3.48	3.57
2.12	3.50	3.62	2.21	1.89	1.63
2.17	2.68	2.88	2.09	1.81	1.74
2.10	2.61	2.77	2.72	3.53	3.57
2.00	3.83	3.67	3.00	3.13	3.13
2.40	2.75	2.75	2.50	2.92	2.83
1.96	2.48	2.83	2.75	3.33	3.62
1.87	2.50	2.63	3.39	2.86	3.07
1.94	2.50	2.57	2.22	3.70	3.74
2.54	2.73	2.54	2.10	3.50	3.50
2.84	2.94	3.19	2.02	2.95	2.98

(concludes on the next page)

WORKLOAD	CRSRTG	INSTRTG
2.36	2.51	2.53
2.62	2.04	2.19
2.09	2.82	2.77
2.07	2.89	2.93
2.17	3.08	3.00
2.85	2.45	2.63
1.96	3.83	3.92
2.11	3.21	3.29
2.06	2.59	2.88
2.08	3.42	3.62
2.37	2.81	3.00
2.30	2.77	3.18
2.37	2.52	3.00
2.43	2.64	2.91
2.02	1.75	1.80
2.35	2.55	2.45
2.52	2.84	2.90
3.74	2.67	2.59
2.18	2.18	1.96
2.03	2.40	3.03
2.09	2.64	3.27
2.17	1.67	2.00
2.30	3.40	3.70
2.22	2.82	3.35
2.68	2.87	3.09
3.04	2.78	3.00
3.29	3.00	2.88
3.25	2.68	3.07
2.00	2.67	2.67
1.73	2.91	2.91
2.14	2.33	2.14
2.26	2.77	3.17
2.18	1.59	1.76
2.32	3.26	3.45
2.22	3.41	3.68
2.17	3.31	3.47
1.33	1.44	1.72

WORKLOAD	CRSRTG	INSTRTG
1.94	3.06	3.21
2.15	2.63	2.81
2.12	2.36	3.07
2.16	3.48	3.68
2.00	3.55	2.72
2.09	2.90	3.03
2.13	1.67	1.33
1.95	2.71	2.43
1.95	2.00	2.10
2.87	2.27	2.59
2.16	3.20	3.40
2.20	3.52	3.73
2.17	2.90	3.13
1.82	2.80	3.09
2.22	3.75	3.62
1.86	2.64	2.41
3.65	2.12	2.19
2.00	3.50	4.00
2.31	3.54	3.46
2.72	.89	1.11
2.26	2.57	2.43
2.37	1.80	1.83
2.38	2.62	3.23
3.47	2.07	2.40
3.65	2.45	2.40
3.00	2.89	3.00
2.36	1.82	1.83
2.10	3.57	3.52
2.07	3.38	3.37
2.12	2.39	2.42
2.43	3.57	3.71
2.29	3.71	3.71
1.97	3.44	3.50
2.17	1.52	1.48
2.50	2.40	2.07
2.78	2.00	1.91

CASE **14**

ROARING FORK TRANSIT

The Roaring Fork Transit Agency (RFTA) operates a multicounty bus system based in Aspen, Colorado, a mountain community known as an upscale resort. The Roaring Fork Valley (named for the Roaring Fork River, which flows from the mountains above Aspen into the Colorado River at Glenwood Springs, some 50 miles away) grew substantially from 1987 to 1990. Combined with lack of available affordable housing (particularly in the winter ski season), increased traffic on state highway 82 ("Killer 82," the only highway into Aspen in winter), and rising gasoline prices ($1.83 per gallon in January 1991), the growth led to more passengers on the "downvalley" routes from Aspen to El Jebel and Carbondale. At times, RFTA had trouble meeting demand for its services on this route.

Monthly ridership figures (total riders) for OUTBOUND (from Aspen in the direction of Carbondale/El Jebel) and INBOUND (from Carbondale/El Jebel to Aspen) routes are given in the data below for 1987–1990 (months are indicated by the variable MONTH, where MONTH=1 for January 1987).

In early 1991, RFTA management asked an analyst to advise them on choices about buying new equipment, leasing extra buses, hiring more drivers, and so forth, for the next few years. Prepare a presentation highlighting any trends or patterns in the data you think the analyst and RFTA management should consider. The data are available in the file ROARING.

This case is based on a real situation. It was described to us and the data were provided by Rebecca R. Horton and Cheryl Cumnock, Personnel Director, Pitkin County, Colorado.

DATA SET

MONTH	OUTBOUND	INBOUND
1	8078	7885
2	7816	7004
3	8025	7058
4	5170	4635
5	3534	3130
6	4449	4242
7	6508	6046
8	5849	5250
9	4262	3759
10	3913	3377
11	4666	4174
12	8534	7995
13	11028	12114
14	11517	11190
15	12012	12109
16	6577	6030
17	4677	4477
18	7583	7144
19	10947	10720
20	11590	10823
21	5652	5086
22	4829	4164
23	7918	7608
24	23536	19697
25	12991	13353
26	11999	12752
27	11986	12151
28	6206	7472
29	6787	6481
30	9432	9463
31	12826	13240
32	12626	12826
33	8223	8059
34	7647	7414
35	10652	10941
36	18767	19889
37	19815	19747
38	17078	17124
39	17662	18030
40	10429	10670
41	8595	8672
42	11478	11968
43	16635	16394
44	16638	16336
45	12500	12644
46	12725	13088
47	15745	17011
48	26362	29890

CASE **15**

STURGEL DIVISION

"Now that I write it all down, I see we have changed a lot!" mused Martha as she put the finishing touches on her annual status report. "In fact," she went on, "the name *information services* indicates most of the changes. We used to be *information systems.*"

Martha was information services (IS) manager for the Sturgel Division of a major manufacturing company. Sturgel developed and manufactured (mostly small) household appliances on a 230-acre site in the southeast part of the United States. While she had managed the IS department for only 15 months, she had been part of the department for nine years, all at the same location.

The late 1990s were a time of many changes for the information systems departments in most companies, including the Sturgel Division. Technological change drove most of the organizational change: the price performance of most computer equipment had improved by compound rates of perhaps 30% per year for several decades. This meant that many companies no longer needed a "computer center." Individual users could simply do whatever they needed to do on a desktop computer in their own offices.

Martha made a rough list of the major influences she'd dealt with over the last few years, calling them "IS Transitions." They included:

- From running the computer to providing information services to the company.
- From "owning" the data to consulting with user departments about use of the company's information resources.
- From emphasizing mainframes to emphasizing terminals, personal computers, and telecommunications. Indeed, sometimes it seemed that the telephone, word processing, clerical, and library departments had all been combined.
- From developing the company's applications to advising users on how to develop their own applications.

Each transition had been difficult, for each required its own combination of hardware, software, data, procedures, and people, most of which were different at the end of the transition.

Currently, the Information Service Department had four sections, each with a manager who reported to Martha:

1. **Systems operations**. The systems operations section ran two mainframe computers and one minicomputer that operated as a telecommunications "node" and link to other divisions and corporate headquarters. They provided operator support for three shifts of operation, as well as systems maintenance, operating system updates, and so forth. Hardware maintenance was handled by the vendor of the mainframes.
2. **Application development**. The application development section really should be renamed, Martha thought. Most of its activities involved data base design and maintenance, though two small groups developed financially oriented applications and maintained several software packages aimed at serving the (fairly small) engineering staff at Sturgel.
3. **PC services**. The PC services section developed division standards for personal computers at Sturgel, maintained the hardware, and consulted with users who wanted PCs of their own. The users purchased PCs out of their own budgets, but they were required to meet certain hardware and software standards as set by IS. This section also evaluated general-purpose software, such as word processing and spreadsheet packages, and recommended to users when they should switch to a new version or release. They also did a fair amount of "hand holding" for users who had difficulty developing their own applications. Recently, the section had developed new procedures for nationwide maintenance (and security) of the company's laptop computers.
4. **Telecommunications.** The telecommunications section maintained the local area network at the Sturgel site, as well as links to corporate headquarters and other company divisions. They also advised users on such activities as links to legal databases and information retrieval services, though that activity, thought Martha, might really fit better in the application development section.

Martha felt good about the organization in general. While it had been a bit of a scramble to develop an organization that could deal with the fast pace of technological change in the computer field, she felt they had more or less done it. One part of it still made her nervous, though. The biggest theme in all the changes, from an IS perspective, was the change from computing and running computers to serving large numbers of people directly.

She felt they were providing good service, but shc had no regular way of knowing how her users perceived it. She thought she should consider a regular survey of her users and their perceptions of service received from the IS department. That way, she figured, she would be in a position to have data to back up her informal sense, and she would (presumably) learn about changes in those perceptions (for better or for worse) more quickly.

What kind of survey should Martha run, or should she? How should it be administered? What kind of forms, questionnaires, or other survey approaches should be developed? Martha would like you to develop an appropriate scheme.

CASE **16**

PHOTOCOPYING ABUSES

Your firm is located in a 10-story building. Each floor has its own copy machine. The firm owns these machines but must pay for paper, toner, and maintenance. Each employee has a key that opens the copy room door on his/her floor only and does not have access to copy machines on other floors.

Because the copy machines are "free goods" right now, you suspect that the firm's photocopying costs could be cut pretty drastically. To test this theory, you have decided to perform an experiment. Each person on the tenth floor has been given a card that operates the tenth-floor machine. These employees have been told that their cards will generate a daily accounting of their photocopying activity. Tenth-floor employees have also been told that they will not be *charged* for their use of the machine, but they certainly know that *someone* will have some sense of individual usage patterns.

To establish a basis of comparison, the group on the third floor has not been converted to the card system. The third-floor machine has an internal mechanism that totals the number of copies made each day, but you do not know *who* is doing *what,* and the third-floor employees have no reason to believe that they are being monitored. The third and tenth floors have been chosen for this experiment because these two floors have had about the same usage rates in the past.

You have collected data from the two machines over the last 50 working days. The data are in the file named COPIERS. This file has 50 observations and three variables: DAY, an indicator of the day; TENTH, the number of copies made on the tenth floor; and THIRD, the number of copies made on the third floor.

Use appropriate graphical or statistical techniques to determine whether the card accounting system will effectively lower photocopying usage if implemented companywide.

The scenario described in this case was related to us by one of our students, who was employed at the firm when the copy machine experiment was undertaken.

DATA SET

DAY	TENTH	THIRD	DAY	TENTH	THIRD
1	500	440	26	250	320
2	420	220	27	250	360
3	440	360	28	230	450
4	480	110	29	240	270
5	450	240	30	220	380
6	460	360	31	190	190
7	450	80	32	150	500
8	420	420	33	170	290
9	410	310	34	120	150
10	405	30	35	150	370
11	380	290	36	140	405
12	360	410	37	130	130
13	360	460	38	150	120
14	370	420	39	130	70
15	350	150	40	110	240
16	320	170	41	90	20
17	350	250	42	80	450
18	360	20	43	90	20
19	310	250	44	70	40
20	320	350	45	20	320
21	290	150	46	50	140
22	290	250	47	40	90
23	270	230	48	20	130
24	250	90	49	30	480
25	240	50	50	30	350

CASE **17**

AGRICOMP

"How do I make sense of all this?" asked Jody, as he prepared to write his report to the dealer relations committee. He stared at the 292 responses to a survey of AgriComp's dealers as he began to ponder the problem. The question was whether to recommend a change to AgriComp's current procedures for settling warranty claim disputes with its dealers.

AgriComp sold computer systems to farmers, who used the systems for such purposes as crop rotation planning and spreadsheet analysis for financial planning. Many also used the systems to provide remote access to agriculture-oriented data bases, market news, and even weather information. The equipment was assembled at company headquarters in southern Minnesota. The software was provided by subcontractors but was distributed under AgriComp's name. Both the hardware and software were sold through some 350 affiliated dealers nationwide, 292 of whom had responded to Jody's survey. It was relations with these dealers that concerned him.

The local dealers handled warranty service for AgriComp products. When hardware or software problems occurred, they arranged for appropriate repairs to be made locally and submitted vouchers to AgriComp headquarters in Minnesota. The headquarters staff reviewed the vouchers and issued reimbursement checks to the dealer. Occasionally claims were denied when the staff found that the particular repair was not covered by the company's warranty or that the warranty had expired. In such cases, the dealer was more or less stuck for the cost of the repairs, and this had caused occasional hard feelings.

The company had an internal appeals process for dealers to follow to protest such denials but, at the last dealers' meeting, Jody had heard a lot of grumbling about that

This case was inspired by an actual survey taken during 1993, but we have disguised the nature of the organization and its products, as well as the particular issue in this survey.

process. More than one dealer had suggested that it was useless, as appeals were always denied. The costs of repairs might correspond to the profits on many systems for the dealer, so their concern was understandable.

Jody knew that clearer warranty instructions would help. Sometimes the dealers couldn't understand exactly what was or wasn't covered by warranty. In that kind of case, they often arranged for the repairs (to keep the customer happy) and took their chances on reimbursement. New documentation currently being developed would probably help with that part of the problem.

In a corridor conversation at the dealers' meeting, one dealer had suggested that perhaps in cases of dispute, an impartial mediator, external to the company, might be called in to settle the matter. In the annual survey of dealers, Jody had included a question about that proposal. As part of a one-page survey, dealers were asked to respond to the statement "The warranty appeals process should be replaced by a process using impartial mediators" on a scale of 1 to 5, where 1 indicated "Strongly agree," 3 indicated "Neither agree nor disagree," and 5 indicated "Strongly disagree." Each dealer was also asked to give the number of times in the last three years in which he or she had used the warranty appeals process. Answers to this question were 0, 1, 2, and "3 or more."

Part of the data from that survey is shown below, and is also available in file AGRICOMP. REP is the dealer's response to the question about a new appeals system and USE is the number of times the dealer used the appeals process, where 3 indicates "3 or more."

Jody was willing to consider changing the warranty appeals process along the lines suggested, if it were important to the dealers. It would cost the company some money, both for the external mediator and (perhaps) for increased costs from appeals the company lost, but keeping the dealers happy was obviously important. Jody wondered, though, just how important it was to the dealers? Was there widespread discontent, or had he just heard from a few malcontents at the dealers' meeting?

DATA SET

REP	USE	REP	USE	REP	USE	REP	USE	REP	USE
5	3	3	0	3	2	4	3	2	1
5	2	5	3	3	1	5	3	4	3
5	1	2	2	2	1	4	2	3	2
5	2	4	3	4	2	5	3	2	2
2	2	5	1	1	1	5	2	3	2
4	3	4	3	3	0	4	3	4	1
1	0	4	3	5	1	2	2	4	3
3	3	4	0	4	3	2	0	4	3
2	2	5	3	5	1	1	0	1	2
4	2	5	1	4	0	2	3	5	2
4	3	3	1	1	0	2	2	4	3
1	0	2	1	1	1	5	3	5	1
2	1	3	1	2	1	4	1	4	3
3	1	5	3	2	1	4	3	4	3
4	2	2	2	1	2	2	2	4	3
3	0	2	3	3	3	3	1	4	1
3	1	2	2	4	3	2	1	2	1

(concludes on the following page)

REP	USE	REP	USE	REP	USE	REP	USE	REP	USE
3	2	3	2	4	3	3	2	2	1
3	3	2	1	1	0	4	3	2	3
2	1	3	3	2	1	2	2	3	3
5	3	2	3	2	2	3	3	2	1
5	3	4	2	3	2	1	0	1	1
4	2	2	1	4	3	1	1	4	0
1	3	5	3	4	3	5	2	1	3
4	3	1	1	4	3	3	3	3	2
5	1	3	2	4	1	4	3	5	3
1	1	5	3	4	1	4	3	5	3
1	0	5	2	3	1	3	1	3	3
4	0	2	1	2	2	4	1	4	3
4	3	4	2	5	3	4	3	3	1
4	3	5	1	2	3	2	0	3	0
3	0	2	3	4	1	4	3	1	1
1	3	5	3	5	3	3	3	5	2
2	3	2	1	2	2	4	3	4	3
4	1	5	3	4	3	5	3	2	3
4	3	2	1	4	3	3	2	4	3
2	2	5	3	2	0	3	2	5	1
3	3	3	2	5	1	5	3	3	0
4	3	3	3	2	2	4	3	1	3
5	3	4	3	1	1	5	2	1	3
3	3	5	3	5	2	5	2	2	2
3	1	5	3	3	3	5	1	5	3
2	1	2	1	3	1	1	1	2	3
2	2	4	3	5	3	4	3	4	3
5	3	3	3	2	2	2	3	4	3
4	3	5	3	1	2	3	2	3	1
4	3	1	1	5	3	1	0	4	3
4	3	5	2	2	1	3	3	5	2
2	0	5	1	4	2	3	1	1	3
4	3	1	0	2	1	5	3	2	3
3	1	1	0	3	0	2	3	1	0
5	3	4	3	3	1	2	1	2	1
4	3	5	3	4	1	5	3	5	3
5	0	5	3	3	3	3	3	3	1
2	1	5	3	3	2	3	0	3	2
1	2	4	3	3	3	2	1	3	1
3	2	3	1	4	3	2	0		
5	3	2	0	1	0	1	1		
2	1	5	3	2	1	1	1		

CASE **18**

TIPS

Foodservers' tips in restaurants may be influenced by many factors, including the nature of the restaurant, size of the party, table location in the restaurant, and so forth. To make appropriate assignments for the foodservers, restaurant managers need to know what these factors are. They must avoid either the substance or appearance of unfair treatment of the foodservers, for whom the tips are a major component of pay. In one restaurant, a foodserver recorded the following data on all customers he had served during a recent interval of two and a half months. The restaurant, located in a suburban shopping mall, was one of a national chain and served a varied menu. Pursuant to local law, the restaurant offered seating in a nonsmoking section to patrons who requested it. The data were recorded on those days and during those times when the foodserver was routinely assigned to work.

The restaurant manager thought the data might provide a means of assessing what sorts of things affected tips and in a way which might be seen as objective. The server had worked at a variety of assignments on different days and times during the data-gathering period and had been happy to cooperate. Are there any patterns of which you think the foodserver or the restaurant manager should be aware, either in the total bills or in the tips? What other variables might be useful, if the experiment were to be repeated?

The data are available in file TIPS.

This case describes a real situation. The data were provided by Thomas J. Kientz, President, Colorado National Bank, Aurora, Colorado, and by Sean Schneider.

DEFINITIONS

TOTBILL	Total bill, including tax, in dollars
TIP	Tip in dollars
SEX	Sex of person paying bill (0 = male, 1 = female)
SMOKER	Smoker in party? (0 = no, 1 = yes)
DAY	3 = Thursday 5 = Saturday 4 = Friday 6 = Sunday
TIME	(0 = day, 1 = night)
SIZE	Size of party

DATA SET

OBS	TOTBILL	TIP	SEX	SMOKER	DAY	TIME	SIZE
1	16.99	1.01	1	0	6	1	2
2	10.34	1.66	0	0	6	1	3
3	21.01	3.50	0	0	6	1	3
4	23.68	3.31	0	0	6	1	2
5	24.59	3.61	1	0	6	1	4
6	25.29	4.71	0	0	6	1	4
7	8.77	2.00	0	0	6	1	2
8	26.88	3.12	0	0	6	1	4
9	15.04	1.96	0	0	6	1	2
10	14.78	3.23	0	0	6	1	2
11	10.27	1.71	0	0	6	1	2
12	35.26	5.00	1	0	6	1	4
13	15.42	1.57	0	0	6	1	2
14	18.43	3.00	0	0	6	1	4
15	14.83	3.02	1	0	6	1	2
16	21.58	3.92	0	0	6	1	2
17	10.33	1.67	1	0	6	1	3
18	16.29	3.71	0	0	6	1	3
19	16.97	3.50	1	0	6	1	3
20	20.65	3.35	0	0	5	1	3
21	17.92	4.08	0	0	5	1	2
22	20.29	2.75	1	0	5	1	2
23	15.77	2.23	1	0	5	1	2
24	39.42	7.58	0	0	5	1	4
25	19.82	3.18	0	0	5	1	2
26	17.81	2.34	0	0	5	1	4
27	13.37	2.00	0	0	5	1	2
28	12.69	2.00	0	0	5	1	2
29	21.70	4.30	0	0	5	1	2
30	19.65	3.00	1	0	5	1	2
31	9.55	1.45	0	0	5	1	2
32	18.35	2.50	0	0	5	1	4
33	15.06	3.00	1	0	5	1	2
34	20.69	2.45	1	0	5	1	4
35	17.78	3.27	0	0	5	1	2
36	24.06	3.60	0	0	5	1	3
37	16.31	2.00	0	0	5	1	3
38	16.93	3.07	1	0	5	1	3
39	18.69	2.31	0	0	5	1	3
40	31.27	5.00	0	0	5	1	3
41	16.04	2.24	0	0	5	1	3
42	17.46	2.54	0	0	6	1	2
43	13.94	3.06	0	0	6	1	2
44	9.68	1.32	0	0	6	1	2
45	30.40	5.60	0	0	6	1	4
46	18.29	3.00	0	0	6	1	2
47	22.23	5.00	0	0	6	1	2
48	32.40	6.00	0	0	6	1	4
49	28.55	2.05	0	0	6	1	3
50	18.04	3.00	0	0	6	1	2
51	12.54	2.50	0	0	6	1	2
52	10.29	2.60	1	0	6	1	2
53	34.81	5.20	1	0	6	1	4
54	9.94	1.56	0	0	6	1	2
55	25.56	4.34	0	0	6	1	4
56	19.49	3.51	0	0	6	1	2
57	38.01	3.00	0	1	5	1	4
58	26.41	1.50	1	0	5	1	2
59	11.24	1.76	0	1	5	1	2
60	48.27	6.73	0	0	5	1	4
61	20.29	3.21	0	1	5	1	2
62	13.81	2.00	0	1	5	1	2
63	11.02	1.98	0	1	5	1	2
64	18.29	3.76	0	1	5	1	4
65	17.59	2.64	0	0	5	1	3
66	20.08	3.15	0	0	5	1	3
67	16.45	2.47	1	0	5	1	2
68	3.07	1.00	1	1	5	1	1

(continues on the following page)

OBS	TOTBILL	TIP	SEX	SMOKER	DAY	TIME	SIZE
69	20.23	2.01	0	0	5	1	2
70	15.01	2.09	0	1	5	1	2
71	12.02	1.97	0	0	5	1	2
72	17.07	3.00	1	0	5	1	3
73	26.86	3.14	1	1	5	1	2
74	25.28	5.00	1	1	5	1	2
75	14.73	2.20	1	0	5	1	2
76	10.51	1.25	0	0	5	1	2
77	17.92	3.08	0	1	5	1	2
78	27.20	4.00	0	0	3	0	4
79	22.76	3.00	0	0	3	0	2
80	17.29	2.71	0	0	3	0	2
81	19.44	3.00	0	1	3	0	2
82	16.66	3.40	0	0	3	0	2
83	10.07	1.83	1	0	3	0	1
84	32.68	5.00	0	1	3	0	2
85	15.98	2.03	0	0	3	0	2
86	34.83	5.17	1	0	3	0	4
87	13.03	2.00	0	0	3	0	2
88	18.28	4.00	0	0	3	0	2
89	24.71	5.85	0	0	3	0	2
90	21.16	3.00	0	0	3	0	2
91	28.97	3.00	0	1	4	1	2
92	22.49	3.50	0	0	4	1	2
93	5.75	1.00	1	1	4	1	2
94	16.32	4.30	1	1	4	1	2
95	22.75	3.25	1	0	4	1	2
96	40.17	4.73	0	1	4	1	4
97	27.28	4.00	0	1	4	1	2
98	12.03	1.50	0	1	4	1	2
99	21.01	3.00	0	1	4	1	2
100	12.46	1.50	0	0	4	1	2
101	11.35	2.50	1	1	4	1	2
102	15.38	3.00	1	1	4	1	2
103	44.30	2.50	1	1	5	1	3
104	22.42	3.48	1	1	5	1	2
105	20.92	4.08	1	0	5	1	2
106	15.36	1.64	0	1	5	1	2
107	20.49	4.06	0	1	5	1	2
108	25.21	4.29	0	1	5	1	2
109	18.24	3.76	0	0	5	1	2
110	14.31	4.00	1	1	5	1	2
111	14.00	3.00	0	0	5	1	2
112	7.25	1.00	1	0	5	1	1
113	38.07	4.00	0	0	6	1	3
114	23.95	2.55	0	0	6	1	2
115	25.71	4.00	1	0	6	1	3
116	17.31	3.50	1	0	6	1	2
117	29.93	5.07	0	0	6	1	4
118	10.65	1.50	1	0	3	0	2
119	12.43	1.80	1	0	3	0	2
120	24.08	2.92	1	0	3	0	4
121	11.69	2.31	0	0	3	0	2
122	13.42	1.68	1	0	3	0	2
123	14.26	2.50	0	0	3	0	2
124	15.95	2.00	0	0	3	0	2
125	12.48	2.52	1	0	3	0	2
126	29.80	4.20	1	0	3	0	6
127	8.52	1.48	0	0	3	0	2
128	14.52	2.00	1	0	3	0	2
129	11.38	2.00	1	0	3	0	2
130	22.82	2.18	0	0	3	0	3
131	19.08	1.50	0	0	3	0	2
132	20.27	2.83	1	0	3	0	2
133	11.17	1.50	1	0	3	0	2
134	12.26	2.00	1	0	3	0	2
135	18.26	3.25	1	0	3	0	2
136	8.51	1.25	1	0	3	0	2
137	10.33	2.00	1	0	3	0	2
138	14.15	2.00	1	0	3	0	2
139	16.00	2.00	0	1	3	0	2
140	13.16	2.75	1	0	3	0	2
141	17.47	3.50	1	0	3	0	2
142	34.30	6.70	0	0	3	0	6
143	41.19	5.00	0	0	3	0	5
144	27.05	5.00	1	0	3	0	6
145	16.43	2.30	1	0	3	0	2
146	8.35	1.50	1	0	3	0	2
147	18.64	1.36	1	0	3	0	3
148	11.87	1.63	1	0	3	0	2
149	9.78	1.73	0	0	3	0	2
150	7.51	2.00	0	0	3	0	2
151	14.07	2.50	0	0	6	1	2
152	13.13	2.00	0	0	6	1	2
153	17.26	2.74	0	0	6	1	3
154	24.55	2.00	0	0	6	1	4
155	19.77	2.00	0	0	6	1	4
156	29.85	5.14	1	0	6	1	5
157	48.17	5.00	0	0	6	1	6
158	25.00	3.75	1	0	6	1	4
159	13.39	2.61	1	0	6	1	2
160	16.49	2.00	0	0	6	1	4
161	21.50	3.50	0	0	6	1	4
162	12.66	2.50	0	0	6	1	2

(concludes on the following page)

OBS	TOTBILL	TIP	SEX	SMOKER	DAY	TIME	SIZE
163	16.21	2.00	1	0	6	1	3
164	13.81	2.00	0	0	6	1	2
165	17.51	3.00	1	1	6	1	2
166	24.52	3.48	0	0	6	1	3
167	20.76	2.24	0	0	6	1	2
168	31.71	4.50	0	0	6	1	4
169	10.59	1.61	1	1	5	1	2
170	10.63	2.00	1	1	5	1	2
171	50.81	10.00	0	1	5	1	3
172	15.81	3.16	0	1	5	1	2
173	7.25	5.15	0	1	6	1	2
174	31.85	3.18	0	1	6	1	2
175	16.82	4.00	0	1	6	1	2
176	32.90	3.11	0	1	6	1	2
177	17.89	2.00	0	1	6	1	2
178	14.48	2.00	0	1	6	1	2
179	9.60	4.00	1	1	6	1	2
180	34.63	3.55	0	1	6	1	2
181	34.65	3.68	0	1	6	1	4
182	23.33	5.65	0	1	6	1	2
183	45.35	3.50	0	1	6	1	3
184	23.17	6.50	0	1	6	1	4
185	40.55	3.00	0	1	6	1	2
186	20.69	5.00	0	0	6	1	5
187	20.90	3.50	1	1	6	1	3
188	30.46	2.00	0	1	6	1	5
189	18.15	3.50	1	1	6	1	3
190	23.10	4.00	0	1	6	1	3
191	15.69	1.50	0	1	6	1	2
192	19.81	4.19	1	1	3	0	2
193	28.44	2.56	0	1	3	0	2
194	15.48	2.02	0	1	3	0	2
195	16.58	4.00	0	1	3	0	2
196	7.56	1.44	0	0	3	0	2
197	10.34	2.00	0	1	3	0	2
198	43.11	5.00	1	1	3	0	4
199	13.00	2.00	1	1	3	0	2
200	13.51	2.00	0	1	3	0	2
201	18.71	4.00	0	1	3	0	3
202	12.74	2.01	1	1	3	0	2
203	13.00	2.00	1	1	3	0	2

OBS	TOTBILL	TIP	SEX	SMOKER	DAY	TIME	SIZE
204	16.40	2.50	1	1	3	0	2
205	20.53	4.00	0	1	3	0	4
206	16.47	3.23	1	1	3	0	3
207	26.59	3.41	0	1	5	1	3
208	38.73	3.00	0	1	5	1	4
209	24.27	2.03	0	1	5	1	2
210	12.76	2.23	1	1	5	1	2
211	30.06	2.00	0	1	5	1	3
212	25.89	5.16	0	1	5	1	4
213	48.33	9.00	0	0	5	1	4
214	13.27	2.50	1	1	5	1	2
215	28.17	6.50	1	1	5	1	3
216	12.90	1.10	1	1	5	1	2
217	28.15	3.00	0	1	5	1	5
218	11.59	1.50	0	1	5	1	2
219	7.74	1.44	0	1	5	1	2
220	30.14	3.09	1	1	5	1	4
221	12.16	2.20	0	1	4	0	2
222	13.42	3.48	1	1	4	0	2
223	8.58	1.92	0	1	4	0	1
224	15.98	3.00	1	0	4	0	3
225	13.42	1.58	0	1	4	0	2
226	16.27	2.50	1	1	4	0	2
227	10.09	2.00	1	1	4	0	2
228	20.45	3.00	0	0	5	1	4
229	13.28	2.72	0	0	5	1	2
230	22.12	2.88	1	1	5	1	2
231	24.01	2.00	0	1	5	1	4
232	15.69	3.00	0	1	5	1	3
233	11.61	3.39	0	0	5	1	2
234	10.77	1.47	0	0	5	1	2
235	15.53	3.00	0	1	5	1	2
236	10.07	1.25	0	0	5	1	2
237	12.60	1.00	0	1	5	1	2
238	32.83	1.17	0	1	5	1	2
239	35.83	4.67	1	0	5	1	3
240	29.03	5.92	0	0	5	1	3
241	27.18	2.00	1	1	5	1	2
242	22.67	2.00	0	1	5	1	2
243	17.82	1.75	0	0	5	1	2
244	18.78	3.00	1	0	3	1	2

CASE **19**

RUBBERGATE

A congressional election is right around the corner. A small but vocal group of Republicans, hoping to unseat Democratic incumbents in the upcoming election, have received a significant amount of attention from the media. They charge ongoing fiscal irresponsibility on the part of the current members of Congress, and their accusations have struck a chord with the American public. The Republicans are using the House Banking Scandal as an example of the type of fiscal improprieties that are "deeply ingrained in the minds of our current elected officials." The House Banking Scandal, also dubbed "Rubbergate," involved the discovery that certain members of the House of Representatives bounced numerous bank checks with impunity in the early 1990s. At the time, Rubbergate received significant press, as, for example, newspapers across the country listed the names and numbers of checks bounced by their representatives. Now the Republican hopefuls are reviving the Rubbergate scandal.

Frank works as a statistician for the Republican party. He has been asked to study the earlier Rubbergate scandal for possible political ammunition. He has collected a random sample of 146 members of the 1992 House of Representatives, which may be found in the file named RUBBERG8. Eight characteristics were collected about each representative. Whether the representative wrote bad checks is indicated by the variable named CHECKS. This variable is coded 1 for a bad-check writer and 0 otherwise. PARTY is the variable indicating political affiliation. Democrats are coded with 1, and Republicans are coded 0. There weren't any Independents in this random sample. SERVICE is the total number of years of service in the House. Service beginning in or after November in a given year is not counted as a year of service in this data set. The representative's age in years is recorded in the variable named AGE. VOTES measures

This case is based on data provided to us by Chris A. Wilson. The data come from three sources: *Politics in America,* Congressional Quarterly, Inc., Washington, D.C., 1991, pp. 9–12; *Vital Statistics on Congress 1991–1992,* Congressional Quarterly, Inc., Washington, D.C., 1992, pp. 214–31; and *Rocky Mountain News,* "The House List," Friday, April 17, 1992, pp. A50–A51.

the returns (in percentages) for the most recent election in which that member participated. DISTRICT measures the representative's home district population rounded to the nearest thousand. STATE is like DISTRICT except it is the population of the representative's home state. If the representative wrote bad checks, Frank also recorded how many he or she wrote. This information is in the variable called BADCHKS.

Frank wants to prepare a report that characterizes the data set. In particular, what should he say about the following issues?

1 What is the general nature of the sample in terms of average and typical fluctuation around the average for the variables named SERVICE, AGE, VOTES, DISTRICT, and STATE?

2 How do these variables differ in a comparison of check-bouncers and non-check-bouncers?

3 How does party affiliation enter into the analysis?

4 For those individuals who did write bad checks, summarize the total number of bad checks written in some appropriate fashion.

DATA SET

CHECKS	PARTY	SERVICE	AGE	VOTES	DISTRICT	STATE	BADCHKS
1	0	6	58	99.6	594	4040	9
1	1	24	69	99.7	574	4040	4
0	1	8	52	92.8	537	4040	0
1	0	4	47	99.5	829	3665	32
0	0	4	48	61.3	747	3665	0
1	1	6	62	60.4	613	2350	1
1	0	0	40	43.0	651	29760	3
0	1	12	48	54.7	758	29760	0
1	1	16	45	60.5	666	29760	99
1	1	28	75	62.7	692	29760	13
1	1	16	59	58.0	566	29760	3
1	1	14	52	74.2	653	29760	12
1	0	17	64	54.6	647	29760	3
0	0	18	68	60.0	638	29760	0
1	1	28	74	70.0	601	29760	11
0	1	12	56	72.7	600	29760	0
1	1	10	64	67.1	599	29760	1
0	1	8	60	60.7	598	29760	0
0	0	8	63	49.7	980	29760	0
0	0	2	38	67.6	691	29760	0
1	0	8	59	68.1	841	29760	4
1	1	18	50	63.7	453	3294	5
0	0	0	47	54.1	529	3294	0
1	1	9	54	71.4	541	3287	60
1	0	3	45	76.5	519	3287	18
1	1	8	43	65.5	666	666	3
1	1	42	80	72.7	569	12938	4
1	0	2	46	59.2	765	12938	9

(continues on the following page)

CHECKS	PARTY	SERVICE	AGE	VOTES	DISTRICT	STATE	BADCHKS
0	0	8	60	58.1	769	12938	0
1	0	8	66	100.0	766	12938	8
0	0	10	51	97.8	544	12938	0
0	0	2	38	60.4	634	12938	0
1	1	10	51	73.0	557	6478	819
1	1	4	50	75.6	540	6478	125
0	1	8	64	68.7	563	6478	0
0	1	14	57	55.8	777	6478	0
0	1	13	63	66.3	594	1108	0
1	1	8	72	93.8	413	11431	716
0	1	2	59	59.2	533	11431	0
1	1	18	59	79.9	445	11431	18
1	0	11	55	67.7	566	11431	1
0	0	6	61	65.8	611	11431	0
1	1	8	39	86.8	467	11431	9
1	1	8	46	66.2	490	11431	12
0	1	6	41	66.0	493	5544	0
1	1	2	38	60.7	573	5544	21
1	0	24	63	57.6	562	5544	61
1	1	24	58	66.4	510	5544	1
1	1	4	47	99.2	462	2777	4
0	0	4	42	71.8	449	2777	0
0	0	6	62	60.1	549	2478	0
1	1	16	53	86.9	527	3685	152
0	0	4	59	69.3	563	3685	0
1	1	6	36	50.8	492	3685	514
1	1	14	43	52.5	467	4212	8
0	1	0	37	60.1	637	1228	0
1	0	12	43	51.0	591	1228	1
0	1	4	47	69.7	541	4782	0
1	1	12	58	65.3	670	4782	6
0	0	32	69	77.5	538	6016	0
0	1	10	50	65.5	545	6016	0
1	1	14	44	99.9	525	6016	92
1	1	18	53	53.4	593	6016	10
1	0	14	58	64.1	536	9295	17
1	0	6	48	75.4	579	9295	20
1	1	17	59	68.6	489	9295	201
1	0	12	58	61.3	521	9295	878
1	1	10	42	63.6	502	9295	547
0	1	8	59	69.7	485	9295	0
1	0	10	38	61.8	480	4375	125
0	1	12	52	72.9	498	4375	0
1	1	16	56	72.9	518	4375	2
0	1	24	70	100.0	530	2573	0
1	1	22	59	60.9	492	5117	328

(continues on the following page)

CHECKS	PARTY	SERVICE	AGE	VOTES	DISTRICT	STATE	BADCHKS
1	1	14	59	61.8	615	5117	9
0	0	2	61	52.1	610	5117	0
1	1	12	53	61.1	417	799	66
0	1	2	49	57.9	560	1578	0
1	0	8	69	63.5	609	1202	2
1	1	0	33	54.3	567	7730	1
0	0	10	37	62.9	615	7730	0
1	0	18	59	74.6	520	7730	8
1	1	2	56	81.5	465	7730	6
1	0	6	47	58.1	613	7730	2
0	0	10	63	100.0	497	7730	0
1	1	16	41	55.8	567	7730	151
1	0	10	48	54.6	513	17991	4
1	1	24	70	72.3	527	17991	133
1	1	8	56	92.9	541	17991	408
1	0	1	32	60.0	536	17991	5
1	1	14	63	80.4	534	17991	3
1	1	2	53	62.8	517	17991	1
1	1	2	43	64.1	520	17991	15
1	0	10	46	100.0	545	17991	8
1	0	28	71	63.0	531	17991	3
0	1	16	51	55.0	499	17991	0
1	1	25	77	64.8	587	6629	63
1	1	4	50	58.1	699	6629	8
0	1	18	51	65.6	601	6629	0
0	0	4	64	61.8	572	6629	0
0	1	0	39	51.1	509	10847	0
1	0	10	46	61.7	511	10847	6
0	0	0	54	62.1	527	10847	0
0	0	24	73	63.2	527	10847	0
0	1	14	59	56.7	521	10847	0
1	0	18	66	58.9	511	10847	14
1	1	8	43	64.8	517	10847	397
0	0	4	56	56.0	518	3146	0
1	1	10	40	73.6	553	3146	8
1	1	16	48	63.1	617	2842	83
0	1	4	43	85.8	539	2842	0
1	0	16	61	57.1	581	11882	4
0	1	22	63	57.0	537	11882	0
1	1	12	44	56.6	578	11882	50
1	1	6	53	100.0	507	11882	7
0	1	10	54	71.8	453	11882	0
0	0	8	60	100.0	534	11882	0
1	1	22	64	65.6	469	11882	3
0	0	12	61	59.4	503	11882	0
0	0	4	63	65.5	616	3487	0

(concludes on the following page)

CHECKS	PARTY	SERVICE	AGE	VOTES	DISTRICT	STATE	BADCHKS
1	1	4	51	61.4	577	3487	2
0	1	4	44	67.6	696	696	0
1	1	16	61	53.0	521	4877	8
1	1	6	41	66.7	607	4877	6
1	1	16	45	58.1	462	4877	388
1	0	8	43	99.6	714	16987	18
0	0	6	41	66.5	705	16987	0
0	1	38	68	57.7	564	16987	0
1	1	2	38	71.3	609	16987	3
1	1	26	63	100.0	658	16987	284
1	1	1	49	99.6	450	16987	3
1	0	4	43	74.8	670	16987	1
0	1	12	48	100.0	635	16987	0
1	1	8	53	100.0	594	16987	18
0	1	0	41	58.3	604	1723	0
1	1	4	60	75.0	654	6187	1
0	1	3	45	99.4	544	6187	0
1	1	0	45	51.7	747	6187	3
1	0	6	52	52.0	657	4867	58
0	0	10	57	70.7	562	4867	0
0	1	2	53	72.3	558	4867	0
0	1	8	39	55.5	488	1794	0
1	1	20	52	99.4	533	4892	6
1	1	7	47	69.2	537	4892	1
1	1	22	52	99.5	544	4892	64
0	0	2	54	55.1	454	454	0

CASE **20**

HITECH ENGINEERING

HiTech Engineering, Inc. (HEI), is a company located in Minnesota that designs and manufactures a variety of industrial products. HEI sells its products directly to existing customers and also uses authorized distributors around the globe to assist in sales and distribution. Although HEI has an edge on the market in terms of engineering and manufacturing expertise, it falls a bit short in knowing the best way to market and sell its latest industrial inventions and innovations.

HEI gets word to the market about a new product in various ways. For instance, sometimes they purchase magazine advertisements and sometimes they issue news releases. Because HEI produces industrial products and not consumer products, it never uses electronic media such as television or radio, but concentrates on print media for the purpose of advertising.

HEI management would like to understand how effective each of the five basic types of advertising they've used has been.

1 Magazines. Advertisements are placed in about 150 different industrial magazines and trade journals. These ads are designed by HEI's advertising department.

2 Postcard decks. Publishers of various magazines and journals assemble postcard announcements of new products from a number of different companies. The postcard decks are then sent out to its list of subscribers. HEI must pay for the postcard deck service.

3 Editorials. HEI's advertising department occasionally produces editorials for magazines or trade publications and hopes that the editorial will be published. In response to the editorial, readers will occasionally write back to HEI requesting more

These data are real even though the name of the company and its location have been changed. Tessa E. Alexander and Edwin C. Peterson have granted us permission to use these data, which they originally collected for study.

information. HEI's expenses for this type of marketing are employee salaries for writing the editorial and preparing a camera-ready version of the editorial.

4 News releases. A news release is a one paragraph description of a new product that typically appears in the back of a magazine. News releases will be published for free by magazines that are already carrying HEI advertisements as a courtesy to their clients. Like requests for literature (below), interested readers send a postcard to the magazine, which is then forwarded to HEI. Costs incurred for this type of advertising are employee salaries for processing requests.

5 Literature. Readers of magazines or trade journals may make a request for literature or more information about HEI by circling a request number on a "bingo card" and sending it back to the magazine or journal. Requests are then sent from the magazine or journal to HEI, which responds to the requests appropriately.

Requests for information by prospective customers are called "leads." By asking customers where they heard about the new product, the advertising group was able to compile information about the number of leads generated by each of these media types; they also know the expenses incurred in each media class. The data cover the last 17 months. These data have been assembled in the file named LEADATA.

The data set contains the following information. MONTH indicates the appropriate month, which takes on values from 1 to 17. RESPONSE is the number of responses for information about the company's product. SPEND indicates expenses incurred in that month for that media type in dollars. Finally, MAGS, POST, EDIT, NEWS, and LIT are binary variables indicating which media type resides in that row of the data set. These correspond to the five media described above.

What is the most effective type of advertising outlet for HEI? Do you have any recommendations on the appropriate mix of the five different types of advertising?

DATA SET

MONTH	RESPONSE	SPEND	MAGS	POST	EDIT	NEWS	LIT
1	150	2505	1	0	0	0	0
2	76	1760	1	0	0	0	0
3	79	3225	1	0	0	0	0
4	82	7527	1	0	0	0	0
5	98	7100	1	0	0	0	0
6	106	6000	1	0	0	0	0
7	81	9330	1	0	0	0	0
8	87	3910	1	0	0	0	0
9	64	7878	1	0	0	0	0
10	111	4233	1	0	0	0	0
11	102	2475	1	0	0	0	0
12	65	1800	1	0	0	0	0
13	42	1689	1	0	0	0	0
14	30	1559	1	0	0	0	0
15	40	865	1	0	0	0	0
16	42	0	1	0	0	0	0
17	10	2635	1	0	0	0	0
1	656	9490	0	1	0	0	0

(continues on the following page)

MONTH	RESPONSE	SPEND	MAGS	POST	EDIT	NEWS	LIT
2	966	10660	0	1	0	0	0
3	503	0	0	1	0	0	0
4	230	5534	0	1	0	0	0
5	379	9517	0	1	0	0	0
6	690	10852	0	1	0	0	0
7	560	6321	0	1	0	0	0
8	498	7263	0	1	0	0	0
9	459	5829	0	1	0	0	0
10	315	4058	0	1	0	0	0
11	226	0	0	1	0	0	0
12	57	0	0	1	0	0	0
13	160	7131	0	1	0	0	0
14	642	3773	0	1	0	0	0
15	153	1436	0	1	0	0	0
16	421	2377	0	1	0	0	0
17	666	4373	0	1	0	0	0
1	60	1779	0	0	1	0	0
2	29	2100	0	0	1	0	0
3	64	1899	0	0	1	0	0
4	105	2200	0	0	1	0	0
5	69	1890	0	0	1	0	0
6	21	1900	0	0	1	0	0
7	24	2112	0	0	1	0	0
8	17	2000	0	0	1	0	0
9	10	1877	0	0	1	0	0
10	50	2200	0	0	1	0	0
11	54	1788	0	0	1	0	0
12	42	2300	0	0	1	0	0
13	56	2166	0	0	1	0	0
14	102	1877	0	0	1	0	0
15	37	1944	0	0	1	0	0
16	36	2000	0	0	1	0	0
17	6	2100	0	0	1	0	0
1	81	189	0	0	0	1	0
2	139	176	0	0	0	1	0
3	174	231	0	0	0	1	0
4	152	179	0	0	0	1	0
5	166	215	0	0	0	1	0
6	98	197	0	0	0	1	0
7	82	241	0	0	0	1	0
8	66	187	0	0	0	1	0
9	74	185	0	0	0	1	0
10	42	216	0	0	0	1	0
11	53	197	0	0	0	1	0
12	26	193	0	0	0	1	0
13	71	216	0	0	0	1	0
14	74	195	0	0	0	1	0

(concludes on the following page)

MONTH	RESPONSE	SPEND	MAGS	POST	EDIT	NEWS	LIT
15	82	208	0	0	0	1	0
16	138	185	0	0	0	1	0
17	91	194	0	0	0	1	0
1	2	375	0	0	0	0	1
2	34	405	0	0	0	0	1
3	121	413	0	0	0	0	1
4	144	378	0	0	0	0	1
5	78	389	0	0	0	0	1
6	116	385	0	0	0	0	1
7	220	407	0	0	0	0	1
8	29	407	0	0	0	0	1
9	24	403	0	0	0	0	1
10	15	399	0	0	0	0	1
11	263	408	0	0	0	0	1
12	117	413	0	0	0	0	1
13	96	401	0	0	0	0	1
14	31	411	0	0	0	0	1
15	84	400	0	0	0	0	1
16	104	420	0	0	0	0	1
17	31	410	0	0	0	0	1

CASE **21**

401(K)

Best Retirement, Inc. (BRI), sells retirement plans to corporations around the country. To capture a market niche, BRI has decided to target smaller firms—those with 500 or fewer employees. BRI is positioned to offer a variety of retirement plans for its clients, from 401(k) plans to individual retirement accounts. BRI also writes group health and group life insurance policies, although the major portion of their revenue comes from retirement packages.

It is extremely important that BRI be able to estimate the dollar amount contributed to each plan by year's end. It would be nice to be able to estimate total contributions before the plan is even written, so that BRI can make internal revenue and cost projections. For this reason, BRI has hired an underwriter who estimates year-end contributions based on information provided by the client on an application form. In the past, the underwriter estimates have not been very accurate. Your task is to prepare more accurate estimates of annual contributions for these retirement plans.

To this end, BRI has provided you with lots of data about 401(k) retirement plans written by BRI representatives from January to March of last year. The file, named 401(k), contains 92 plans sold over this time period with the ultimate contributions to each plan at year's end. This variable, labeled CONTRIB, is the one you are attempting to understand. The underwriter's estimate (ESTIMATE) is also given; this is the variable you would like to improve upon.

A number of factors are likely to influence employee contributions to a retirement plan. Clearly employees' salaries would be an important factor. "Matching-contribution" provisions would also be a potential determinant; we might expect that employer contributions matching employee contributions would induce greater employee participation

These data are real, but the names of the company and of the person who supplied the data to us have been withheld upon their request.

in the retirement plan. Firms with high employee turnover rates might have low employee participation rates, all else the same. BRI also suspects that it is to their advantage to offer "package deals" and that 401(k) contributions are more likely if BRI also writes group health and group life insurance policies for that client.

Several other provisions of the policy might influence contribution levels. For instance, some plans have fail-safe provisions. This means that if the plan fails to meet government antidiscrimination rules, the employer will make contributions to the lower-paid employees' accounts. Also, a plan might have immediate vesting of employer contributions; employees who withdraw their money also receive all of the employer's matching contributions at the time of withdrawal if the plan is vested.

Finally, BRI has one sales representative, Susan Shepard, who has been specifically trained to deal exclusively with 401(k) plans. Other sales representatives can write 401(k) plans, but don't have quite the in-depth knowledge that Ms. Shepard does. BRI would like to know if Susan's expertise is a factor that influences year-end contributions to 401(k) plans. If so, BRI would like to consider training some of its other sales representatives.

In the data set shown below, CONTRIB is the variable indicating contribution to the plan in dollars at the end of its first year. SUSAN is a categorical variable taking on the value of 1 if Susan Shepard sold the policy and 0 if she did not. GROUP takes on the value of 1 if the client also has a group life or group health insurance policy with Best Retirement; if not, GROUP is 0. TURNOVER is the employee turnover rate. It measures the percentage of employees terminating employment each year. ESTIMATE is the estimated end-of-year contributions in dollars and SALARY is the average annual employee salary in dollars. ELIGIBLE is the number of employees eligible to participate in the 401(k) plan. VEST and FAILSAFE are both binary variables: does the plan have immediate vesting of employer contributions (0 = no, 1 = yes) and does the plan have a fail-safe provision (0 = no, 1 = yes)? MATCH is the percentage of the contributions matched by the employer.

Can you do better than the underwriter at estimating year-end contributions?

DATA SET

CONTRIB	GROUP	TURNOVER	ELIGIBLE	VEST	FAILSAFE	MATCH	SALARY	ESTIMATE	SUSAN
36675	1	14.00	69	1	0	25	26296.3	75432	0
63733	0	10.00	33	1	0	50	14133.7	50000	0
25560	1	8.00	21	1	0	50	24000.0	45000	0
177970	0	10.00	67	1	0	25	36833.3	235000	0
86873	1	10.00	47	1	0	50	41140.1	146965	0
39051	0	10.00	12	1	0	0	66463.8	95000	0
131449	1	12.00	85	1	0	25	25929.3	155045	0
30711	0	10.00	40	1	0	50	17198.7	54392	1
13691	0	10.00	30	1	1	0	28788.6	50000	0
49587	1	10.00	63	1	0	25	32124.4	100000	0
37898	0	10.00	25	1	1	100	35288.1	86000	0
41686	1	0.10	30	0	0	100	23266.7	56000	0
107657	0	14.00	21	1	0	50	54798.0	115650	0
39811	0	14.00	29	1	0	50	18617.9	50000	0
148274	0	14.00	99	1	0	0	25516.6	252316	0

(continues on the following page)

CONTRIB	GROUP	TURNOVER	ELIGIBLE	VEST	FAILSAFE	MATCH	SALARY	ESTIMATE	SUSAN
91496	0	10.80	21	1	1	25	41250.0	113152	1
126988	0	10.00	28	1	0	0	21599.3	77280	0
158941	0	7.50	283	1	0	25	20405.2	321750	0
13691	0	12.00	50	0	0	100	23778.9	35772	0
118396	0	14.00	250	1	0	0	28626.9	243750	0
102670	0	7.00	30	1	0	0	37906.4	101431	0
21012	0	12.00	28	1	0	100	33941.2	78330	0
58448	0	14.00	36	1	0	25	23492.0	60304	0
39787	0	12.00	23	1	0	50	28222.1	41458	0
73614	0	14.00	66	0	0	0	35874.8	144316	0
218158	0	12.00	150	0	0	100	21626.1	200000	1
41335	0	12.00	121	1	0	20	57182.2	141481	0
19120	1	14.00	69	1	0	0	21503.3	103171	0
73053	1	11.00	175	1	0	0	25102.2	152646	0
166110	1	14.00	146	1	0	25	37219.7	267268	0
141009	1	14.00	70	1	0	0	20380.5	154135	0
98052	0	13.00	176	1	0	0	34764.0	276222	0
42293	1	14.00	38	0	0	0	34182.8	90074	0
169983	0	12.00	144	1	0	25	26809.3	255126	0
71849	0	14.00	69	0	0	0	67391.1	119856	0
28429	1	12.00	110	1	0	0	30397.0	144000	0
24154	1	14.00	21	1	0	75	33374.8	50088	0
12038	1	14.00	49	1	0	0	21133.9	51592	0
289886	1	11.80	391	1	0	0	27309.8	432562	0
150329	1	11.00	80	0	1	0	37927.6	173846	0
69020	0	14.00	32	1	0	0	43603.4	90000	1
25749	1	18.00	281	1	0	0	18943.7	120000	0
32147	1	14.00	63	1	0	0	30144.4	103320	0
101665	1	11.80	45	0	0	25	35661.8	126625	0
224721	0	8.21	437	1	0	25	24986.9	679972	0
249063	1	14.00	400	1	0	25	20511.8	425040	0
17865	0	12.00	9	1	0	100	59528.6	40500	0
28912	0	14.00	31	1	0	50	27289.5	59271	0
42703	0	14.00	25	1	0	0	34137.7	51026	0
120940	0	14.00	50	1	0	50	35888.3	106000	0
149654	0	14.00	230	1	0	40	26807.3	200000	0
277688	0	12.00	244	1	0	0	36729.4	263676	1
25214	1	12.00	54	1	0	10	18039.7	73031	0
141359	0	12.00	234	0	0	0	29970.1	217420	0
57282	1	12.00	23	1	0	100	26647.8	50000	0
76269	0	12.00	23	1	0	25	37104.9	64488	0
75572	0	12.00	63	1	0	10	42790.9	175000	0
71330	1	13.00	85	1	0	50	25932.4	192000	0
28748	1	14.00	69	1	0	100	21250.7	123084	0
151405	1	14.00	141	1	0	25	30506.1	262005	0
24285	0	14.00	49	1	1	10	15423.5	46104	1
58131	1	14.00	130	0	0	0	38793.3	140576	0

(concludes on the next page)

CONTRIB	GROUP	TURNOVER	ELIGIBLE	VEST	FAILSAFE	MATCH	SALARY	ESTIMATE	SUSAN
44323	1	14.00	41	1	0	0	37380.2	97680	0
318022	0	14.00	154	0	0	100	51865.4	524000	0
34052	0	14.00	17	0	0	100	30398.9	61222	0
82837	0	12.00	28	1	0	25	41309.2	70275	0
86394	0	14.00	100	1	0	0	30265.6	264075	1
26810	1	14.00	126	1	0	0	29346.0	281954	0
37516	1	12.00	62	1	0	50	24814.9	65052	0
44228	1	14.00	135	1	0	25	19549.9	165102	0
24541	1	12.00	60	1	0	0	39189.3	121135	0
80602	1	14.00	96	1	0	50	26440.6	165240	0
56209	1	14.00	64	1	0	0	45428.3	170928	0
187159	1	12.50	350	1	0	0	23659.6	394980	0
34878	1	14.00	60	1	0	25	32408.3	135936	0
126924	1	14.00	86	1	0	0	37579.8	150000	0
197015	0	14.00	27	1	0	50	55804.3	123405	1
228169	0	14.00	84	0	0	20	51481.8	399431	1
147450	1	13.00	176	1	0	25	25424.2	251064	0
39691	1	14.00	68	0	0	25	27399.6	134133	0
52618	0	14.00	37	1	0	20	36806.7	113010	0
25530	1	14.00	37	1	0	0	43253.4	87000	0
39311	1	15.00	70	1	0	0	22034.8	72680	0
45969	1	14.00	75	1	0	25	18465.4	82320	0
13604	1	14.00	36	1	0	0	30352.5	50930	0
304864	0	14.00	141	1	1	25	42847.9	271153	0
66685	0	12.00	29	1	0	0	47012.6	118145	0
26192	1	20.00	49	1	0	25	2234.4	67340	0
67353	0	14.00	52	1	0	100	37295.0	220153	1
22954	0	14.00	150	1	0	25	18352.1	147000	0
109116	1	14.00	246	0	0	0	31887.7	506530	0
18022	0	14.00	51	1	0	25	71320.0	98076	0

CASE **22**

LATE CHARTS

Organizations that serve their customers by appointments at the company's offices must have some mechanism for ensuring that appropriate customer records are in the right office in time for the appointment. Misplaced records cause delays and hard feelings; in extreme cases, appointments are canceled for lack of appropriate records. Small organizations may have a simple "system" to provide records: the agent assigned may retrieve the file from a cabinet. If the file isn't there, she may look around the office until she finds it. In larger organizations, though, the large number of records involved usually means that some more formal arrangement is needed to make records available to the appropriate agent for the appointment. The records delivery system (if we may call it that) is often connected to the system that schedules the appointments. Tax audits, counseling sessions at social agencies, and insurance planning sessions share this need, as do many medical offices.

At a large medical center in the southwest, patients fill out forms and doctors dictate reports that are kept in paper form in file folders in a central filing room in the basement of the building. The center has considered converting medical records to a computerized form, but the conversion process is remarkably complicated when all its details and implications are considered, so while such a conversion is an active project, it is not yet complete.

Patients may schedule appointments with any of the more than 60 physicians either personally or by telephone. In either case, an operator consults a computer, which displays available appointment times and any restrictions on the kinds of appointments for which they may be scheduled. The operator works with the patient to find a mutually agreeable time and indicates that time to the computer system, which updates the reservation database. If the patient is physically present when the reservation is made, the

This case is an amalgam of several situations with which the authors have dealt.

operator gives him or her a card with the scheduled time on it; otherwise, the operator confirms the time orally with the patient.

Each night, the computer produces a list of patients who are scheduled for the following day for each physician and a corresponding list of charts to be pulled from the central filing area and delivered to the physicians' offices. A medical records clerk works in the evening to be sure the files are delivered. First, he collects and re-files any records returned from that day's appointments or from the centralized dictation area. Then he pulls the files needed for the next day and delivers them along with the list of appointments for each physician to each physician's office.

Occasionally, charts are unavailable. Usually this happens because they are still in use by a physician who saw the patient recently, but occasionally they are misplaced, left in someone's desk drawer or waiting in some other pile of folders somewhere in the building. The clerk enters the patient names for such files on a separate list and leaves the list for the first-shift records staff. They attack the problem as their first order of business in the morning. Usually, the computerized records of which physician last saw the patient provide a clue they can use to track down the file, but in perhaps 10 or 15 percent of the cases, they can't deliver the file in time for the scheduled appointment.

As part of a quality management initiative at the medical center, the medical records staff sampled the appointment records to assess the frequency of late charts. They sampled 100 patient appointments at random for each of 20 successive weeks. By consulting the physicians' staffs, they determined the number of appointments for which patient charts had been delivered late. The records staff knew that "late chart" was to some extent a subjective judgment on the part of the nurses and receptionists in the physicians' offices, since they kept no record or time-stamp of when the chart was delivered. Still, they figured, it was their (internal) customer's impression of quality, and it mattered. The data they collected are given below (and in file CHARTS). WEEK is the week number (1 through 20); NLATE is the number of late charts.

What do you conclude about the chart delivery process?

DATA SET

WEEK	NLATE	WEEK	NLATE	WEEK	NLATE	WEEK	NLATE
1	14	6	11	11	8	16	10
2	10	7	10	12	12	17	8
3	12	8	12	13	9	18	12
4	13	9	13	14	10	19	10
5	9	10	10	15	11	20	16

CASE **23**

MONEY SUPPLY AND INTEREST RATES

As Kellie sat in her economics class, she wondered whether her professor spent too much time in the ivory tower. She thought her question to the professor was quite straightforward, but he seemed flustered and uncertain in his response.

It all started innocently enough. The lecture was about monetary policy and the methods by which the Federal Reserve can control monetary instruments. At some point during the hour, Professor Hicks said, "When the Feds increase the money supply, by printing more money or reducing the reserve requirement, for example, the interest rate will increase." He then went into a long discussion of the mechanisms that cause this positive relationship between money and interest rates.

Although it was all very interesting, Kellie thought the ideas were too nebulous to be of any practical value. So she raised her hand. "Professor Hicks, I certainly enjoy your lively lectures, but there is one point on which I am very confused. What do you mean when you say *the* interest rate and *the* money supply? For instance, last week you told us about three different measures of money supply—M1, the narrow measure; M2, an intermediate quantity; and M3, the broadest measure of money. Are you referring to all three of these measures when you say *the* money supply?"

Professor Hicks replied that he meant the money supply in a generic sense and that the general theory applied equally to all three measures.

"Well, then, Professor Hicks, what is *the* interest rate? I mean, I have an interest rate on my new car loan, my parents have an interest rate on their mortgage, my credit card has an interest rate, and they're all different. Plus, last week, you were talking about the prime rate, the discount rate, a commercial paper rate, and a whole bunch of other things. Are you referring to only one of these interest rates, or all of them?"

Professor Hicks said, "You know, you make an interesting point. We would expect the different interest rates to react somewhat differently to changes in the stock of money, but I haven't really spent much time thinking about the issue you raise. Give me time to collect some data, and I'll report back to you during our next class meeting."

The file named MONEY contains a variety of interest rate and money stock variables measured on a monthly basis beginning in January 1980. These data are not seasonally adjusted. Asterisks at the end of the data indicate that the data were not available at the time Professor Hicks's student assistant prepared the data set for him. Updated monthly observations are available on the Federal Reserve Board's website: the money data from *www.bog.frb.us/Releases/H6/hist/* and the interest rate data from *www.bog.frb.fed.us/Releases/H15/data.HTM#FN3.*

Are some of the interest rates more responsive to certain measures of money supply than others? Is there any evidence to suggest that certain interest rates are unrelated to money supply? Professor Hicks wants to be fully prepared for the next class.

DEFINITIONS

M1	Currency in circulation, travelers checks, demand deposits, and other checkable deposits, in billions of dollars
M2	M1 plus retail money market mutual funds, savings accounts, and small time deposits, in billions of dollars
M3	M2 plus large time deposits, Eurodollars, and institutions-only money market mutual funds, in billions of dollars
PRIME	The interest rate banks charge their best customers for short-term loans
DISCOUNT	The interest rate charged on loans made by the New York Federal Reserve Bank to its member banks
HOME	The interest rate on 30-year, fixed-rate, conventional mortgage loans
T-BILL	Yield on three-month U.S. Treasury Bills, new issues

DATA SET

M1	M2	M3	PRIME	DISCOUNT	HOME	T-BILL
390.5	1487.5	1822.8	15.25	12.00	12.88	12.04
380.7	1487.4	1831.7	15.63	12.52	13.04	12.82
382.2	1498.4	1844.7	18.31	13.00	15.28	15.53
386.9	1509.1	1853.2	19.77	13.00	16.33	14.00
377.6	1505.2	1855.0	16.57	12.94	14.26	9.15
387.4	1528.7	1876.6	12.63	11.40	12.71	7.00
394.6	1550.3	1902.2	11.48	10.87	12.19	8.13
398.3	1561.9	1919.5	11.12	10.00	12.56	9.26
404.7	1574.3	1931.2	12.23	10.17	13.20	10.32
410.8	1589.8	1952.2	13.79	11.00	13.79	11.58
415.8	1600.4	1975.4	16.06	11.47	14.21	13.89
419.5	1606.2	1997.0	20.35	12.87	14.79	15.66
416.2	1611.8	2019.0	20.16	13.00	14.90	14.73
405.5	1611.9	2027.2	19.43	13.00	15.13	14.91
412.3	1634.5	2048.9	18.05	13.00	15.40	13.48
431.1	1665.8	2080.4	17.15	13.00	15.58	13.63
418.4	1656.8	2085.8	19.61	13.87	16.40	16.29
422.8	1668.5	2104.6	20.03	14.00	16.70	14.56

(continues on the following page)

M1	M2	M3	PRIME	DISCOUNT	HOME	T-BILL
427.6	1686.4	2131.7	20.39	14.00	16.83	14.70
425.9	1696.0	2153.4	20.50	14.00	17.29	15.61
427.0	1707.9	2172.0	20.08	14.00	18.16	14.95
429.6	1727.8	2195.9	18.45	14.00	18.45	13.87
435.0	1742.4	2218.5	16.84	13.03	17.83	11.27
447.0	1761.9	2245.1	15.75	12.10	16.92	10.93
448.5	1774.3	2265.8	15.75	12.00	17.40	12.41
432.4	1766.8	2263.9	16.56	12.00	17.60	13.78
435.6	1783.3	2286.2	16.50	12.00	17.16	12.49
451.1	1809.3	2316.4	16.50	12.00	16.89	12.82
440.9	1806.3	2319.5	16.50	12.00	16.68	12.15
446.3	1824.3	2338.8	16.50	12.00	16.70	12.11
449.4	1839.3	2356.8	16.26	11.81	16.82	11.92
449.8	1851.7	2383.6	14.39	10.68	16.27	9.01
456.2	1865.6	2398.2	13.50	10.00	15.43	8.20
465.8	1882.3	2423.1	12.52	9.68	14.61	7.75
474.4	1896.4	2438.0	11.85	9.35	13.83	8.04
485.8	1919.9	2450.4	11.50	8.73	13.62	8.02
482.8	1969.4	2479.3	11.16	8.50	13.25	7.81
474.1	1991.4	2495.3	10.98	8.50	13.04	8.13
482.7	2015.0	2518.0	10.50	8.50	12.80	8.30
498.5	2038.7	2543.5	10.50	8.50	12.78	8.25
493.9	2036.8	2547.9	10.50	8.50	12.63	8.19
503.5	2056.8	2569.2	10.50	8.50	12.87	8.82
510.5	2074.6	2585.7	10.50	8.50	13.42	9.12
508.3	2077.4	2598.8	10.89	8.50	13.81	9.39
511.3	2086.0	2614.4	11.00	8.50	13.73	9.05
517.1	2106.9	2637.2	11.00	8.50	13.54	8.71
521.9	2122.8	2667.3	11.00	8.50	13.44	8.71
533.2	2139.1	2695.5	11.00	8.50	13.42	8.96
530.2	2150.0	2709.1	11.00	8.50	13.37	7.14
516.9	2152.9	2720.9	11.00	8.50	13.23	9.03
523.3	2174.2	2752.7	11.21	8.50	13.39	9.08
539.0	2199.9	2786.9	11.93	8.87	13.65	9.69
530.7	2196.4	2804.2	12.39	9.00	13.94	9.90
541.4	2219.6	2832.5	12.60	9.00	14.42	9.94
543.4	2233.3	2857.8	13.00	9.00	14.67	10.13
539.0	2234.1	2872.5	13.00	9.00	14.47	10.49
542.5	2247.9	2892.2	12.97	9.00	14.35	10.41
542.2	2265.1	2920.5	12.58	9.00	14.13	9.97
549.8	2292.0	2954.7	11.77	8.83	13.64	8.79
564.6	2324.2	2992.9	11.06	8.37	13.18	8.16
561.1	2345.3	3014.0	10.61	8.00	13.08	7.76
551.9	2348.9	3019.7	10.50	8.00	12.92	8.17
558.4	2365.9	3039.9	10.50	8.00	13.17	8.57
575.1	2384.9	3051.1	10.50	8.00	13.20	8.00
569.4	2381.8	3056.2	10.31	7.81	12.91	7.56

(continues on the following page)

M1	M2	M3	PRIME	DISCOUNT	HOME	T-BILL
585.2	2419.2	3090.2	9.78	7.50	12.22	7.01
592.0	2440.4	3102.8	9.50	7.50	12.03	7.05
595.0	2448.2	3120.2	9.50	7.50	12.19	7.18
602.0	2458.5	3138.3	9.50	7.50	12.19	7.08
605.4	2472.2	3158.2	9.50	7.50	12.14	7.17
615.0	2487.8	3181.5	9.50	7.50	11.78	7.20
633.4	2510.1	3212.2	9.50	7.50	11.26	7.07
626.7	2516.7	3232.4	9.50	7.50	10.88	7.04
612.9	2507.9	3231.5	9.50	7.50	10.71	7.03
624.4	2533.2	3262.3	9.10	7.10	10.08	6.59
647.1	2570.0	3299.0	8.83	6.83	9.94	6.06
645.8	2576.7	3307.0	8.50	6.50	10.14	6.12
663.0	2611.1	3336.4	8.50	6.50	10.68	6.21
673.5	2638.7	3370.6	8.16	6.16	10.51	5.84
678.5	2652.3	3396.2	7.90	5.82	10.20	5.57
684.7	2668.3	3421.3	7.50	5.50	10.01	5.19
692.3	2691.2	3442.0	7.50	5.50	9.97	5.18
708.9	2711.9	3463.8	7.50	5.50	9.70	5.35
740.0	2747.3	3501.4	7.50	5.50	9.31	5.49
737.4	2760.1	3523.3	7.50	5.50	9.20	5.45
717.3	2741.8	3509.4	7.50	5.50	9.08	5.59
723.3	2754.3	3522.3	7.50	5.50	9.04	5.56
752.2	2784.8	3551.8	7.75	5.50	9.83	5.76
739.5	2765.6	3550.1	8.14	5.50	10.60	5.75
744.0	2779.5	3573.1	8.25	5.50	10.54	5.69
746.3	2791.7	3585.9	8.25	5.50	10.28	5.78
744.3	2793.9	3608.0	8.25	5.50	10.33	6.00
744.7	2799.8	3628.7	8.70	5.95	10.89	6.32
753.3	2817.4	3653.2	9.07	6.00	11.26	6.40
755.7	2828.3	3673.8	8.78	6.00	10.65	5.81
765.5	2845.4	3685.8	8.75	6.00	10.65	5.80
764.4	2864.6	3708.9	8.75	6.00	10.43	5.90
744.7	2866.2	3714.8	8.51	6.00	9.89	5.69
751.8	2894.1	3746.3	8.50	6.00	9.93	5.69
778.2	2932.3	3785.0	8.50	6.00	10.20	5.92
763.7	2917.3	3786.4	8.84	6.00	10.46	6.27
778.8	2942.1	3815.6	9.00	6.00	10.46	6.50
785.7	2961.1	3839.3	9.29	6.00	10.43	6.73
781.2	2957.8	3849.8	9.84	6.37	10.60	7.02
779.9	2958.2	3858.0	10.00	6.50	10.48	7.23
781.0	2970.4	3876.9	10.00	6.50	10.30	7.34
787.1	2991.0	3906.3	10.05	6.50	10.27	7.68
803.3	3008.7	3925.3	10.50	6.50	10.61	8.09
792.3	3007.2	3927.8	10.50	6.50	10.73	8.29
771.8	2990.7	3918.4	10.93	6.59	10.65	8.48
774.8	3006.9	3946.9	11.50	7.00	11.03	8.83
790.4	3030.5	3965.3	11.50	7.00	11.05	8.70
766.4	3003.5	3945.0	11.50	7.00	10.77	8.40

(continues on the following page)

M1	M2	M3	PRIME	DISCOUNT	HOME	T-BILL
773.1	3030.0	3976.8	11.07	7.00	10.20	8.22
781.0	3063.1	4007.1	10.98	7.00	9.88	7.92
776.8	3078.8	4016.3	10.50	7.00	9.99	7.91
778.0	3094.1	4019.3	10.50	7.00	10.13	7.72
783.6	3117.5	4032.2	10.50	7.00	9.95	7.63
790.5	3143.6	4059.3	10.50	7.00	9.77	7.65
811.0	3173.0	4077.9	10.50	7.00	9.74	7.64
801.3	3177.9	4075.1	10.11	7.00	9.90	7.64
787.1	3177.1	4068.2	10.00	7.00	10.20	7.76
795.0	3199.6	4084.3	10.00	7.00	10.27	7.87
816.8	3226.9	4097.5	10.00	7.00	10.37	7.78
796.0	3193.3	4072.0	10.00	7.00	10.48	7.78
809.5	3217.1	4091.6	10.00	7.00	10.16	7.74
811.5	3231.4	4104.0	10.00	7.00	10.04	7.66
813.1	3245.6	4118.8	10.00	7.00	10.10	7.44
817.6	3255.4	4117.3	10.00	7.00	10.18	7.38
816.6	3260.2	4119.2	10.00	7.00	10.18	7.19
825.1	3272.6	4125.1	10.00	7.00	10.01	7.07
843.2	3291.7	4138.5	10.00	6.79	9.67	6.81
832.3	3296.4	4151.1	9.52	6.50	9.64	6.30
822.6	3303.3	4164.1	9.05	6.00	9.37	5.95
834.2	3334.0	4183.0	9.00	6.00	9.50	5.91
852.4	3356.0	4193.8	9.00	5.98	9.49	5.67
840.9	3335.6	4167.7	8.50	5.50	9.47	5.51
857.3	3355.2	4176.0	8.50	5.50	9.62	5.60
861.5	3361.1	4170.4	8.50	5.50	9.58	5.58
863.7	3357.3	4169.8	8.50	5.50	9.24	5.39
866.5	3352.8	4153.6	8.20	5.20	9.01	5.25
874.5	3359.0	4158.7	8.00	5.00	8.86	5.03
892.8	3376.4	4181.2	7.58	4.58	8.71	4.60
916.0	3393.2	4195.2	7.21	4.11	8.50	4.12
916.9	3390.3	4190.5	6.50	3.50	8.43	3.84
915.5	3396.6	4202.7	6.50	3.50	8.76	3.84
930.0	3412.8	4212.1	6.50	3.50	8.94	4.05
954.1	3426.8	4210.4	6.50	3.50	8.85	3.81
943.3	3395.9	4181.7	6.50	3.50	8.67	3.66
951.4	3398.0	4183.3	6.50	3.50	8.51	3.70
962.4	3402.4	4183.0	6.02	3.02	8.13	3.28
970.4	3405.1	4199.2	6.00	3.00	7.98	3.14
983.0	3410.2	4194.4	6.00	3.00	7.92	2.97
1001.2	3426.9	4196.0	6.00	3.00	8.09	2.84
1021.8	3440.4	4209.1	6.00	3.00	8.31	3.14
1046.0	3450.4	4207.8	6.00	3.00	8.22	3.25
1040.2	3433.8	4178.5	6.00	3.00	8.02	3.06
1022.3	3410.2	4165.9	6.00	3.00	7.68	2.95
1030.8	3420.7	4174.8	6.00	3.00	7.50	2.97
1057.8	3441.6	4194.8	6.00	3.00	7.47	2.89
1057.2	3435.7	4198.4	6.00	3.00	7.47	2.96

(continues on the following page)

M1	M2	M3	PRIME	DISCOUNT	HOME	T-BILL
1072.2	3450.2	4206.1	6.00	3.00	7.42	3.10
1083.6	3453.8	4203.2	6.00	3.00	7.21	3.05
1088.6	3453.1	4207.7	6.00	3.00	7.11	3.05
1099.2	3455.8	4209.6	6.00	3.00	6.92	2.96
1112.0	3464.5	4225.3	6.00	3.00	6.83	3.04
1129.5	3486.2	4254.7	6.00	3.00	7.16	3.12
1153.7	3506.6	4274.8	6.00	3.00	7.17	3.08
1142.2	3494.2	4263.6	6.00	3.00	7.06	3.02
1124.1	3475.1	4231.6	6.00	3.00	7.15	3.21
1131.4	3492.8	4248.5	6.06	3.00	7.68	3.52
1152.5	3520.7	4275.2	6.45	3.00	8.32	3.74
1132.4	3490.9	4252.0	6.99	3.24	8.60	4.19
1141.9	3491.7	4262.1	7.25	3.50	8.40	4.18
1150.7	3504.7	4282.3	7.25	3.50	8.61	4.39
1144.1	3495.7	4282.6	7.51	3.76	8.51	4.50
1146.6	3491.3	4283.7	7.75	4.00	8.64	4.64
1148.0	3492.9	4298.2	7.75	4.00	8.93	4.96
1156.1	3506.1	4323.3	8.15	4.40	9.17	5.25
1174.4	3522.5	4347.4	8.50	4.75	9.20	5.64
1159.3	3510.2	4353.8	8.50	4.75	9.15	5.81
1135.1	3489.7	4342.4	9.00	5.25	8.83	5.80
1139.1	3506.6	4369.7	9.00	5.25	8.46	5.73
1160.1	3536.9	4407.1	9.00	5.25	8.32	5.67
1133.8	3520.9	4412.0	9.00	5.25	7.96	5.70
1140.8	3560.8	4460.1	9.00	5.25	7.57	5.50
1145.6	3586.6	4491.7	8.80	5.25	7.61	5.47
1139.3	3603.6	4524.4	8.75	5.25	7.86	5.41
1138.5	3613.5	4539.3	8.75	5.25	7.64	5.26
1132.9	3621.5	4561.5	8.75	5.25	7.48	5.30
1138.7	3643.2	4587.1	8.75	5.25	7.38	5.35
1152.8	3675.3	4612.0	8.65	5.25	7.20	5.16
1130.1	3673.3	4626.5	8.50	5.24	7.03	5.02
1105.7	3668.6	4641.0	8.25	5.00	7.08	4.87
1117.8	3715.7	4691.9	8.25	5.00	7.62	4.96
1131.3	3741.6	4715.7	8.25	5.00	7.93	4.99
1105.3	3709.1	4712.7	8.25	5.00	8.07	5.02
1114.2	3739.2	4748.7	8.25	5.00	8.32	5.11
1109.8	3753.1	4767.5	8.25	5.00	8.25	5.17
1096.5	3761.0	4788.1	8.25	5.00	8.00	5.09
1090.2	3762.5	4805.4	8.25	5.00	8.23	5.15
1076.9	3769.4	4845.4	8.25	5.00	7.92	5.01
1085.3	3801.2	4884.2	8.25	5.00	7.62	5.03
1103.1	3837.7	4935.0	8.25	5.00	7.60	4.87
1085.9	3836.3	4945.6	8.25	5.00	7.82	5.05
1066.5	3834.0	4968.2	8.25	5.00	7.65	5.00
1067.4	3870.0	5015.2	8.30	5.00	7.90	5.14
1071.8	3900.1	5055.1	8.50	5.00	8.14	5.17

(concludes on the following page)

M1	M2	M3	PRIME	DISCOUNT	HOME	T-BILL
1051.8	3863.6	5036.3	8.50	5.00	7.94	5.13
1062.4	3892.1	5070.1	8.50	5.00	7.69	4.92
1063.7	3910.3	5115.3	8.50	5.00	7.50	5.07
1067.5	3944.0	5170.8	8.50	5.00	7.48	5.13
1057.7	3952.4	5197.3	8.50	5.00	7.43	4.97
1054.6	3965.4	5244.3	*	*	*	4.93

CASE **24**

SAT SCORES

Many administrators, teachers, and parents perceive a "crisis in education" in the United States. Learned skills and test scores suggest lower achievement by American students both over time and compared to their counterparts around the world. Much of the concern is whether our children will be able to compete in a global economy if they are underprepared to meet the challenges of the coming decades.

Suggested solutions to the crisis in educational achievements are varied. Some argue that instilling a sense of achievement, goal-orientation, and hard work is crucial for motivating our students. Others suggest that the key is to commit more financial resources to the educational system.

Using the SAT scores of high school students as a measure of educational achievement, does it appear that the level of financial resources committed to schools in the U.S. influences educational success? If not, what does influence SAT scores of U.S. students? The data file named SAT contains some information that may assist you in answering these questions. This data set contains 51 observations and 10 variables; these represent SAT scores and other variables for each of the 50 U. S. states and the District of Columbia.

DEFINITIONS

SAT	Average combined verbal/math scores on the Scholastic Assessment Test (ranges from 400 to 1600) for students taking the exam during the 1994–95 school year

An earlier version of these data were compiled by Lisa Radovich, who used the data set in a course project. The data come from U.S. Department of Education, National Center for Education Statistics, Digest of Education Statistics, NCES 96-122, Washington D.C. 1996; and College Entrance Examination Board, "College-Bound Seniors: 1995 Profile of SAT Program Test Takers," 1995.

EXPEND	Current expenditures for public elementary and secondary education for the 1995–96 school year, in billions of dollars
TAKERS	Percentage of 1994–95 school year graduates who took the SAT. This number is calculated using the number of high school graduates in 1995 and the number of 1995 seniors who took the SAT exam.
ENROLL	Total enrollment in public elementary and secondary schools in the United States; fall 1995 figures in thousands of students
TEACHERS	Total number of public elementary and secondary teachers in the United States; fall 1995 figures in thousands of full-time equivalents (FTEs)
RATIO	Student-to-teacher ratio (the average number of students taught per teacher) in the United States, fall 1995. Calculated as ENROLL ÷ TEACHERS
SALARY	Estimated average annual salary of teachers in public elementary and secondary schools for the 1995–96 school year, in thousands of current dollars
STATE	1 = Alabama 2 = Alaska 3 = Arizona 4 = Arkansas 5 = California 6 = Colorado 7 = Connecticut 8 = Washington, D.C. 9 = Delaware 10 = Florida 11 = Georgia 12 = Hawaii 13 = Idaho 14 = Illinois 15 = Indiana 16 = Iowa 17 = Kansas 18 = Kentucky 19 = Louisiana 20 = Maine 21 = Maryland 22 = Massachusetts 23 = Michigan 24 = Minnesota 25 = Mississippi 26 = Missouri 27 = Montana 28 = Nebraska 29 = Nevada 30 = New Hampshire 31 = New Jersey 32 = New Mexico 33 = New York 34 = North Carolina 35 = North Dakota 36 = Ohio 37 = Oklahoma 38 = Oregon 39 = Pennsylvania 40 = Rhode Island 41 = South Carolina 42 = South Dakota 43 = Tennessee 44 = Texas 45 = Utah 46 = Vermont 47 = Virginia 48 = Washington 49 = West Virginia 50 = Wisconsin 51 = Wyoming

DATA SET

STATE	SAT	EXPEND	TAKERS	ENROLL	TEACHERS	RATIO	SALARY
1	1029	3.162	8	735.95	44.1	16.7	31.3
2	934	1.173	47	125.26	7.4	16.9	49.6
3	944	3.394	27	766.50	39.6	19.4	32.5

(concludes on the following page)

STATE	SAT	EXPEND	TAKERS	ENROLL	TEACHERS	RATIO	SALARY
4	1005	1.497	6	454.28	29.2	15.6	29.3
5	902	26.842	45	5447.85	225.0	24.2	42.5
6	980	3.338	29	656.28	35.2	18.6	35.4
7	908	4.094	81	514.63	36.0	14.3	50.4
8	897	0.753	68	108.46	6.5	16.7	40.5
9	857	0.503	53	79.80	6.4	12.5	43.7
10	889	11.767	48	2172.79	116.8	18.6	33.3
11	854	6.271	65	1311.13	80.2	16.3	34.3
12	889	0.808	57	186.57	10.2	18.3	35.8
13	979	1.031	15	243.10	12.8	19.0	30.9
14	1048	10.419	13	1927.52	113.0	17.1	41.0
15	882	5.670	58	980.20	55.6	17.6	37.8
16	1099	2.728	5	502.30	32.5	15.5	32.4
17	1060	2.588	9	464.09	30.9	15.0	35.5
18	999	3.683	11	638.63	37.2	17.2	33.0
19	1021	3.499	9	781.14	47.6	16.4	26.8
20	896	1.282	68	219.23	14.6	15.0	32.9
21	909	5.238	64	805.58	46.2	17.4	41.2
22	907	6.302	80	910.02	61.6	14.8	43.8
23	1033	10.140	11	1643.10	80.8	20.3	49.2
24	1085	5.788	9	835.42	48.4	17.3	36.9
25	1036	2.015	4	503.60	28.8	17.5	27.7
26	1045	3.879	9	873.64	57.7	15.1	33.3
27	1009	0.909	21	165.50	10.1	16.4	29.4
28	1050	1.498	9	289.73	20.0	14.5	31.5
29	917	1.247	30	265.04	13.7	19.3	36.2
30	935	1.195	70	190.45	12.2	15.6	35.8
31	898	11.720	70	1197.56	86.9	13.8	47.9
32	1015	2.065	11	328.46	19.3	17.0	29.3
33	892	24.600	74	2830.00	185.9	15.2	48.1
34	865	6.114	60	1165.39	71.8	16.2	30.6
35	1107	0.552	5	119.09	7.5	15.9	27.0
36	975	10.260	23	1838.41	104.2	17.6	37.8
37	1027	2.331	9	616.50	39.3	15.7	28.9
38	947	3.028	51	527.91	26.7	19.8	39.7
39	880	12.916	70	1801.97	104.6	17.2	46.9
40	888	1.098	70	148.08	10.2	14.5	42.2
41	844	2.970	58	637.52	39.2	16.3	31.6
42	1068	0.658	5	144.11	9.3	15.5	26.3
43	1040	3.763	12	880.96	50.7	17.4	33.5
44	893	19.330	47	3740.26	240.2	15.6	32.0
45	1076	1.738	4	473.67	20.1	23.6	30.5
46	901	0.682	68	105.96	7.7	13.8	36.3
47	896	5.957	65	1079.85	73.3	14.7	34.7
48	937	5.585	48	951.70	45.3	21.0	38.0
49	932	1.863	17	306.45	20.9	14.7	32.2
50	1073	5.755	9	869.17	56.2	15.5	38.6
51	1001	0.580	10	99.86	6.7	14.9	31.6

CASE **25**

EMERGENCY ADMISSIONS

The emergency department at "General Hospital" is a very busy place, particularly at certain times of the day, certain days of the week, and different months of the year—or so it seems to Cindy, the director of the emergency department at General Hospital. Among other tasks, preparing the weekly staffing schedule is one of Cindy's responsibilities. This is a challenging job, because she needs to make sure that enough nurses, physicians, and other staff members are available to handle incoming emergencies. Yet one of the obvious problems facing all hospitals these days is the budget crunch; Cindy does not want to overstaff the emergency department and pay personnel to stand around with nothing to do.

To understand general trends in admissions to the emergency department, Cindy has collected some historical data from hospital records. In the file named EDADMITS are the data on daily admissions to the emergency department over the period January 1 (a Wednesday) to August 28. (Records are missing for a couple of days, as indicated by an asterisk in the data set.) Cindy has hired you to study the data and provide a forecasting model of the number of admissions to be expected on a daily basis.

Your task is twofold. First, derive a forecasting model and use your model to forecast the number of admissions for August 29, 30, and 31. Explain your model to Cindy in nontechnical terms. Second, prepare a worksheet that Cindy might use to create her own forecasts.

These are real data collected from an emergency department, although we have disguised the name of the institution. These data come from a study project originally prepared by Carolyn Briggs, Christopher Jordan, Christie Martinez, and Joann Schauer.

DATA SET

Read across the row

115	121	118	111	109	103	75	114	103
117	128	76	97	106	78	108	114	124
83	99	122	98	81	128	141	101	121
110	108	106	131	147	104	125	124	121
123	132	145	102	116	114	143	108	140
111	118	117	134	130	104	128	132	122
108	112	112	*	141	129	125	136	134
107	118	137	121	87	112	132	*	111
135	114	111	130	121	99	*	*	105
88	*	*	109	116	112	120	119	118
*	116	134	149	127	110	129	135	137
116	142	128	108	135	126	133	132	164
134	92	127	129	133	148	*	139	*
138	137	129	121	133	135	137	149	130
124	146	151	150	104	127	107	125	133
162	139	96	118	134	149	130	128	146
134	117	143	113	118	150	152	121	143
146	137	121	134	157	128	121	128	133
139	158	156	148	126	157	160	134	143
159	126	138	164	137	132	147	151	104
138	152	127	129	148	149	143	152	126
107	156	138	138	147	147	137	134	140
125	135	102	141	135	136	158	141	146
101	148	143	134	102	127	127	121	179
145	158	141	158	158	128	148	137	127
134	140	136	105	146	135	129	120	129
138	129	146	112	106	105	128		

CASE **26**

DAVE'S TRUCK WASHING SERVICE

Dave is the owner of the Truck Washing Service (TWS). TWS provides cleaning services to truckers and trucking operations within a moderately sized metropolitan area. Dave is interested in understanding the factors that influence monthly fluctuations in total sales of TWS.

TWS offers two types of services. Type 1 sales are from Dave's mobile truck washing service. Dave also provides a mobile chemical treatment service for removing decals and logos from trucks, called type 2 sales. Most of TWS revenues come from the type 1 operations. Occasionally, there will be an upcoming auction of used trucks in town. Trucks are generally cleaned up prior to auctions, and Dave can pick up extra business at auction time.

Dave does all of the marketing for TWS and also handles the administrative and financial side of the business. When necessary, he occasionally works in the field. Dave started out with only one other full-time employee. Later, he was able to add an additional part-time employee to the payroll.

Dave has recently become sensitive to the old adage, "While the cat's away, the mice will play." Sales seem to drop pretty dramatically when Dave is on vacation. Dave has also recently enrolled in an MBA program, which reduces the amount of time that he can commit to his business. During the semester, Dave suspects that business drops off.

The data file is named TWS. This file contains 46 monthly time-series observations and eight variables. The time-series begins in May. The definitions of the variables on that file are SALES, the total monthly sales of TWS in dollars; SALES2, the monthly sales of type 2 in dollars; TEMP, the average monthly temperature (degrees Fahrenheit);

These data are real, but the name of the company has been changed. The data were made available to us by David S. Larson and Carmin Hunter-Siegert. Carmin is with the Colorado Student Obligation Bond Authority. We would also like to thank Dave and Carmin for coming to several of our classes to discuss the data set and their study with our students.

and PREC, the total monthly precipitation in the metro area in inches. CAT and SCHOOL are binary variables indicating whether Dave was unavailable to work at TWS that month. CAT takes on a value of 0 if Dave was not vacationing that month, and 1 if he was; SCHOOL is 0 if Dave was not in school that month, and 1 if he was. EMPLOY represents the number of employees on the payroll that month. If there was an auction in town, the variable AUCTION is 1 (0 for no auction).

Dave has hired you to look at the data for him. He thinks the data can shed some light on two general areas:

1 Factors that influence fluctuations in revenues of TWS. Provide an overview of the data. Does it appear that Dave is right about his absenteeism hurting the business? What other factors influence fluctuations in revenues of TWS?

2 Future values of TWS revenues. Dave would like to try his hand at forecasting future values of TWS revenues. If he is successful at doing this, budgeting for employee hours, equipment, and chemicals will be much easier.

For the second task, construct a forecasting model of total sales for TWS. Use this model to predict total sales for the upcoming months of March and April.

Write a business memo to Dave that addresses the issues raised above. Without relying on any jargon, explain to Dave how he might use your forecasting model to predict revenues later in the year.

DATA SET

SALES	SALES2	TEMP	PREC	CAT	SCHOOL	EMPLOY	AUCTION
4214	262	59.0	4.26	0	0	2.0	0
5595	560	71.9	1.28	0	0	2.0	0
3274	175	74.2	2.19	1	0	2.0	0
6075	420	73.6	1.83	0	0	2.0	1
4354	259	62.3	9.90	0	0	2.0	0
3994	200	51.3	0.06	1	0	2.0	0
6256	335	42.8	0.47	0	0	2.0	1
4150	540	27.3	1.04	0	0	2.0	0
4277	23	33.5	1.14	0	0	2.0	0
3955	812	22.4	0.66	0	0	2.0	0
6986	1486	43.3	0.56	0	0	2.0	0
5940	939	51.1	1.00	0	0	2.0	0
2973	261	59.0	3.83	0	0	2.0	0
5687	333	65.4	2.04	0	0	1.5	0
4779	95	75.9	1.64	1	0	1.0	0
5321	367	71.7	1.28	0	0	1.0	0
4849	1085	62.5	1.55	0	0	1.0	0
6484	525	51.3	0.81	0	0	1.0	0
5569	89	42.8	0.15	1	0	1.0	0
3834	1245	27.3	0.81	0	0	2.0	1
5469	188	36.4	0.74	1	0	2.0	0
5863	580	33.3	0.55	0	0	2.0	0
6148	175	39.5	3.10	0	0	2.0	0
5578	353	49.1	1.01	0	0	2.0	0

(concludes on the next page)

SALES	SALES2	TEMP	PREC	CAT	SCHOOL	EMPLOY	AUCTION
8318	1377	56.6	1.51	0	0	2.0	0
7132	663	72.6	0.21	0	0	2.0	1
6888	95	70.8	3.57	0	0	2.0	0
8301	200	71.3	1.96	0	0	2.0	1
6838	432	66.9	1.46	0	0	2.0	0
6145	230	52.3	1.03	0	0	2.0	0
6470	325	44.0	1.28	1	0	2.0	0
4580	0	25.7	0.27	1	0	2.0	1
6555	100	27.9	0.76	0	0	2.5	0
6540	205	40.3	0.08	0	0	2.5	0
5874	230	42.8	0.76	0	0	2.5	0
5692	0	47.8	1.94	1	0	2.5	0
5922	475	60.3	2.43	1	0	2.5	0
5189	1100	69.4	2.20	1	0	2.5	0
5458	347	73.1	4.11	0	0	2.5	0
10165	2657	71.5	3.69	0	0	2.5	1
5829	631	62.5	0.79	0	1	2.5	0
4210	460	50.3	0.70	1	1	2.5	0
5066	784	34.7	2.67	0	1	2.5	1
6960	885	33.1	0.19	0	0	2.5	0
5484	378	31.8	1.19	0	1	2.5	0
5892	1020	40.1	0.10	0	1	2.5	0

CASE **27**

HOUSING PRICES—I

You currently own a house in Eastville, Oregon, and are planning on moving sometime in the next year. A friend of yours has suggested that you try to sell the house on your own. The decision is a difficult one. Real estate commissions can be quite hefty, but brokers bring a great deal of expertise to the process of selling a house, particularly the ability to set a reasonable asking price. Real estate professionals can quantify a variety of unobservable or unmeasurable characteristics of a house, such as that shiny purple wallpaper in the upstairs bathroom and the overall desirability of the neighborhood.

To get an initial feel for the types of houses in your area, you have collected the data in the file named HOUSES. This file contains information such as the selling price of the house, square feet, numbers of bedrooms and bathrooms, and the presence of a fireplace for 108 homes sold in your neighborhood last winter. In the data set:

SQ_FT is the variable measuring the total square feet in the house. BEDS and BATHS are number of bedrooms and bathrooms, respectively. HEAT and STYLE are categorical variables. HEAT takes on the value of 0 for gas forced air heating and 1 for electric heat. STYLE is the architectural style of the home: 0 indicates a trilevel, 1 indicates a two-story house, and 2 indicates that the house is a ranch-styled home. GARAGE is the number of cars that can fit into the garage. We don't know whether it is an attached garage. AGE is the age of the home in years. FIRE and BASEMENT indicate the presence (1) or absence (0) of at least one fireplace or basement. PRICE is the selling price of the house in thousands of dollars and SCHOOL is the school district (0 = Eastville school district; 1 = Apple Valley school district). All else the same, Apple Valley is viewed as the preferred school district.

These data were used by Ellen L. Chilikas, David S. Abelson, and Brian R. Landry to price a house owned by one of the group members as part of a student project. The name of the suburb from which the data were taken has been changed.

Your task is to use these data to answer a couple of questions.

- What is the nature of the selling prices of these houses?
- What is the breakout of selling prices in terms of architectural style?
- How does the existence of a basement contribute to selling price?
- How does the existence of a fireplace add to the selling price?
- What other interesting or useful information do you see in this data set?

DATA SET

SQ_FT	BEDS	BATHS	HEAT	STYLE	GARAGE	BASEMENT	AGE	FIRE	PRICE	SCHOOL
1238	3	2	0	0	1	1	12	1	59.900	1
1707	3	2	1	0	2	0	13	1	64.000	0
1296	4	2	0	0	2	1	17	0	66.500	0
1320	3	2	0	0	2	1	11	1	66.500	0
1210	3	2	0	0	1	0	6	1	66.900	0
1296	3	2	0	0	2	1	17	1	68.000	0
1765	3	2	0	0	2	1	20	0	68.500	0
1725	4	3	0	0	2	1	12	0	69.000	0
1794	4	2	0	0	2	1	18	0	70.950	0
1294	3	2	0	0	2	0	13	1	71.000	0
1372	3	2	0	0	2	1	9	0	72.692	1
1162	3	2	0	0	1	0	8	1	72.801	0
1996	4	2	0	0	2	0	13	1	75.207	1
1764	4	2	1	0	2	1	13	1	76.000	0
1416	3	2	0	0	2	0	8	0	76.000	1
1730	4	2	0	0	2	0	15	1	77.500	0
1392	3	2	0	0	2	1	8	1	79.900	1
1664	3	3	0	0	2	0	11	1	79.900	0
1332	3	2	0	0	2	1	14	0	81.000	1
1752	3	3	0	0	2	0	18	1	82.800	0
2167	3	3	1	0	2	1	13	1	84.900	0
1664	3	2	0	0	2	0	9	1	85.000	0
1973	4	3	0	0	2	0	13	1	86.000	0
1384	3	2	0	0	2	0	5	1	89.280	0
1431	3	2	0	0	2	1	7	1	89.900	1
1950	5	3	0	0	2	0	13	1	90.000	0
1452	3	2	0	0	2	1	4	1	92.000	0
1829	3	2	0	0	2	0	10	1	92.439	1
1652	4	3	0	0	2	1	7	1	94.646	1
1516	3	2	0	0	2	1	10	1	97.293	1
1998	4	3	0	0	2	1	17	1	98.100	1
1984	4	3	0	0	2	1	9	1	98.149	0
1840	3	3	0	0	2	1	9	1	105.000	0
1823	4	3	0	0	2	1	3	1	110.000	1
2150	5	3	0	0	2	1	12	1	111.900	0
2096	3	3	0	0	2	1	9	1	113.000	1
2212	4	3	0	0	3	1	17	1	124.000	1

(continues on the following page)

SQ_FT	BEDS	BATHS	HEAT	STYLE	GARAGE	BASEMENT	AGE	FIRE	PRICE	SCHOOL
2375	4	3	0	0	2	1	11	1	133.000	1
2809	4	3	0	0	3	1	6	1	195.000	0
912	4	2	0	1	2	1	19	1	59.000	0
816	3	2	0	1	2	1	19	0	61.500	0
1008	5	2	0	1	2	1	17	0	63.500	0
912	4	2	0	1	2	1	16	1	66.950	1
1008	5	2	0	1	2	1	17	1	68.000	0
838	4	2	0	1	1	1	11	0	68.694	1
1008	4	3	0	1	2	1	19	1	69.000	0
1120	3	2	0	1	2	1	11	1	70.452	1
1242	4	3	0	1	2	1	12	0	75.000	0
912	4	2	0	1	2	1	18	1	76.900	1
1316	3	2	0	1	2	1	8	1	83.000	0
1490	3	2	0	1	2	1	13	1	88.000	0
1118	4	3	0	1	2	1	16	1	88.879	1
1278	3	2	0	1	2	1	11	1	89.347	0
1490	3	2	0	1	2	1	10	1	89.700	1
1418	4	2	0	1	2	1	14	1	90.000	1
1250	4	3	0	1	2	1	17	1	90.800	1
1881	4	3	0	1	2	1	9	1	91.500	1
1355	3	2	0	1	2	1	7	1	95.000	1
1518	3	2	0	1	2	1	22	0	98.000	1
1606	4	3	0	1	2	1	13	1	99.500	0
2124	3	2	0	1	2	1	14	1	107.000	0
2099	4	3	0	1	2	1	13	1	113.750	1
2232	5	3	0	1	2	1	9	1	125.000	1
1608	5	3	0	1	3	1	13	1	132.000	1
1320	3	2	0	2	2	1	13	1	76.000	1
1680	3	3	0	2	2	0	7	1	79.900	1
1434	3	3	0	2	2	1	12	1	79.900	1
1664	3	2	0	2	2	1	7	1	85.000	1
1379	3	2	0	2	2	1	8	1	87.158	1
1634	3	3	0	2	2	1	10	1	87.200	1
1400	3	3	0	2	2	0	1	1	91.500	1
1664	4	4	0	2	2	1	16	1	91.500	1
1548	3	3	0	2	2	1	8	1	92.500	1
2040	4	3	0	2	2	1	15	1	93.500	0
1792	4	3	0	2	2	1	9	1	94.000	1
1865	3	3	0	2	2	1	12	1	96.700	0
1833	3	3	0	2	2	1	7	1	97.400	1
1627	3	3	0	2	2	1	3	1	98.500	1
2198	4	3	0	2	2	1	12	1	99.000	0
1707	3	3	0	2	2	1	12	1	99.000	0
2096	3	3	0	2	2	1	7	1	102.400	1
2004	4	3	0	2	2	1	13	1	104.000	0
1818	4	3	0	2	2	1	19	1	107.000	1

(concludes on the following page)

SQ_FT	BEDS	BATHS	HEAT	STYLE	GARAGE	BASEMENT	AGE	FIRE	PRICE	SCHOOL
1935	3	3	0	2	2	1	11	1	107.900	1
2381	3	3	0	2	2	0	17	1	109.000	1
1977	3	3	0	2	2	1	9	1	110.000	1
2317	5	3	0	2	2	1	9	1	110.500	1
2072	4	3	0	2	2	1	11	1	112.000	1
2108	3	3	0	2	2	1	7	1	112.900	0
2060	4	4	0	2	2	1	9	1	113.600	1
2064	3	3	0	2	2	1	5	1	114.500	1
2148	5	4	0	2	2	1	8	1	114.700	1
2170	6	4	0	2	2	1	21	1	115.000	1
1984	4	3	1	2	2	1	11	1	121.000	1
2130	4	3	0	2	2	1	20	1	121.000	1
2049	4	3	0	2	2	1	19	1	125.000	1
2035	3	3	0	2	3	1	6	1	128.000	1
2264	3	3	0	2	3	1	7	1	130.000	1
2264	3	3	0	2	3	1	10	1	131.000	1
2362	4	3	0	2	2	1	6	1	132.000	0
2262	5	3	0	2	2	1	12	1	132.500	1
2353	4	3	0	2	2	1	5	1	132.500	0
2298	3	3	0	2	3	1	2	1	144.500	1
2282	4	3	0	2	2	1	9	1	150.577	1
2542	4	3	0	2	2	1	6	1	153.800	1
2380	4	3	0	2	2	1	6	1	155.000	1
2505	4	3	0	2	2	1	7	1	156.900	1
2804	4	3	0	2	3	1	1	1	192.000	1

CASE **28**

HOUSING PRICES—II

You currently own a home in Eastville, Oregon, and want to put your house on the market. You're not in any particular hurry to get rid of the house, and would like to try selling it yourself. One way to determine a reasonable asking price of a house is to call one or more real estate agents and seek their advice. Another is to hire an appraiser—this approach would cost several hundred dollars. You've been wondering if there might be an easier and cheaper way to understand what determines selling prices in the area.

On your daily after-dinner stroll through the neighborhood, you've passed several houses on the market. At first, you hoped that you could get a feel for the market by simply studying these few houses. But the more you think about this problem, the more confused you get. You're pretty sure that bigger homes sell for more, but you don't know how *much* more. Also, it seems as if fireplaces would be a desirable amenity in Oregon; your house doesn't have a fireplace, and you're not sure how much to lower the asking price because of this. You've also heard that location is the most important thing in the real estate market, and you wonder if prices tend to vary by school district, one indication of the quality of the neighborhood.

A friend has collected some information for you about selling prices and other characteristics of houses sold within the last few months in your neighborhood. These include square feet, number of bedrooms and bathrooms, type of heating, and school district. This data set is shown below. At first, you decided that your problem was solved—you would only need to find a house just like yours in the data set, and use its price as your guess.

Of course, it's never that easy. Although you know the specific characteristics of your house, such as square feet, number of bedrooms and bathrooms, and so forth, *not*

These data were used by Ellen L. Chilikas, David S. Abelson, and Brian R. Landry to price a house owned by one of the group members as part of a student project. The name of the suburb from which the data were taken has been changed.

one of the houses in the data set is a perfect match for your own home. Some of the houses in the data set come *close* to yours, but every comparable house is different from yours in at least three aspects. You're wondering if there might not be another way to use the data to come up with an estimate.

The data are contained in the file named HOUSES, which includes information on 108 houses. In this data set, SQ_FT is the variable measuring the total square feet in the house. BEDS and BATHS are number of bedrooms and bathrooms, respectively. HEAT and STYLE are categorical variables. HEAT takes on the value of 0 for gas forced air heating and 1 for electric heat. STYLE is the architectural style of the home: 0 indicates a trilevel, 1 indicates a two-story house, and 2 indicates that the house is a ranch-styled home. GARAGE is the number of cars that might fit in the garage, although you don't know whether the garage is attached. AGE is the age of the house in years. FIRE indicates the presence (FIRE = 1) or absence (FIRE = 0) of a fireplace and BASEMENT indicates the presence (BASEMENT = 1) or absence (BASEMENT = 0) of a basement. PRICE is the selling price of the house in thousands of dollars and SCHOOL is the school district (0 = Eastville school district; 1 = Apple Valley school district). All else the same, Apple Valley is viewed as being the more desirable of the two school districts.

Use these data to uncover the important determinants of selling prices of houses in your neighborhood. Then prepare a description of how your findings might be used as a general method for estimating the selling price of *any* house in your neighborhood, such as yours.

DATA SET

SQ_FT	BEDS	BATHS	HEAT	STYLE	GARAGE	BASEMENT	AGE	FIRE	PRICE	SCHOOL
1238	3	2	0	0	1	1	12	1	59.900	1
1707	3	2	1	0	2	0	13	1	64.000	0
1296	4	2	0	0	2	1	17	0	66.500	0
1320	3	2	0	0	2	1	11	1	66.500	0
1210	3	2	0	0	1	0	6	1	66.900	0
1296	3	2	0	0	2	1	17	1	68.000	0
1765	3	2	0	0	2	1	20	0	68.500	0
1725	4	3	0	0	2	1	12	0	69.000	0
1794	4	2	0	0	2	1	18	0	70.950	0
1294	3	2	0	0	2	0	13	1	71.000	0
1372	3	2	0	0	2	1	9	0	72.692	1
1162	3	2	0	0	1	0	8	1	72.801	0
1996	4	2	0	0	2	0	13	1	75.207	1
1764	4	2	1	0	2	1	13	1	76.000	0
1416	3	2	0	0	2	0	8	0	76.000	1
1730	4	2	0	0	2	0	15	1	77.500	0
1392	3	2	0	0	2	1	8	1	79.900	1
1664	3	3	0	0	2	0	11	1	79.900	0
1332	3	2	0	0	2	1	14	0	81.000	1
1752	3	3	0	0	2	0	18	1	82.800	0
2167	3	3	1	0	2	1	13	1	84.900	0
1664	3	2	0	0	2	0	9	1	85.000	0

(continues on the next page)

SQ_FT	BEDS	BATHS	HEAT	STYLE	GARAGE	BASEMENT	AGE	FIRE	PRICE	SCHOOL
1973	4	3	0	0	2	0	13	1	86.000	0
1384	3	2	0	0	2	0	5	1	89.280	0
1431	3	2	0	0	2	1	7	1	89.900	1
1950	5	3	0	0	2	0	13	1	90.000	0
1452	3	2	0	0	2	1	4	1	92.000	0
1829	3	2	0	0	2	0	10	1	92.439	1
1652	4	3	0	0	2	1	7	1	94.646	1
1516	3	2	0	0	2	1	10	1	97.293	1
1998	4	3	0	0	2	1	17	1	98.100	1
1984	4	3	0	0	2	1	9	1	98.149	0
1840	3	3	0	0	2	1	9	1	105.000	0
1823	4	3	0	0	2	1	3	1	110.000	1
2150	5	3	0	0	2	1	12	1	111.900	0
2096	3	3	0	0	2	1	9	1	113.000	1
2212	4	3	0	0	3	1	17	1	124.000	1
2375	4	3	0	0	2	1	11	1	133.000	1
2809	4	3	0	0	3	1	6	1	195.000	0
912	4	2	0	1	2	1	19	1	59.000	0
816	3	2	0	1	2	1	19	0	61.500	0
1008	5	2	0	1	2	1	17	0	63.500	0
912	4	2	0	1	2	1	16	1	66.950	1
1008	5	2	0	1	2	1	17	1	68.000	0
838	4	2	0	1	1	1	11	0	68.694	1
1008	4	3	0	1	2	1	19	1	69.000	0
1120	3	2	0	1	2	1	11	1	70.452	1
1242	4	3	0	1	2	1	12	0	75.000	0
912	4	2	0	1	2	1	18	1	76.900	1
1316	3	2	0	1	2	1	8	1	83.000	0
1490	3	2	0	1	2	1	13	1	88.000	0
1118	4	3	0	1	2	1	16	1	88.879	1
1270	3	2	0	1	2	1	11	1	89.347	0
1490	3	2	0	1	2	1	10	1	89.700	1
1418	4	2	0	1	2	1	14	1	90.000	1
1250	4	3	0	1	2	1	17	1	90.800	1
1881	4	3	0	1	2	1	9	1	91.500	1
1355	3	2	0	1	2	1	7	1	95.000	1
1518	3	2	0	1	2	1	22	0	98.000	1
1606	4	3	0	1	2	1	13	1	99.500	0
2124	3	2	0	1	2	1	14	1	107.000	0
2099	4	3	0	1	2	1	13	1	113.750	1
2232	5	3	0	1	2	1	9	1	125.000	1
1608	5	3	0	1	3	1	13	1	132.000	1
1320	3	2	0	2	2	1	13	1	76.000	1
1680	3	3	0	2	2	0	7	1	79.900	1
1434	3	3	0	2	2	1	12	1	79.900	1
1664	3	2	0	2	2	1	7	1	85.000	1
1379	3	2	0	2	2	1	8	1	87.158	1

(concludes on the next page)

SQ_FT	BEDS	BATHS	HEAT	STYLE	GARAGE	BASEMENT	AGE	FIRE	PRICE	SCHOOL
1634	3	3	0	2	2	1	10	1	87.200	1
1400	3	3	0	2	2	0	1	1	91.500	1
1664	4	4	0	2	2	1	16	1	91.500	1
1548	3	3	0	2	2	1	8	1	92.500	1
2040	4	3	0	2	2	1	15	1	93.500	0
1792	4	3	0	2	2	1	9	1	94.000	1
1865	3	3	0	2	2	1	12	1	96.700	0
1833	3	3	0	2	2	1	7	1	97.400	1
1627	3	3	0	2	2	1	3	1	98.500	1
2198	4	3	0	2	2	1	12	1	99.000	0
1707	3	3	0	2	2	1	12	1	99.000	0
2096	3	3	0	2	2	1	7	1	102.400	1
2004	4	3	0	2	2	1	13	1	104.000	0
1818	4	3	0	2	2	1	19	1	107.000	1
1935	3	3	0	2	2	1	11	1	107.900	1
2381	3	3	0	2	2	0	17	1	109.000	1
1977	3	3	0	2	2	1	9	1	110.000	1
2317	5	3	0	2	2	1	9	1	110.500	1
2072	4	3	0	2	2	1	11	1	112.000	1
2108	3	3	0	2	2	1	7	1	112.900	0
2060	4	4	0	2	2	1	9	1	113.600	1
2064	3	3	0	2	2	1	5	1	114.500	1
2148	5	4	0	2	2	1	8	1	114.700	1
2170	6	4	0	2	2	1	21	1	115.000	1
1984	4	3	1	2	2	1	11	1	121.000	1
2130	4	3	0	2	2	1	20	1	121.000	1
2049	4	3	0	2	2	1	19	1	125.000	1
2035	3	3	0	2	3	1	6	1	128.000	1
2264	3	3	0	2	3	1	7	1	130.000	1
2264	3	3	0	2	3	1	10	1	131.000	1
2362	4	3	0	2	2	1	6	1	132.000	0
2262	5	3	0	2	2	1	12	1	132.500	1
2353	4	3	0	2	2	1	5	1	132.500	0
2298	3	3	0	2	3	1	2	1	144.500	1
2282	4	3	0	2	2	1	9	1	150.577	1
2542	4	3	0	2	2	1	6	1	153.800	1
2380	4	3	0	2	2	1	6	1	155.000	1
2505	4	3	0	2	2	1	7	1	156.900	1
2804	4	3	0	2	3	1	1	1	192.000	1

CASE **29**

BOOKS

The retail book market has become very competitive. These days bookstores offer more than just books: magazines, international and domestic newspapers, notecards, stationery, maps, books on tape, reading lights, and gifts are just some of the items found on bookstore shelves. Large retailers such as Barnes & Noble are able to offer low prices and large inventories to their customers, much to the chagrin of local bookstore owners. The large chain bookstores have recently added coffee shops or delicatessens to entice buyers to stay and browse.

A client of the consulting firm for which you work is considering investment into the retail book sales industry, and has hired your firm to undertake a financial study. Your job is to examine the data in the file named BOOKS. This data set contains general financial characteristics of five retail bookstores (Standard Industrial Classification [SIC] code 5942). The firms were chosen to include different-sized business entities: two large corporations (Barnes & Noble and Borders), two medium size firms (Books-A-Million and Crown) and, for comparison, one of the smaller stores—Village Green Bookstore.

Barnes & Noble, in addition to selling hardcover and paperback books, publishes books for exclusive sale by Barnes & Noble outlets. The firm also operates a retail book mail-order business. Borders Group, Inc. is a holding company which owns and operates book "superstores" and smaller stores such as those found in shopping malls. Borders bookstores offer prerecorded music and videotapes, and claim to specialize in hard-to-find items. Borders is one of the largest firms in its SIC group, employing almost 23,000 people in 1997.

Books-A-Million is a relatively new corporation offering the full line of hardcover and softcover books and magazines. Books-A-Million also distributes books to wholesale clubs, supermarkets, department stores, and mass merchandisers. Crown Books

These data and company descriptions come from various company reports.

offers the usual range of books, audiotapes, videotapes, and newspapers, and also some computer software. Headquartered in Columbus, Ohio, Crown has recently experienced financial difficulties: net income for Crown was −$860,000 in 1997.

Village Bookstore, out of Rochester, New York, sells much more than books. According to corporate reports, Village offers "books, newspapers, magazines, periodicals, greeting cards, boxed stationery, writing papers, books-on-tape, selective video tape rentals and multimedia products, humorous and traditional gift items and food items such as fresh and packaged desserts, gourmet chocolates, coffees and ice creams as well as other food items; and operates educational retail systems which sell children's educational toys and products."

The data set contains the longest time-series available for each of these five firms. Incomplete financial information is indicated with an asterisk. Use these data to undertake a study of this industry. What do the historical trends have to say about these firms? Use an appropriate statistical method to characterize net operating assets of these firms. What factors influence net operating assets? How? Prepare a summary of these matters for presentation at next week's meeting with your client.

DEFINITIONS

Firm	1 = Barnes & Noble, Inc. 2 = Books-A-Million, Inc. 3 = Borders Group, Inc. 4 = Crown Books Corp. 5 = Village Green Bookstore, Inc.
Year	Year, 19XX
PPE	Property, plant, and equipment - Total (net), in millions of $
Assets	Current assets - Total, in millions of $
Liab	Current liabilities - Total, in millions of $
OpAssets	Net operating assets - Total, in millions of $

DATA SET

Firm	Year	PPE	Assets	Liab	OpAssets
1	91	84.101	306.168	332.338	57.933
1	92	92.638	336.898	364.789	64.747
1	93	135.985	415.416	300.739	250.662
1	94	190.187	546.342	363.939	372.590
1	95	243.635	613.872	457.896	399.611
1	96	319.880	848.188	621.688	546.380
1	97	434.774	866.955	654.263	647.466
2	91	5.338	30.812	*	*
2	92	6.157	34.758	*	*
2	93	10.353	57.239	24.455	43.137
2	94	15.599	99.333	28.261	86.671
2	95	35.864	125.843	29.112	132.595
2	96	49.253	140.003	40.151	149.105
2	97	63.147	168.755	75.650	156.252
3	94	208.200	521.900	439.700	290.400
3	95	244.200	816.200	558.200	502.200

(concludes on the following page)

Firm	Year	PPE	Assets	Liab	OpAssets
3	96	243.500	740.200	542.900	440.800
3	97	289.200	846.400	634.500	501.100
4	88	7.541	85.032	31.177	61.396
4	89	7.318	96.213	34.588	68.943
4	90	8.091	100.562	37.631	71.022
4	91	9.637	127.097	49.495	87.239
4	92	10.714	149.432	62.622	97.524
4	93	15.018	147.809	56.492	106.335
4	94	25.228	180.210	94.981	110.457
4	95	23.004	182.331	105.553	99.782
4	96	23.762	150.574	77.606	96.730
4	97	25.379	141.583	84.515	82.447
5	88	1.462	2.907	1.651	2.718
5	89	1.380	2.713	1.757	2.336
5	90	0.496	3.060	2.127	1.429
5	91	0.499	3.335	2.061	1.773
5	92	0.515	3.394	2.002	1.907
5	93	0.615	5.004	4.137	1.482
5	94	0.670	4.815	3.304	2.181
5	95	1.237	5.760	5.478	1.519
5	96	1.934	7.112	5.191	3.855
5	97	1.331	3.101	3.042	1.390

CASE **30**

ELISA PLATE UNIFORMITY

ELISAs, Inc., manufactures plates used by hospital laboratories to test for the presence of antibodies to viruses such as lupus or HIV. The specific test process, called the ELISA method, consists of coating the bottom of plastic wells (like little test tubes) with a substance that contains molecules with which the specific antibodies will react. Serums from the patient are placed in the well, and if they contain the specific antibodies, those antibodies will bind to the bottom of the well, too. The test produces a blue color if antibodies are present. The amount of blueness depends on the number of antibodies present. The degree of blueness is then measured by a machine that produces readings of "optical densities," the final value of interest, on a scale of 0.000 to 3.000. A low reading on this scale indicates low numbers of antibodies present.

The plastic wells are manufactured in lots of 500 plates. Each plate contains 96 wells, arranged in a pattern of 8 rows and 12 columns, as shown below:

Column number

	1	2	3	4	5	6	7	8	9	10	11	12
Injector 1 ->	O	O	O	O	O	O	O	O	O	O	O	O
Injector 2 ->	O	O	O	O	O	O	O	O	O	O	O	O
Injector 3 ->	O	O	O	O	O	O	O	O	O	O	O	O
Injector 4 ->	O	O	O	O	O	O	O	O	O	O	O	O
Injector 5 ->	O	O	O	O	O	O	O	O	O	O	O	O
Injector 6 ->	O	O	O	O	O	O	O	O	O	O	O	O
Injector 7 ->	O	O	O	O	O	O	O	O	O	O	O	O
Injector 8 ->	X	O	O	O	O	O	O	O	O	O	O	O

This case describes a real situation, though the name of the company has been changed. (The ELISA process is real.) Information and data about the situation are used here with permission. They were provided by a team of students, including Susilo Hertanto, Michael Tigges, and several others.

The machine that fills the wells with the substance to which the antibodies bind has a single bank of eight injectors, which correspond to the rows in the diagram above. The machine first fills the eight wells in column 1 simultaneously: each cell is filled by a squirt of solution from the injector directly over it. The bank of injectors then moves to column 2 and fills those eight wells simultaneously, and so forth. After filling the wells in column 12, the entire bank of injectors returns to the positions over column 1, in preparation for the next plate. Differences in the resulting wells, then, may be attributed to variations between the columns and/or variations between the injectors, as well as to variations in the other parts of the total measurement process.

In principle, if the wells are filled with identical serums (from the same patient) and if the resulting optical densities are measured by using the approved method, the results should not vary. In practice, of course, they do.

ELISAs, Inc., has provided the data below, which consists of the readings for 95 observations on each of five plates taken sequentially from a single lot. The same patient's serums were used throughout. The specific plates tested were designed to detect antibodies of the disease lupus. Note that for each plate, the well in row 8, column 1 (marked with an X in the diagram), is used as a control. Its value in the data below is "missing." ELISAs, Inc., wants to know how well (i.e., how uniformly) its process works and to understand any ways in which it might be improved.

Based on the data below (also available in file ELISA), how would you advise ELISAs, Inc.?

DEFINITIONS

PLATE	Plate number (1 through 5)
ROW	Row number (injector number, 1 through 8)
COL	Column number (1 through 12)
OPTDEN	Optical density (0.000 through 3.000)

DATA SET

OBS	PLATE	ROW	COL	OPTDEN
1	1	1	1	1.421
2	1	2	1	1.446
3	1	3	1	1.614
4	1	4	1	1.578
5	1	5	1	1.452
6	1	6	1	1.790
7	1	7	1	1.991
8	1	8	1	*
9	1	1	2	1.266
10	1	2	2	1.270
11	1	3	2	1.106
12	1	4	2	1.159
13	1	5	2	1.141
14	1	6	2	1.002
15	1	7	2	1.207
16	1	8	2	1.259
17	1	1	3	1.804
18	1	2	3	1.334
19	1	3	3	1.287
20	1	4	3	1.084
21	1	5	3	1.199
22	1	6	3	1.120
23	1	7	3	1.532
24	1	8	3	1.389
25	1	1	4	1.466
26	1	2	4	1.298
27	1	3	4	1.288
28	1	4	4	1.165
29	1	5	4	1.293
30	1	6	4	1.392
31	1	7	4	1.523
32	1	8	4	1.657
33	1	1	5	1.454
34	1	2	5	1.356

(continues on the following page)

OBS	PLATE	ROW	COL	OPTDEN
35	1	3	5	1.522
36	1	4	5	1.392
37	1	5	5	1.483
38	1	6	5	1.633
39	1	7	5	1.744
40	1	8	5	1.550
41	1	1	6	1.578
42	1	2	6	1.370
43	1	3	6	1.788
44	1	4	6	1.442
45	1	5	6	1.383
46	1	6	6	1.786
47	1	7	6	1.844
48	1	8	6	1.778
49	1	1	7	1.503
50	1	2	7	1.604
51	1	3	7	1.771
52	1	4	7	1.500
53	1	5	7	1.586
54	1	6	7	1.803
55	1	7	7	1.679
56	1	8	7	2.013
57	1	1	8	1.645
58	1	2	8	1.695
59	1	3	8	1.891
60	1	4	8	1.628
61	1	5	8	1.840
62	1	6	8	1.611
63	1	7	8	1.847
64	1	8	8	1.802
65	1	1	9	1.717
66	1	2	9	1.814
67	1	3	9	1.875
68	1	4	9	1.778
69	1	5	9	1.424
70	1	6	9	2.001
71	1	7	9	1.885
72	1	8	9	1.652
73	1	1	10	1.714
74	1	2	10	1.874
75	1	3	10	1.957
76	1	4	10	1.884
77	1	5	10	1.762
78	1	6	10	1.873
79	1	7	10	1.753
80	1	8	10	1.708
81	1	1	11	1.569
82	1	2	11	1.799

OBS	PLATE	ROW	COL	OPTDEN
83	1	3	11	1.705
84	1	4	11	1.586
85	1	5	11	1.531
86	1	6	11	1.778
87	1	7	11	1.675
88	1	8	11	1.889
89	1	1	12	1.742
90	1	2	12	1.806
91	1	3	12	1.803
92	1	4	12	1.553
93	1	5	12	1.282
94	1	6	12	1.568
95	1	7	12	1.505
96	1	8	12	1.752
97	2	1	1	1.610
98	2	2	1	1.701
99	2	3	1	1.946
100	2	4	1	1.815
101	2	5	1	1.838
102	2	6	1	2.061
103	2	7	1	1.665
104	2	8	1	*
105	2	1	2	1.479
106	2	2	2	1.340
107	2	3	2	1.207
108	2	4	2	1.218
109	2	5	2	1.160
110	2	6	2	1.093
111	2	7	2	1.112
112	2	8	2	1.272
113	2	1	3	1.243
114	2	2	3	1.198
115	2	3	3	1.241
116	2	4	3	1.111
117	2	5	3	1.210
118	2	6	3	1.239
119	2	7	3	1.218
120	2	8	3	1.245
121	2	1	4	1.197
122	2	2	4	1.125
123	2	3	4	1.246
124	2	4	4	1.210
125	2	5	4	1.074
126	2	6	4	1.196
127	2	7	4	1.267
128	2	8	4	1.303
129	2	1	5	1.020
130	2	2	5	1.083

(continues on the following page)

OBS	PLATE	ROW	COL	OPTDEN
131	2	3	5	1.087
132	2	4	5	1.148
133	2	5	5	1.190
134	2	6	5	1.184
135	2	7	5	1.298
136	2	8	5	1.487
137	2	1	6	1.148
138	2	2	6	1.133
139	2	3	6	1.069
140	2	4	6	1.137
141	2	5	6	1.116
142	2	6	6	1.190
143	2	7	6	1.368
144	2	8	6	1.416
145	2	1	7	1.137
146	2	2	7	1.223
147	2	3	7	1.226
148	2	4	7	1.282
149	2	5	7	1.040
150	2	6	7	1.318
151	2	7	7	1.251
152	2	8	7	1.625
153	2	1	8	1.176
154	2	2	8	1.327
155	2	3	8	1.207
156	2	4	8	1.260
157	2	5	8	1.112
158	2	6	8	1.332
159	2	7	8	1.419
160	2	8	8	1.547
161	2	1	9	1.209
162	2	2	9	1.302
163	2	3	9	1.309
164	2	4	9	1.543
165	2	5	9	1.259
166	2	6	9	1.423
167	2	7	9	1.448
168	2	8	9	1.578
169	2	1	10	1.255
170	2	2	10	1.319
171	2	3	10	1.269
172	2	4	10	1.356
173	2	5	10	1.195
174	2	6	10	1.305
175	2	7	10	1.434
176	2	8	10	1.439
177	2	1	11	1.249
178	2	2	11	1.292
179	2	3	11	1.184
180	2	4	11	1.400
181	2	5	11	1.220
182	2	6	11	1.276
183	2	7	11	1.350
184	2	8	11	1.476
185	2	1	12	1.391
186	2	2	12	1.511
187	2	3	12	1.233
188	2	4	12	1.284
189	2	5	12	1.220
190	2	6	12	1.230
191	2	7	12	1.363
192	2	8	12	1.366
193	3	1	1	1.531
194	3	2	1	1.946
195	3	3	1	1.662
196	3	4	1	1.799
197	3	5	1	1.596
198	3	6	1	1.504
199	3	7	1	1.415
200	3	8	1	*
201	3	1	2	1.575
202	3	2	2	1.675
203	3	3	2	1.298
204	3	4	2	1.179
205	3	5	2	1.253
206	3	6	2	1.221
207	3	7	2	1.019
208	3	8	2	0.943
209	3	1	3	1.915
210	3	2	3	1.582
211	3	3	3	1.538
212	3	4	3	1.577
213	3	5	3	1.444
214	3	6	3	1.276
215	3	7	3	1.467
216	3	8	3	1.398
217	3	1	4	1.199
218	3	2	4	1.407
219	3	3	4	1.267
220	3	4	4	1.170
221	3	5	4	1.257
222	3	6	4	1.119
223	3	7	4	1.252
224	3	8	4	1.290
225	3	1	5	1.213
226	3	2	5	1.310

(continues on the following page)

OBS	PLATE	ROW	COL	OPTDEN
227	3	3	5	1.425
228	3	4	5	1.327
229	3	5	5	1.268
230	3	6	5	1.279
231	3	7	5	1.344
232	3	8	5	1.272
233	3	1	6	1.152
234	3	2	6	1.427
235	3	3	6	1.465
236	3	4	6	1.509
237	3	5	6	1.381
238	3	6	6	1.338
239	3	7	6	1.435
240	3	8	6	1.446
241	3	1	7	1.437
242	3	2	7	1.574
243	3	3	7	1.503
244	3	4	7	1.722
245	3	5	7	1.506
246	3	6	7	1.394
247	3	7	7	1.446
248	3	8	7	1.695
249	3	1	8	1.493
250	3	2	8	1.535
251	3	3	8	1.457
252	3	4	8	1.592
253	3	5	8	1.565
254	3	6	8	1.476
255	3	7	8	1.423
256	3	8	8	1.552
257	3	1	9	1.514
258	3	2	9	1.554
259	3	3	9	1.527
260	3	4	9	1.761
261	3	5	9	1.553
262	3	6	9	1.618
263	3	7	9	1.623
264	3	8	9	1.503
265	3	1	10	1.341
266	3	2	10	1.591
267	3	3	10	1.650
268	3	4	10	1.517
269	3	5	10	1.435
270	3	6	10	1.607
271	3	7	10	1.338
272	3	8	10	1.636
273	3	1	11	1.449
274	3	2	11	1.581

OBS	PLATE	ROW	COL	OPTDEN
275	3	3	11	1.668
276	3	4	11	1.584
277	3	5	11	1.175
278	3	6	11	1.446
279	3	7	11	1.684
280	3	8	11	1.284
281	3	1	12	2.057
282	3	2	12	1.801
283	3	3	12	1.562
284	3	4	12	2.054
285	3	5	12	1.636
286	3	6	12	1.769
287	3	7	12	1.759
288	3	8	12	1.922
289	4	1	1	1.425
290	4	2	1	1.867
291	4	3	1	1.625
292	4	4	1	1.460
293	4	5	1	1.491
294	4	6	1	1.222
295	4	7	1	1.371
296	4	8	1	*
297	4	1	2	1.416
298	4	2	2	1.192
299	4	3	2	1.246
300	4	4	2	1.130
301	4	5	2	1.283
302	4	6	2	1.116
303	4	7	2	1.156
304	4	8	2	1.295
305	4	1	3	1.185
306	4	2	3	1.149
307	4	3	3	1.174
308	4	4	3	1.198
309	4	5	3	1.163
310	4	6	3	1.096
311	4	7	3	1.268
312	4	8	3	1.278
313	4	1	4	1.107
314	4	2	4	0.968
315	4	3	4	1.306
316	4	4	4	1.364
317	4	5	4	1.200
318	4	6	4	1.175
319	4	7	4	1.304
320	4	8	4	1.286
321	4	1	5	1.333
322	4	2	5	1.287

(continues on the following page)

OBS	PLATE	ROW	COL	OPTDEN
323	4	3	5	1.198
324	4	4	5	1.379
325	4	5	5	1.442
326	4	6	5	1.524
327	4	7	5	1.458
328	4	8	5	1.375
329	4	1	6	1.480
330	4	2	6	1.340
331	4	3	6	1.280
332	4	4	6	1.508
333	4	5	6	1.399
334	4	6	6	1.513
335	4	7	6	1.608
336	4	8	6	1.001
337	4	1	7	1.296
338	4	2	7	1.279
339	4	3	7	1.426
340	4	4	7	1.338
341	4	5	7	1.312
342	4	6	7	1.212
343	4	7	7	1.477
344	4	8	7	1.423
345	4	1	8	1.374
346	4	2	8	1.693
347	4	3	8	1.504
348	4	4	8	1.414
349	4	5	8	1.523
350	4	6	8	1.378
351	4	7	8	1.680
352	4	8	8	1.293
353	4	1	9	1.596
354	4	2	9	1.646
355	4	3	9	1.647
356	4	4	9	1.728
357	4	5	9	1.601
358	4	6	9	1.654
359	4	7	9	1.689
360	4	8	9	1.369
361	4	1	10	1.384
362	4	2	10	1.723
363	4	3	10	1.569
364	4	4	10	1.654
365	4	5	10	1.599
366	4	6	10	1.607
367	4	7	10	1.597
368	4	8	10	1.611
369	4	1	11	1.592
370	4	2	11	1.408

OBS	PLATE	ROW	COL	OPTDEN
371	4	3	11	1.530
372	4	4	11	1.432
373	4	5	11	1.491
374	4	6	11	1.418
375	4	7	11	1.646
376	4	8	11	1.536
377	4	1	12	1.793
378	4	2	12	1.742
379	4	3	12	1.970
380	4	4	12	1.591
381	4	5	12	1.992
382	4	6	12	1.815
383	4	7	12	1.692
384	4	8	12	1.380
385	5	1	1	1.794
386	5	2	1	2.190
387	5	3	1	1.981
388	5	4	1	2.348
389	5	5	1	2.011
390	5	6	1	2.024
391	5	7	1	2.000
392	5	8	1	*
393	5	1	2	1.913
394	5	2	2	1.012
395	5	3	2	1.683
396	5	4	2	1.908
397	5	5	2	1.690
398	5	6	2	1.610
399	5	7	2	1.656
400	5	8	2	1.684
401	5	1	3	1.669
402	5	2	3	1.571
403	5	3	3	1.751
404	5	4	3	1.689
405	5	5	3	1.538
406	5	6	3	1.356
407	5	7	3	1.597
408	5	8	3	1.796
409	5	1	4	1.321
410	5	2	4	1.712
411	5	3	4	1.821
412	5	4	4	1.728
413	5	5	4	1.615
414	5	6	4	1.426
415	5	7	4	1.610
416	5	8	4	1.751
417	5	1	5	1.709
418	5	2	5	1.708

(concludes on the following page)

OBS	PLATE	ROW	COL	OPTDEN
419	5	3	5	1.465
420	5	4	5	1.781
421	5	5	5	1.524
422	5	6	5	1.773
423	5	7	5	1.833
424	5	8	5	1.707
425	5	1	6	1.366
426	5	2	6	1.654
427	5	3	6	1.402
428	5	4	6	1.642
429	5	5	6	1.540
430	5	6	6	1.680
431	5	7	6	1.904
432	5	8	6	1.980
433	5	1	7	1.192
434	5	2	7	1.765
435	5	3	7	1.765
436	5	4	7	1.932
437	5	5	7	1.696
438	5	6	7	1.894
439	5	7	7	2.098
440	5	8	7	2.107
441	5	1	8	1.148
442	5	2	8	1.891
443	5	3	8	1.775
444	5	4	8	2.028
445	5	5	8	1.779
446	5	6	8	1.976
447	5	7	8	2.321
448	5	8	8	2.041
449	5	1	9	1.232
450	5	2	9	1.662
451	5	3	9	1.729
452	5	4	9	1.992
453	5	5	9	1.877
454	5	6	9	1.936
455	5	7	9	2.109
456	5	8	9	2.208
457	5	1	10	1.210
458	5	2	10	1.673
459	5	3	10	1.668
460	5	4	10	1.911
461	5	5	10	1.803
462	5	6	10	2.079
463	5	7	10	1.796
464	5	8	10	1.951
465	5	1	11	1.832
466	5	2	11	1.837
467	5	3	11	1.797
468	5	4	11	1.855
469	5	5	11	1.824
470	5	6	11	1.832
471	5	7	11	2.022
472	5	8	11	1.722
473	5	1	12	1.370
474	5	2	12	2.034
475	5	3	12	2.246
476	5	4	12	1.926
477	5	5	12	1.835
478	5	6	12	2.253
479	5	7	12	2.314
480	5	8	12	2.281

CASE **31**

VIOLENCE ON TV—I

As the general manager of KTDS, the NBC affiliate in Tidusville, Oklahoma, Chris has a range of responsibilities, including programming, personnel, advertising, and public relations. His least favorite activity is responding to customer complaints. Unfortunately, there's been an unusually large number of complaints in the past few months from viewers and advertisers alike.

Most of the recent comments are objections to the level of violence on KTDS programs. Chris is sensitive to this issue because he has observed a gradual increase in violence on TV over the past 20 years. Chris really prefers the old-time movies in which dirty deeds were neatly sanitized and violent crimes occurred behind the scenes. He is sympathetic to the recent callers.

Nonetheless, he's in a tight spot. Chris knows that small, vocal groups do not necessarily represent the population at large. People who feel strongly about an issue are likely to speak out, while those who are content tend to remain silent. While the recent callers have denounced the level of violence on KTDS shows, Chris knows he must understand and serve *all* of the KTDS viewers.

Chris has a suspicion about the source of the recent calls. Four months ago, a flamboyant politician announced his candidacy for mayor. This candidate has received a great deal of air time on the local news due, in part, to his impassioned outbursts. Some people love him, others despise him, but almost everyone tunes in to the evening news in hopes of catching the latest controversy. One continuing theme of his platform is violence in America in general and violence on KTDS in particular. Over the past four months, the candidate has suggested that those opposed to violence in the media "let

This case is based on an idea of Ann Lee Bailey, an MBA student at the University of Colorado at Denver.

their voices be heard." Chris suspects that this fellow has inspired a large portion of the recent complaints to KTDS.

The situation has become so volatile that Chris is convinced he needs to take some kind of action. To sort out the hype from any real concerns of his constituents, Chris has commissioned you to help him get some structured feedback. He would like to undertake a survey (and has the staff to do the legwork), but he would like advice on the specifics. In particular, he needs to know:

- What questions should be asked, and what is the best way to ask them?
- How should survey participants be chosen and how many should be chosen?
- What sort of confidentiality guarantees should be made to participants, if any?
- What is the best way to contact potential participants?
- Should the station offer some kind of free gift to induce higher participation rates? If so, what?
- Should one survey be used to address the concerns of the viewers and another survey to appease the station's advertisers? If so, how should the surveys differ?

Prepare a report for Chris that addresses these and any other issues you think he should be aware of. Include in your report a copy of the questionnaire.

CASE **32**

OVERDUE BILLS

Quick Stab Collection Agency (QSCA) collects bills in an eastern town. The company specializes in small accounts and avoids risky collections, such as those in which the debtor tends to be chronically late in payments or is known to be hostile.

The business can be very profitable. OSCA buys the rights to collect debts from their original owners at a substantial discount. For example, QSCA might pay $10 for the right to collect a $60 debt. QCSA takes the risk of not collecting the debt *at all,* of course, but often a single official-looking letter yields full or nearly full payment, particularly for small debts.

Profitability at QSCA depends critically on the number of days to collect the payment and on the size of the bill, as well as on the discount rate offered.

A random sample of accounts closed out during the months of January through June yielded the data set below (and in file OVERDUE). Write a brief memo to QSCA management advising them on the relationship, if any, between size of bill and number of days to collect. In this data set, the variable DAYS is the number of days to collect the payment, and BILL is the amount of the overdue bill in dollars, while TYPE =1 for residential accounts and 0 for commercial accounts.

DATA SET

DAYS	BILL	TYPE	DAYS	BILL	TYPE	DAYS	BILL	TYPE
41	215	1	52	302	1	83	98	0
60	205	0	60	197	0	55	225	0
86	79	0	47	288	0	69	150	0
81	97	0	26	150	1	48	273	1
37	201	1	71	158	0	25	146	1
90	50	0	59	215	0	60	210	0
94	46	0	31	158	1	42	210	1
83	95	0	30	149	1	36	205	1
84	100	0	70	154	0	50	302	0
79	140	0	34	180	1	68	187	0
47	299	0	38	205	1	22	95	1
33	187	1	42	220	1	11	46	1
47	264	1	29	162	1	44	301	0
69	180	0	83	97	0	47	289	0
19	97	1	50	311	1	19	98	1
36	179	1	49	250	0	67	199	0
30	154	1	25	153	1	73	149	0
39	310	0	16	80	1	91	70	0
63	205	0	43	225	1	82	90	0
17	110	1	51	310	1	63	211	0
85	75	0	71	179	0	74	153	0
21	100	1	74	150	0	24	150	1
49	301	1	67	201	0	92	80	0
83	95	0	22	97	1	65	146	0
13	75	1	53	273	0	99	60	0
16	79	1	5	90	1	47	288	1
53	240	0	57	220	0	51	264	0
40	197	1	10	50	1	39	211	1
47	311	0	80	110	0	27	140	1
48	299	1	47	289	1	44	250	1
70	162	0	15	70	1	35	199	1
43	240	1	11	60	1	6	95	1

CASE **33**

THE BRAZING OPERATION

In one of the phases in manufacturing a precision flow meter, a vacuum brazing process is used to attach hardware to the flow meter tube. The process consists of placing the "fixtured" assembly in a sealed vacuum furnace, pumping a hard vacuum on the furnace, slowly heating the assemblies to approximately 2050 degrees Fahrenheit, and then rapidly cooling them down to the ambient temperature.

The company wants the assembly to have a metallurgical structure that resists corrosion. One aspect of such a structure is a good grain size, a measure of the size of the actual crystals in the crystalline structure of the flow wall. The grain size is measured by a standard specification called the grain count, a numerical reading between 1 and 10. The larger the grain count, the finer the crystalline structure (and the more corrosion resistant).

The following data (and in file BRAZING) were gathered from process control charts. The run numbers are not consecutive because a variety of operations, not just the brazing operation, are performed in the furnace.

Based on the data beginning on the next page, what recommendations do you make concerning the process?

DEFINITIONS

RUN	Run number
TEMPAVE	Average maximum temperature (degrees Fahrenheit)
TIMEAVE	Average time at that temperature (minutes)
VACLEVEL	Vacuum level (in torr units; higher numbers indicate a poorer vacuum)
COOLRATE	Cooling rate, in degrees Fahrenheit per minute
GRNCOUNT	Grain count, in standard units, computed from a sample of the resulting product

David R. Berge provided the data on which this case is based and described the manufacturing situation.

DATA SET

RUN	TEMPAVE	TIMEAVE	COOLRATE	VACLEVEL	GRNCOUNT
326	2045.9	12.4	342	2.6	6.6
327	2046.0	12.6	352	6.0	6.4
332	2046.1	12.4	331	5.0	7.4
335	2047.0	12.4	304	6.1	8.0
336	2044.5	12.6	376	7.5	6.4
337	2047.4	12.2	347	5.7	6.1
341	2046.6	12.6	391	4.7	5.8
346	2047.5	12.7	354	5.5	6.0
348	2046.3	12.2	310	5.0	7.7
349	2046.1	12.4	345	6.0	6.6
351	2046.6	12.3	305	5.0	8.1
353	2046.8	12.4	349	2.9	6.4
355	2047.2	12.3	311	6.2	7.6
358	2046.0	12.8	363	2.3	6.2
361	2045.2	12.7	354	2.2	6.8
365	2042.9	12.4	346	5.0	7.4
367	2048.6	12.3	302	3.7	7.2
369	2044.5	12.8	361	3.7	6.9
371	2046.6	12.4	326	5.0	7.2
373	2050.9	12.6	389	3.4	4.6
375	2050.8	12.7	390	6.5	3.8
377	2050.1	12.3	336	5.0	5.9
381	2048.3	12.4	306	3.7	7.0
385	2049.3	12.8	353	1.7	5.9
389	2048.3	12.6	357	4.9	5.9
391	2049.3	12.3	310	4.9	6.9
394	2049.6	12.6	363	2.0	5.1
397	2050.9	12.3	330	6.4	6.1
399	2050.5	12.6	351	2.2	5.0
403	2049.0	12.5	392	5.0	4.9
407	2048.0	12.4	337	3.4	6.6
412	2050.2	12.2	326	3.0	6.5
415	2050.1	12.7	363	4.4	5.1
417	2049.0	12.6	391	3.9	4.2
419	2048.7	12.7	385	5.0	4.9
421	2048.2	12.3	347	3.9	6.1
425	2048.4	12.6	372	7.4	6.1
427	2050.0	12.6	398	2.7	3.9
429	2049.6	12.3	311	6.0	6.4
430	2051.3	12.7	352	3.3	4.9
433	2046.8	12.6	350	7.0	6.1
437	2046.4	12.4	342	3.2	6.8
440	2048.3	12.4	338	6.3	6.4
449	2052.7	12.6	354	3.2	4.3
453	2050.2	12.7	355	6.7	5.1
456	2048.2	12.3	311	7.9	7.0

(concludes on the next page)

RUN	TEMPAVE	TIMEAVE	COOLRATE	VACLEVEL	GRNCOUNT
458	2050.1	12.4	315	4.4	5.8
463	2050.1	12.5	348	4.6	5.5
471	2050.1	12.6	352	7.0	5.8
473	2050.3	12.8	360	5.4	5.2
477	2049.7	12.7	353	7.0	5.9
481	2047.2	12.6	374	4.8	6.2
486	2048.8	12.2	347	6.9	6.3

CASE **34**

LEADING INDICATORS

Decision makers need reliable forecasts of key business activities to efficiently run an organization. Failure to anticipate future demand for a product can lead to foregone revenue, customer dissatisfaction, or lost market share. On the other hand, overestimation of future demand may lead to surplus inventory, with associated costs. Accurate projections of *all* costs of doing business, from personnel to capital, are needed to prepare budgets for an upcoming fiscal year. Although predictions about future business needs rarely turn out to be perfectly reliable, all firms do some kind of forecasting to assist in planning.

Some businesses rely on leading indicators for preparation of forecasts. Leading indicators are *current* levels of economic activity believed to foreshadow the direction and magnitude of *future* activities. For instance, since durable goods such as refrigerators and cars are time-consuming and costly to produce, manufacturers are more likely to step up production only when future demand is expected to be strong. Demand for large-ticket items seems to depend on consumers' confidence in their economic future, since consumers can postpone purchases of durable goods when times are hard and expected to remain so. Consumer confidence is one of the many leading indicators watched closely by business executives.

Inventories, wholesale trade, shipments of durable goods, housing starts, and the producer price index are other important leading indicators. Economic indicators are collected and published by organizations such as the Federal Reserve Board, the Bureau of the Census, and Dismal Sciences. See, for instance, www.dismal.com or www.census.gov/briefrm/esbr/www/brief.html.

Leading indicators can help prepare company forecasts if the firm can isolate which indicators apply to their industry. For instance, we might expect that mortgage applications and the interest rate on mortgages might foreshadow future activities in the construction industry.

Can you prepare a model to forecast the need for new orders for materials and supplies in the construction industry using the leading indicators available in the data in the file named LEADING? The definitions below include citations from the World Wide Web that will assist you in updating and revising the data set. Make use of explanatory material about these data on the web if necessary.

DEFINITIONS

Yr.Month	Year and month of the time series; 91.01 is January 1991, and so forth
Pop	U.S. resident population in thousands (source: U.S. Census Bureau, *Current Population Reports* www.census.gov/main/www/subjects.html)
PermitSA	Building permits (seasonally adjusted annual rate in thousands) for new, privately owned housing units; the data are based on 17,000 permit-issuing places through 1993, and 19,000 places beginning in 1994 (source: U.S. Census Bureau, *Housing Units Authorized by Building Permits,* C 40 www.census.gov/const/www/index.html)
PermitNA	Building permits (unadjusted monthly estimates in thousands) for new, privately owned housing units; the data are based on 17,000 permit-issuing places through 1993, and 19,000 places beginning in 1994 (source: U.S. Census Bureau, *Housing Units Authorized by Building Permits,* C 40 www.census.gov/const/www/index.html)
Const_SA	New orders for construction materials and supplies in the U.S., in millions of dollars, seasonally adjusted (source: U.S. Census Bureau, *Current Industrial Reports* www.census.gov/ftp/pub/indicator/www/m3/index.htm)
Const_NA	New orders for construction materials and supplies in the U.S., in millions of dollars, not seasonally adjusted (source: U.S. Census Bureau, *Current Industrial Reports* www.census.gov/ftp/pub/indicator/www/m3/index.htm)
MortRate	Interest rate on 30-year, fixed rate, conventional mortgage loans (source: Federal Home Loan Mortgage Corporation www.bog.frb.fed.us)

DATA SET

Yr.Month	Pop	PermitSA	Const_SA	Const_NA	PermitNA	MortRate
91.01	250660	786	14043	12320	51.5	9.64
91.02	250848	853	13695	13131	54.1	9.37
91.03	251028	911	13523	13826	78.7	9.50
91.04	251246	916	14199	14333	90.9	9.49
91.05	251524	991	14385	14735	98.3	9.47
91.06	251861	964	14389	15685	93.2	9.62
91.07	252106	973	15298	14845	91.3	9.58
91.08	252379	944	14566	15262	84.9	9.24

(continues on the following page)

Yr.Month	Pop	PermitSA	Const_SA	Const_NA	PermitNA	MortRate
91.09	252661	974	14645	15637	80.6	9.01
91.10	252932	991	14761	15445	89.8	8.86
91.11	253177	984	14553	14173	67.1	8.71
91.12	253385	1061	14457	13229	68.4	8.50
92.01	253589	1077	14790	12964	67.9	8.43
92.02	253780	1146	14653	14132	72.6	8.76
92.03	253993	1082	15085	15384	97.7	8.94
92.04	254238	1054	15305	15462	103.9	8.85
92.05	254488	1056	15185	15582	97.1	8.67
92.06	254749	1057	15860	17198	109.2	8.51
92.07	255011	1089	15102	14611	101.1	8.13
92.08	255304	1075	15275	16052	93.0	7.98
92.09	255577	1114	15864	16913	99.2	7.92
92.10	255843	1132	15649	16351	97.1	8.09
92.11	256082	1118	15523	15164	76.4	8.31
92.12	256297	1176	15961	14572	79.5	8.22
93.01	256512	1177	15872	13926	67.1	8.02
93.02	256700	1148	16766	16290	73.1	7.68
93.03	256889	1056	16307	16628	99.1	7.50
93.04	257100	1104	16248	16372	107.3	7.47
93.05	257318	1112	16566	17005	101.7	7.47
93.06	257551	1130	16284	17676	117.8	7.42
93.07	257795	1174	16079	15521	104.5	7.21
93.08	258056	1230	16305	17214	111.9	7.11
93.09	258307	1251	16639	17722	111.5	6.92
93.10	258555	1287	16660	17428	107.0	6.83
93.11	258766	1357	17185	16858	98.1	7.16
93.12	258965	1461	17426	15823	100.1	7.17
94.01	259159	1392	17231	15180	80.6	7.06
94.02	259312	1279	17421	16994	81.9	7.15
94.03	259494	1331	18476	18745	126.3	7.68
94.04	259723	1377	17985	18016	128.0	8.32
94.05	259918	1383	17971	18444	132.0	8.60
94.06	260140	1336	18201	19589	138.8	8.40
94.07	260372	1347	18143	17424	114.8	8.61
94.08	260610	1386	18342	19355	131.5	8.51
94.09	260853	1426	18384	19515	127.2	8.64
94.10	261087	1401	18747	19596	117.0	8.93
94.11	261297	1358	19049	18766	100.5	9.17
94.12	261496	1420	19350	17681	94.2	9.20
95.01	261687	1293	19750	17560	79.1	9.15
95.02	261838	1282	19474	18948	81.3	8.83
95.03	262024	1235	19030	19292	113.4	8.46
95.04	262251	1243	18548	18016	110.2	8.32
95.05	262451	1243	18807	18444	122.7	7.96
95.06	262661	1275	18529	19589	129.3	7.57
95.07	262890	1355	18902	17424	116.4	7.61

(concludes on the following page)

Yr.Month	Pop	PermitSA	Const_SA	Const_NA	PermitNA	MortRate
95.08	263123	1368	19382	19355	132.3	7.86
95.09	263366	1405	19105	19515	121.6	7.64
95.10	263609	1384	19169	19596	121.1	7.48
95.11	263804	1450	18926	18766	107.8	7.38
95.12	263988	1487	18731	17681	97.4	7.20
96.01	264162	1378	18927	16807	87.3	7.03
96.02	264295	1417	18639	18115	94.9	7.08
96.03	264485	1423	18676	18862	119.9	7.62
96.04	264701	1459	19265	19209	139.4	7.93
96.05	264878	1452	19652	20161	140.5	8.07
96.06	265072	1415	19653	21093	130.6	8.32
96.07	265284	1457	20269	19538	135.7	8.25
96.08	265498	1423	20053	21237	129.8	8.00
96.09	265723	1399	20195	21493	121.0	8.23
96.10	265943	1349	20057	21159	123.7	7.92
96.11	266130	1391	19903	19603	100.7	7.62
96.12	266313	1405	20264	18500	99.2	7.60
97.01	266490	1395	20471	18123	88.9	7.82
97.02	266620	1438	20507	19749	92.8	7.65
97.03	266789	1457	20421	20371	120.1	7.90
97.04	266993	1442	21342	21498	137.0	8.14
97.05	267177	1432	20613	21094	131.8	7.94
97.06	267368	1402	20896	22442	131.6	7.69
97.07	267575	1414	21574	21058	132.5	7.50
97.08	267792	1397	20873	21970	124.2	7.48
97.09	268014	1460	21213	22752	134.6	7.43
97.10	268234	1476	*	*	134.0	*

CASE **35**

INTERRUPTING TINA

Tina is a graduate student adviser in the College of Business at the University of Michigan at Trenton. Her job is sometimes hectic, but she likes its flexibility. She sets her own schedule, which can vary from week to week if needed, as long as she gets the job done and works an average of 20 hours per week.

Tina is one of five advisers in the College of Business. Advisers have a myriad of job responsibilities that include counseling students, processing admissions, maintaining and updating student records, responding to requests from professors, and providing information to prospective students. Each adviser is trained to respond to all of these needs but also has a specific area of expertise. Tina is solely responsible for verifying student eligibility for graduation. She has to pull each student file to check whether all required courses have been completed and the student has a minimum grade point average of 3.0. This is a time-consuming process—about 200 students apply for graduation each semester.

Tina's office is located in the middle of the advising center, which is in the middle of the College of Business. She suspects that a quieter office in a corner somewhere would eliminate an awful lot of the noise, confusion, and extraneous requests. On the other hand, she needs to be available to answer quick questions for her colleagues and students. Tina is the senior student adviser with a great deal of knowledge. She is also an expert in the Student Information Computer System (SICS), which handles all of the student-related administrative record keeping. Often she's the only one on the floor who knows the ins and outs of the SICS.

Student advising is a big part of her job. Tina can schedule appointments in advance, or she might see students on a walk-in basis. It's tough to know from day to day how

These data were collected and provided to us by T. J. Gaub. The name of the institution has been changed.

many students will need to be served on a walk-in basis or how long their appointments will last. Nonetheless, the college has recently instituted a total quality management (TQM) program, which requires an eye toward customer (in this case, student) satisfaction. This means that students must now be seen on demand. Tina also advises students over the phone; advising sessions by phone can be either prescheduled or unscheduled.

In addition to these chores, Tina interacts with her supervisor, answers questions for professors, trains newly hired advisers, processes administrative paperwork, and answers phone calls.

Tina had more difficulty than usual last semester completing all of the graduation checks. It wasn't that there were more students graduating last semester. The problem seemed to be an unusually large number of interruptions. Even though the job calls for her do a variety of tasks beyond just graduation checks, things seem to be getting out of hand! Tina's supervisor is unable to pay any overtime, so the only solution for this semester is to identify and eliminate interruptions where possible.

To study this problem in more detail, Tina decided to perform an experiment. She created a worksheet on which she recorded every incident that took her away from graduation checks over a six-week period. The worksheet includes the time she clocked in and out, the type of incident, and the starting and ending time of the incident. The worksheet is shown below. At the end of the day, Tina calculated the length of each incident in minutes and categorized each incident by general type:

A = SICS problems
B = Work with coworkers
C = Work with supervisor
D = Answer phone (unscheduled)
E = Clerical
F = Student advising: walk-in
G = Student advising: scheduled appointment
H = Meetings
I = Student advising by phone: scheduled
M = Other

As the weeks went on, Tina's coworkers asked what she was doing with the clipboard and worksheet. Tina was mysteriously unresponsive to their questions.

Tina has asked you to analyze and interpret the data for her. Use whatever statistical technique you feel is appropriate to prepare a report.

WORKSHEET

DATE	TIME IN	TIME OUT	INTERRUPTION	START TIME	END TIME	LENGTH (mins)	TYPE
10/12	11:40	12:50	computer problem	11:52	11:53	1	A
			counsel Rick	11:55	11:56	1	B
			where is Betty?	12:09	12:10	1	C
			proof memo	12:13	12:14	1	C

(continues on the following page)

DATE	TIME IN	TIME OUT	INTERRUPTION	START TIME	END TIME	LENGTH (mins)	TYPE
			phone calls	12:20	12:23	3	D
			find file	12:23	12:50	27	E
10/13	3:15	4:55	check address	3:22	3:23	1	E
			phone calls	3:27	3:29	2	D
			give Rick message	3:29	3:29	1	B
			phone calls	3:50	3:52	2	D
			computer help	4:02	4:04	2	A
			phone call	4:26	4:26	1	D
			walk-in	4:33	4:35	2	F
			walk-in	4:36	4:40	4	F
10/14	1:45	5:00	question about answering machine	1:50	1:51	1	E
			question about file	1:52	1:54	2	E
			get memo for Betty	2:01	2:03	2	C
			walk-in	2:10	2:23	13	F
			SICS problem	2:25	2:26	1	A
			walk-in	2:59	3:15	16	F
			dean's signature	3:42	3:44	2	E
			walk-in	3:46	3:47	1	F
10/18	4:15	5:50	counsel Rick	4:16	4:20	4	B
			walk-in	4:27	4:28	1	F
			telephone	4:42	4:56	14	D
			walk-in	5:31	5:50	19	F
10/19	10:40	12:40	phone call	10:40	10:42	2	D
			phone appointment	10:47	11:02	15	I
			phone call	11:23	11:26	3	D
			phone call	11:38	11:39	1	D

(continues on the following page)

DATE	TIME IN	TIME OUT	INTERRUPTION	START TIME	END TIME	LENGTH (mins)	TYPE
			phone call	11:55	11:56	1	D
			phone call	11:56	11:58	2	D
			walk-in	12:05	12:14	9	F
			phone call	12:20	12:22	2	D
			walk-in	12:25	12:31	6	F
			phone call	12:34	12:35	1	D
			phone call	12:37	12:38	1	D
			phone call	12:38	12:40	2	D
10/27	2:30	5:00	counsel Rick	2:39	2:40	1	B
			forgot password	2:43	2:52	9	A
			appointment	2:55	3:05	10	G
			computer problem	3:05	3:21	16	A
			appointment	4:00	4:40	40	G
			appointment	4:45	5:00	15	G
11/2	10:30	12:45	SICS problem	10:31	10:38	7	A
			Mark steals the computer	10:38	10:41	3	B
			phone call	10:44	10:55	11	D
			check on professor	11:55	12:00	5	E
	2:10	6:00	walk-in	2:45	3:21	36	F
			phone appointment	3:33	3:40	7	I
			appointment	5:00	5:15	15	G
11/3	3:00	5:15	appointment	3:15	3:25	10	G
			appointment	4:15	4:30	15	G
			appointment	4:45	4:50	5	G
11/4	10:15	12:45	appointment	10:50	11:00	10	G
	2:00	5:00	walk-in	2:45	3:00	15	F
			walk-in	3:10	3:26	16	F

(concludes on the following page)

DATE	TIME IN	TIME OUT	INTERRUPTION	START TIME	END TIME	LENGTH (mins)	TYPE
			walk-in	3:58	4:05	7	F
11/8	4:00	6:30	appointment	4:05	4:30	25	G
			appointment	5:05	5:30	25	G
			training Shelly	5:35	5:40	5	B
			walk-in	5:50	5:56	6	F
11/9	9:45	12:30	appointment	9:50	10:20	30	G
			appointment	11:00	11:30	30	G
	2:00	6:30	personal call	3:10	3:20	10	M
			walk-in	3:50	3:55	5	F
			walk-in	3:55	4:10	15	F
			walk-in	4:30	4:35	5	F
			appointment	5:00	5:25	25	G
11/11	11:00	5:30	walk-in	1:15	2:30	75	F
			walk-in	3:06	3:08	2	F
			walk-in	4:24	4:26	2	F
11/12	3:00	5:30	Betty	4:00	4:15	15	C
			walk-in	4:45	4:50	5	F
11/15	4:00	6:30	appointment	4:00	4:15	15	G
			meeting	4:55	6:15	80	H
11/16	10:45	6:30	walk-in	11:15	11:16	1	F
			move furniture	12:30	12:45	15	M
			records	2:30	3:10	40	E
			phone call	4:40	4:55	15	D
11/18	10:15	12:15					
	1:15	5:15	walk-in	2:45	3:15	30	F
			walk-in	3:15	4:00	45	F
			walk-in	4:00	4:45	45	F

CASE **36**

VIOLENCE ON TV—II

As the general manager of KTDS, the NBC affiliate in Tidusville, Oklahoma, Chris has a range of responsibilities that include programming, personnel, advertising, and public relations. His least favorite activity is responding to customer complaints. Unfortunately, there's been an unusually large number of complaints in the past few months from viewers and advertisers alike.

Most of the recent comments are objections to the level of violence on KTDS programs. Chris is sensitive to this issue because he has observed a gradual increase in violence on TV over the past 20 years. Chris really prefers the old-time movies in which dirty deeds were neatly sanitized and violent crimes occurred behind the scenes. He is sympathetic to the recent callers.

Nonetheless, he's in a tight spot. Chris knows that small, vocal groups do not necessarily represent the population at large. People who feel strongly about an issue are likely to speak out, while those who are content tend to remain silent. While the recent callers have denounced the level of violence on KTDS shows, Chris knows he must understand and serve *all* of the KTDS viewers.

Chris has a suspicion about the source of the recent calls. Four months ago, a flamboyant politician announced his candidacy for mayor. This candidate has received a great deal of air time on the local news due, in part, to his impassioned outbursts. Some people love him, others despise him, but almost everyone tunes in to the evening news in hopes of catching the latest controversy. One continuing theme of his platform is violence in America in general and violence on KTDS in particular. Over the past four months, the candidate has suggested that those opposed to violence in the media "let their voices be heard." Chris suspects that this fellow has inspired a large portion of the recent complaints to KTDS.

The survey design and data set compilation were undertaken by Ann Lee Bailey, an MBA student at the University of Colorado at Denver. The scenario has been altered to preserve the anonymity of the survey respondents.

Chris needs to sort all of this out. To understand the views of all KTDS patrons, he has commissioned an independent consulting firm to devise and administer a survey. The survey results are in, and he wants you to analyze the data (which reside in the file named VIOLENCE). Below is a copy of the questionnaire.

The survey was done over a three-week period in October. Two hundred phone calls were made and 106 people declined the invitation to participate in the survey. Of the 94 participating respondents, two people were offended by the question of income and refused to answer that particular question. Nonresponses are indicated in the data set by an asterisk. A random selection of phone numbers from the Tidusville phone book was used to select the sample.

What is Tidusville's perception of the level of violence on KTDS? Does one's perception of violence on TV vary with gender, age, marital status, income, or education? Do parents with children at home have a different tolerance for violence than those without children at home? Do viewers who spend a lot of time watching TV become desensitized to violence? Chris needs a report as soon as possible.

SURVEY

1. Gender

(0)	(1)
male	female

2. Age

(1)	(2)	(3)	(4)	(5)
under 20	20-30	31-40	41-50	over 50

3. Marital status

(0)	(1)
married	single or divorced

4. Do you have children at home?

(0)	(1)
yes	no

5. Household Income

(1)	(2)	(3)	(4)
under $20,000	$20,000 -40,000	$40,000 -60,000	over $60,000

6. Education

(1)	(2)	(3)	(4)
high school	some college	college graduate	graduate school

7. How many hours per week do you watch TV?

(1)	(2)	(3)	(4)	(5)	(6)	(7)
0-7	8-14	15-21	22-28	29-35	36-42	43 or more

8. In your opinion, how violent are most TV programs?

(1)	(2)	(3)	(4)	(5)
much too violent	somewhat too violent	violent	a little violent	not very violent

DATA SET

GENDER	AGE	MARRIED	CHILDREN	INCOME	SCHOOL	HOURS	VIOLENT
0	3	0	1	2	4	2	5
1	2	1	1	1	2	2	3
1	4	0	1	3	1	3	1
1	3	0	1	3	3	4	2
1	4	0	0	4	4	1	2
1	5	0	0	*	4	1	1
1	3	1	0	2	3	1	1
0	3	0	0	3	2	3	5
1	3	0	0	2	4	5	1
1	4	0	1	4	4	5	4
1	5	0	1	2	1	2	3
1	4	0	0	4	2	1	1
1	3	1	1	2	2	1	3
1	3	1	0	2	3	2	3
1	2	0	1	3	3	4	4
1	4	0	1	2	2	2	2
1	2	1	1	2	4	1	4
1	3	0	1	2	4	2	2
1	2	1	1	1	2	3	3
0	3	1	0	2	2	3	3
1	2	1	1	1	2	3	2
0	4	1	0	2	3	4	2
1	3	1	1	2	3	1	4
1	3	1	0	2	2	3	2
1	3	0	0	3	3	4	1
1	2	1	1	1	3	1	1
0	2	0	1	2	4	1	1
0	2	1	1	2	4	2	2
1	2	0	1	4	3	1	2
1	2	0	1	2	4	2	3
1	2	1	1	2	4	1	4
1	2	0	0	3	3	1	1
1	4	1	0	2	2	3	1
1	3	1	0	3	4	3	1
1	4	0	0	3	4	1	1
0	2	1	1	1	2	4	5
1	2	1	1	2	2	1	3
0	4	1	1	2	3	1	2
1	5	1	1	1	2	3	3
1	4	1	0	2	3	1	2
1	2	1	1	2	2	1	1
0	3	1	1	2	3	2	3
1	5	1	1	2	4	1	1
1	4	0	1	3	2	5	1
1	3	1	1	2	3	4	3
1	3	1	0	2	3	4	3

(concludes on the next page)

GENDER	AGE	MARRIED	CHILDREN	INCOME	SCHOOL	HOURS	VIOLENT
1	3	1	1	2	3	4	4
1	3	1	1	2	4	1	1
1	2	0	1	1	4	4	2
1	3	1	1	*	2	2	1
0	5	0	1	4	2	4	4
0	5	1	1	1	4	2	1
0	4	0	0	3	3	3	3
0	2	0	0	2	2	3	4
1	2	0	0	3	2	2	2
0	3	0	1	3	3	2	1
1	4	0	0	4	4	1	1
0	2	1	1	1	1	4	4
0	3	0	0	3	3	2	2
1	4	0	0	2	3	2	3
1	2	1	1	1	3	3	1
1	4	1	1	1	2	2	1
0	3	0	1	2	2	4	4
0	1	1	1	1	1	4	5
0	3	0	0	1	2	1	2
1	4	1	1	2	3	1	2
1	2	1	0	1	2	1	3
1	2	0	1	2	2	2	4
0	2	0	1	2	2	2	2
0	1	1	1	1	1	5	5
1	2	1	1	2	3	1	3
1	4	1	1	1	3	2	1
1	3	1	1	1	3	2	1
1	3	0	0	3	3	2	2
1	3	0	1	3	4	2	2
0	3	0	0	3	3	3	2
0	3	0	0	3	3	3	4
0	2	1	1	1	1	4	5
0	4	1	1	2	2	3	4
0	4	1	0	2	3	1	3
0	3	0	0	4	3	1	1
1	3	0	0	2	2	4	2
0	2	1	1	2	3	3	3
1	2	0	0	2	3	2	3
0	2	1	1	1	2	5	5
1	2	0	1	1	3	1	3
1	2	0	0	1	1	2	3
0	2	0	1	2	1	4	4
0	2	0	1	1	4	3	1
1	2	0	1	1	4	2	4
1	2	1	1	1	2	4	4
1	3	1	1	2	2	5	4
0	2	1	1	2	2	3	4
0	2	0	1	1	4	2	2

CASE **37**

COMMERCIAL PROPERTY

In early 1992, the Arapahoe Assessor's office made local headlines. Arapahoe County, adjacent to Denver County in Colorado, had just undergone its annual property assessments. Many people thought the assessed values were much too low. If this were true, it would have implications for funding of schools, the perceived value of the properties and property tax revenues for the county. The Arapahoe County Assessor was in the center of all of this—he was responsible for the low assessed values and he owned one of the properties.

Is there any way to determine whether the assessments were fair? To understand how buildings might be assessed, you have collected information on recent selling prices of commercial property in Arapahoe County. Of course, the selling price need not equal the assessed value of the property, but in most jurisdictions the assessed value is supposed to be proportional to the fair market value of the property. Selling price is an obvious measure of that value.

Understanding the determinants of selling prices of commercial property is difficult. Private residences typically range in size from several hundred to several thousand square feet, but commercial property can be the corner drug store or the high-rise downtown. The expected *use* of the commercial property may influence its selling price; buildings designed to be restaurants, warehouses, or barbershops probably sell for different amounts. Like residential property, though, it might be expected that older commercial property sells for less, all else the same.

The data file, named COMMPROP, contains data on selling prices and physical characteristics of 75 commercial properties sold in Arapahoe County from July 1988 to June 1990. Do these data suggest which factors determine selling prices of

These data were compiled and analyzed by George Carr, a tax appraiser and consultant for the Colorado Division of Property taxation; Rochelle Hrigora Hass; Lisa E. Pieper, Research Associate; and Mark Truitt. The data come from "Arapahoe County 1992 Property Assessment Study Final Report," by Thimgan & Associates, Inc., conducted for the Colorado Legislative Council, Denver, Colorado, September 15, 1992, p. 110.

commercial property and, hence, which factors should be included in a fair assessment process? Is there seasonality to selling prices of commercial property? If so, how might this seasonality influence the ability to use selling price as a proxy for assessed value? Prepare a report of your findings.

DEFINITIONS

CLASS	Primary use of the property: 0 = Merchandising 1 = Lodging 2 = Offices 3 = Special purpose 4 = Warehousing 5 = Multi-use 6 = Manufacturing
LANDSIZE	Size of the land parcel in square feet
TOWNSHIP	Location by township within Arapahoe County: 0 = Greenwood Village 1 = Aurora 2 = Englewood 3 = Littleton 4 = Unincorporated
SQFEET	Number of square feet in the building
FLOORS	Number of floors in the building
DESIGN	Specific property characteristics impacting its use: 0 = Store/Retail 1 = Hotel 2 = General office 3 = Medical office 4 = Restaurant 5 = Garage/Auto 6 = Warehouse 7 = Industrial 8 = Other 9 = Commercial condominiums
QUALCON	Quality of construction as assessed by an independent audit firm: 0 = Fair 1 = Average 2 = Good
BUILT	Year the building was built (19XX)
SYEAR	Year the building was sold (19XX)
SMONTH	Month the building was sold (1 = January, etc.)
PRICE	Selling price of the property in dollars

DATA SET

CLASS	LANDSIZE	TOWNSHIP	SQFEET	FLOORS	DESIGN	QUALCON	BUILT	SYEAR	SMONTH	PRICE
6	24393	2	11000	1	7	0	76	88	12	295000
3	8235	2	2400	1	7	0	75	89	5	120000
6	35912	2	23668	1	7	0	73	89	3	395000
6	28000	2	11567	1	7	0	70	89	6	280000
4	18229	2	6885	1	6	0	79	90	2	195000
6	14840	2	10000	1	7	0	71	90	2	205000
6	13160	2	8550	1	7	0	74	90	4	201000
6	43065	2	10240	1	7	0	73	88	12	350000
4	21587	2	11280	1	6	0	78	89	10	300000
4	45000	2	25224	1	6	0	78	89	2	425000
4	17480	2	6900	1	6	0	85	89	8	207500
2	6250	2	3300	1	2	0	80	89	10	230000
1	48260	1	43228	3	1	1	72	89	11	600000
0	15000	1	7450	1	0	0	42	90	3	250000
3	25000	1	3864	1	4	1	70	88	11	200000
3	41136	1	3478	1	5	1	73	89	7	350000
3	37635	1	3544	1	4	1	60	90	2	217000
5	20000	1	5640	1	0	1	78	89	7	262500
0	13983	1	2400	1	0	1	77	90	1	241250
2	79933	1	38924	3	2	1	82	89	12	1300000
1	267737	1	187528	7	1	2	82	89	3	13700000
0	17550	1	5040	1	0	0	76	89	12	182250
0	9828	4	2490	2	2	0	65	89	2	80000
2	49648	1	8776	1	2	0	77	89	3	400000
3	60364	1	8641	2	4	1	79	89	6	618000
2	1024	1	1390	2	9	1	80	90	4	92500
2	1024	1	1408	2	9	1	80	90	1	85000
2	1152	1	1440	2	9	1	84	88	12	99500
2	1152	1	1440	2	9	1	84	89	11	87000
2	1152	1	1408	2	9	1	83	90	4	83000
2	1152	1	1408	2	9	1	83	90	5	84000
3	233438	1	44562	1	0	0	86	89	8	2100000
2	11644	1	3812	1	9	1	84	90	6	102500
2	25655	1	9352	2	2	1	80	89	3	759000
2	54320	1	32130	3	2	1	83	89	4	1450000
4	30381	1	12980	2	6	0	81	88	7	300000
2	5502	1	1625	1	9	1	82	89	5	58100
4	28070	1	8918	1	6	1	85	89	12	175000
4	102148	1	11603	1	6	1	86	90	6	285000
5	107157	1	34122	1	0	0	79	90	4	850000
0	106252	1	35836	1	0	1	78	89	12	1075000
2	4800	1	4170	1	2	1	81	89	11	165000
2	6300	1	5571	1	2	1	81	89	9	350000
4	44012	4	12843	2	6	1	82	89	8	400000

(concludes on the next page)

CLASS	LANDSIZE	TOWNSHIP	SQFEET	FLOORS	DESIGN	QUALCON	BUILT	SYEAR	SMONTH	PRICE
4	41528	4	8400	1	6	0	84	89	1	232000
5	23460	0	5480	1	0	1	71	89	8	310000
2	104548	0	59854	4	2	1	74	88	8	3350000
2	87991	0	14764	1	2	1	65	90	4	810000
2	85410	0	10492	1	2	1	68	88	10	600000
2	193321	0	77792	3	2	1	80	89	7	3325900
3	174240	3	48887	3	3	2	85	89	1	1440000
2	191403	4	111231	3	2	1	82	89	12	4683300
2	87120	4	19174	2	2	1	85	89	9	652350
4	266108	4	81190	1	6	1	82	90	5	2100000
4	6400	4	6292	1	2	1	80	88	11	227000
4	8000	4	7880	1	8	1	79	89	8	300000
4	4800	4	4704	1	2	1	79	90	2	189900
4	8000	4	7892	1	2	0	79	89	11	200000
3	20034	2	8352	2	3	1	65	88	9	475000
3	81295	2	39218	6	3	1	78	89	1	700000
5	15625	2	6084	1	0	1	85	89	9	513000
0	4625	2	1921	2	0	0	30	89	7	85000
0	25000	2	6080	1	0	0	60	89	2	260000
4	24000	2	10260	1	6	0	83	88	7	550000
3	21978	2	6080	1	5	1	88	89	8	600000
0	28828	3	910	1	0	0	65	90	3	90000
3	24700	3	5365	1	0	0	75	89	2	300000
0	5625	2	4500	1	0	0	53	89	1	105000
3	84040	3	31461	3	3	1	66	89	11	650000
0	4000	3	4489	1	0	0	60	89	4	105000
3	16800	3	3478	1	3	1	71	88	8	215000
5	267300	3	89820	1	0	1	56	90	6	1900000
4	217800	3	74550	1	6	1	85	90	5	1900000
0	24137	3	9775	1	0	1	81	88	9	440000
5	91500	3	18850	1	0	1	78	89	11	640000

CASE **38**

AIDS RATES

Our intuitive impressions of rates are often erratic or wrong, particularly about issues that are sensitive, emotional, or scary. For example, ask 10 or 20 people to estimate the rate of acquired immunodeficiency syndrome (AIDS) cases in the United States, first for males, and then for females, and record their answers in units of, say, cases per 1000 people. Prepare an appropriate summary of their responses. Record your own estimates, too.

Now use the World Wide Web to find the latest data from, say, the Centers for Disease Control and Prevention, about AIDS rates in the United States, and prepare an appropriate summary of *that* data. How accurate were the opinions of the people in your sample? How consistent?

CASE **39**

SOFTWARE SERVICE CALLS

"What can you do?" asked Sam. "The number of service calls we get in any week or month jumps around so much, it's no wonder that sometimes we seem to be overstaffed, while at other times we are completely swamped and can't get to all the reported problems anywhere near as quickly as we ought to." Sam knew his supervisor wouldn't like his explanation much, but he still felt it was an accurate description of the problem he faced.

Sam managed a department in the national service center for a software development company. As part of the customer service and warranty operation, the national service center staffed a hotline, to which customers could report discrepancies or problems. Customers could file formal, written trouble reports (STR), but Sam was concerned about telephoned trouble reports, not STRs.

His department received direct service requests by phone (SCALLs) for one specific product in the firm's catalog, a fairly large package for computer-aided design and drafting. The package had been sold for 18 months. His problem was to have the appropriate number of service representatives on hand to provide decent response time to customers. "Decent" was not defined in any specific terms (at least not yet), but Sam agreed with his supervisor that performance lately had certainly been unsatisfactory, at least in spurts. Sam knew that the allocation of service representatives to different products could in principle be changed from time to time. The difficulty was how to know in advance how many would be needed for which product.

Sam stared at the numbers on his desk and remembered the tone in his supervisor's voice. He had to figure out something. The numbers on his desk are also available in

This case describes a real business situation. The data are used here with the permission of the students who originally provided them to us.

file SERVCALL. Q is the quantity of the product sold in each month, and MONTH is a count of calendar months elapsed since the product's introduction. The other variables are defined above.

DATA SET

MONTH	Q	STR	SCALL
1	20	28	76
2	19	18	63
3	28	10	79
4	37	53	73
5	2	47	89
6	91	41	97
7	4	38	68
8	11	34	118
9	45	66	85
10	4	28	88
11	9	63	99
12	14	58	68
13	10	40	70
14	8	43	84
15	7	41	85
16	9	38	72
17	11	32	74
18	8	33	72

CASE **40**

TITANIC SHIPPING COMPANY

Titanic Shipping Company depends on its nationwide system of computer terminals for entering orders, tracking shipments, and other routine functions. Some 1500 terminals and personal computers in more than 70 locations are connected through a state-of-the-art telecommunications network to central computing facilities in Newark, New Jersey. Thanks to good backup procedures and steady improvements in communications technology, the system has become reliable, and failures (in the strict sense of the term) have been rare.

The performance of the system has troubled the users, though. At any point during a normal day, several hundred users may be entering or retrieving information about the company's shipments or updating customer records. When response time for these transactions deteriorates, users can't work effectively. Recently, they've been letting the computer center manager know their feelings—forcefully.

The computer center manager understands their concerns, but isn't sure how to respond. To some extent, of course, improving response time requires things like new facilities or better performance tuning of their systems, both of which take money. That can be negotiated in the normal company budget process. Right now, he is more concerned about losing his credibility with the users he is supposed to serve. Several years ago, the users' representatives agreed that response times of two seconds would be adequate. Since then, the computer center has tracked average response time, as measured by an internal tool in the computer's operating system. According to this measure, performance has been generally satisfactory, at least for large portions of the time. Still, the users complain that the system is not meeting their needs. The manager realizes that part of the problem is one of "translation": the users don't know how to specify exactly

The company name and the specific data for this case are simulated, but the situation corresponds closely to a number of real situations with which we are familiar.

how the computer center ought to measure response time (at least not in any precise, quantifiable terms), but when performance is bad, they know it, and they complain. As long as the computer center continues to track performance by using a measure that (apparently) is unrelated to the users' perceptions of their needs, relations between the users and computer center management will probably continue to deteriorate.

During the last quarter the computer center has developed a system that asks each user to rate the system performance as "OK" or "Bad" before signing off. Thirteen weeks' worth of data from the 10:00–11:00 hour for each workday are given below and in file TITANIC. The 10:00 hour is usually a busy one, and generates many of the user complaints. Data from two holidays have been omitted.

The manager wonders if these data can provide some insight that will allow him to develop a better measure of response time. He really wants to know two things:

- How should he measure response time?
- What target or standard performance should he use?

The measurements in the data set come from the software in the operating system and could be obtained routinely if it were helpful to do so. In setting the standard, the manager wants to use a value that corresponds as closely as possible to the users' perceptions, so that when the computer center's own measurements indicate acceptable performance, the users will agree. Can you advise him?

DEFINITIONS

AVERESP	Average response time during that hour (seconds)
RESP50	50th percentile of the response times during that hour (seconds)
RESP75	75th percentile of the response times during that hour (seconds)
RESP90	90th percentile of the response times during that hour (seconds)
RESP95	95th percentile of the response times during that hour (seconds)
NBAD	Number of users who rated performance "bad" when signing off
NOK	Number of users who rated performance "OK" when signing off

DATA SET

DAY	AVERESP	RESP50	RESP75	RESP90	RESP95	NBAD	NOK
1	2.223	1.559	2.982	4.944	6.426	322	5
2	1.253	0.989	1.825	3.008	3.951	2	276
3	1.758	1.157	2.257	3.734	4.882	144	155
4	1.703	1.287	2.632	4.238	5.593	310	4
5	1.395	1.119	2.174	3.682	4.764	104	191
6	2.380	1.928	3.757	6.228	8.017	343	7
7	2.486	1.694	3.384	5.695	7.366	338	1
8	2.049	1.556	3.039	5.034	6.523	326	3
9	2.771	1.903	3.822	6.293	8.147	351	1
10	2.132	1.734	3.367	5.482	7.182	337	2
11	2.566	1.785	3.621	5.948	7.633	343	3
12	2.090	1.488	2.888	4.856	6.277	314	10
13	3.032	1.819	3.686	6.044	7.879	346	2
14	1.029	0.776	1.556	2.504	3.270	15	247
15	1.465	1.080	2.028	3.342	4.314	5	283
16	2.144	1.558	3.083	5.090	6.612	317	13
17	1.211	0.951	1.928	3.211	4.168	2	281
18	0.963	0.797	1.618	2.639	3.364	5	261
19	2.800	2.011	4.043	6.736	8.708	356	1
20	2.507	1.629	3.327	5.430	7.027	315	23
21	2.387	1.661	3.327	5.479	7.132	328	10
22	3.118	2.085	4.078	6.697	8.696	357	1
23	2.182	1.851	3.590	5.981	7.743	310	35
24	1.812	1.807	3.556	5.934	7.699	343	1
25	0.916	0.861	1.603	2.603	3.347	7	258
26	2.754	2.053	4.188	6.813	8.907	355	6
27	2.592	1.768	3.533	5.758	7.571	341	3
28	3.135	2.085	4.061	6.755	8.692	340	18
29	3.076	2.121	4.213	6.868	8.974	353	8
30	1.795	1.207	2.421	3.960	5.179	263	43
31	2.795	2.089	4.090	6.813	8.807	355	3
32	1.803	1.186	2.211	3.710	4.811	128	169
33	1.786	1.374	2.662	4.283	5.562	312	3
34	2.733	2.090	4.160	6.935	9.043	352	8
35	2.880	2.086	4.037	6.662	8.663	353	4
36	1.589	1.081	2.052	3.432	4.444	31	258
37	1.682	1.346	2.580	4.192	5.436	306	6
38	0.902	0.785	1.551	2.493	3.269	4	257
39	1.045	0.898	1.701	2.841	3.657	4	267
40	1.194	1.009	1.963	3.208	4.091	20	265
41	2.306	1.696	3.291	5.421	7.020	334	3
42	1.329	1.111	2.125	3.474	4.432	7	286
43	3.016	1.937	3.724	6.147	8.060	348	1
44	2.477	1.668	3.377	5.547	7.185	330	9
45	1.567	1.187	2.282	3.797	4.916	185	115

(concludes on the next page)

DAY	AVERESP	RESP50	RESP75	RESP90	RESP95	NBAD	NOK
46	1.617	1.183	2.355	3.864	5.005	221	82
47	2.440	1.713	3.455	5.699	7.405	339	2
48	1.741	1.402	2.821	4.574	6.007	315	6
49	2.028	1.369	2.730	4.392	5.776	317	1
50	1.776	1.464	3.008	4.956	6.401	303	25
51	1.116	0.831	1.633	2.675	3.508	2	265
52	2.912	1.924	3.857	6.409	8.249	351	1
53	1.276	0.898	1.768	2.904	3.756	3	271
54	1.392	1.312	2.523	4.186	5.416	303	7
55	2.142	1.863	3.767	6.175	7.997	336	14
56	1.533	1.156	2.148	3.590	4.648	64	230
57	2.311	1.704	3.486	5.714	7.395	335	7
58	2.464	1.851	3.713	6.054	7.867	329	20
59	2.551	2.126	4.186	6.887	8.882	358	3
60	1.396	0.853	1.536	2.516	3.355	1	259
61	2.876	2.030	4.060	6.628	8.601	350	8
62	2.451	1.723	3.273	5.468	7.102	318	18
63	1.924	1.446	2.743	4.625	5.942	316	2

CASE **41**

CAMPAIGN STATISTICS

"That's good!" said the congressman. "We can use that to show the voters just what happens when our opponents stay in office for four years." He was referring to data that showed that the effective income tax rate had increased from 14.1% to 15.2% of adjusted gross income during the period 1974 to 1978.

"What I'd like, though," he went on, "is to break it down into smaller categories, like, say, $10,000 to $14,999. That way, we can kill two birds with one stone. First, we can choose the category that looks the worst for our opponents. Second, the people in that category will know that we're talking about them, not just some 'average' or someone a lot richer or poorer than they are. See what you can work up for me!"

With that, he was off to a campaign appearance. The data he referred to are given below. Provide the breakdown he wants.

TABLE

TOTAL INCOME AND TOTAL TAX (IN THOUSANDS OF DOLLARS)

Adjusted gross income	1974 income	1974 tax	1978 income	1978 tax
Under $5,000	41,651,643	2,244,467	19,879,622	689,318
$5,000–$9,999	146,400,740	13,646,348	122,853,315	8,819,461
$10,000–$14,999	192,688,922	21,449,597	171,858,024	17,155,758
$15,000–$99,999	470,010,790	75,038,230	865,037,814	137,860,951
$100,000 or more	29,427,152	11,311,672	62,806,159	24,051,698
Total	880,179,247	123,690,314	1,242,434,934	188,577,186

This case is based on real data, originally reported by C. H. Wagner in *The American Statistician,* Volume 36, No. 1, February 1982, pp. 46–48. Reprinted with permission from *The American Statistician.*

The effective tax rate for a given category is the tax paid divided by the total income reported in that category. For example, for 1974 the effective tax rate in total was 123,690,314/880,179,247=.141 and in 1978 the corresponding number was 188,577,186/1,242,434,934=.152. These are the numbers to which the congressman was referring before he left for the campaign trail.

CASE **42**

HOSPITAL CHARGES

When administrative duty's to be done, the hospital manager's lot is not a happy one. For example:

- Physicians have control over a substantial fraction of hospital costs, including the length of time a patient spends in the hospital and which drugs and other supplies are used in the patient's treatment.
- On the other hand, the physicians are usually *not* employees of the hospital, so while the hospital is responsible for managing the costs, its administration has rather less control over those costs than it does in many other business settings. Hospitals *do* generally grant "admitting privileges" to individual physicians and review their performance on a periodic basis, but short of rescinding those privileges (or threatening to do so), the hospital has little direct influence over many expenditures.
- Further, the hospital's revenues are determined largely by the patient's insurance coverage, which may pay either a more or less fixed amount based upon the "Diagnosis Related Group" (DRG, a classification system used by many federal and other programs), or a percentage of the bill (in more "classical" kinds of insurance schemes). To what extent differences in insurance coverage may or may not influence the patient's treatment or quality of care is a matter of occasional controversy.

Being sure that analyses of physician performance data are fair and balanced is important in any such performance review. The 289 patient records in the data for this case (available in file HOSPITAL) are for normal deliveries of babies. DAYS is the number of days the patient spent in the hospital, CHRGS is the total expenses charged to that patient, PHYS is a code identifying the physician, and PAYOR indicates the type of insurance the patient carried: 1 for managed care insurance and 0 for commercial insur-

These data are real, but the name of the hospital and of the person who collected them are confidential.

ance. Other things being equal, do there appear to be any substantial differences in charges among the various physicians? Write a brief report summarizing your findings.

DATA SET

DAYS	CHRGS	PHYS	PAYOR	DAYS	CHRGS	PHYS	PAYOR
2	1701	10	0	2	929	13	1
2	3500	4	1	3	3464	14	1
2	3165	10	0	2	3173	10	0
1	1953	14	1	2	3034	14	1
2	2607	2	1	2	3716	12	1
2	2944	12	1	2	2137	6	1
4	5063	2	1	3	3430	6	1
2	2396	13	1	3	3041	6	1
4	4903	2	1	2	2247	12	1
2	2280	7	1	3	4357	12	1
2	2939	7	1	2	3050	12	1
3	3585	4	1	2	2779	12	1
2	3073	13	1	2	2357	4	1
2	2880	14	1	2	2620	7	1
2	2358	10	0	3	3369	13	1
3	3418	2	1	2	3090	7	1
3	3604	2	0	4	6340	11	1
3	3480	6	1	2	2900	10	1
2	3306	10	0	3	3709	2	1
3	3047	4	1	3	3503	4	1
2	2157	11	1	2	2991	12	1
2	2439	4	1	2	2941	14	1
3	2647	11	1	2	2668	14	1
2	1874	6	1	2	2138	2	1
1	2139	10	1	2	2227	12	0
2	2424	14	1	2	2838	10	0
2	2450	12	1	2	2681	2	0
2	2198	14	0	3	3945	7	1
2	3355	10	1	2	2026	4	0
2	2609	4	0	3	3160	10	1
2	2324	2	1	2	3270	12	0
2	2207	12	1	3	4146	6	1
2	2649	11	0	2	3757	11	1
3	3375	10	1	2	2435	7	1
2	3906	6	1	2	2343	10	1
3	4049	14	1	3	3932	2	1
2	2564	13	1	2	2285	10	1
2	2745	11	1	2	3001	14	1
3	1254	6	1	2	1864	7	0
2	2953	2	1	3	4711	12	1
2	2334	7	1	3	3283	2	0
2	2809	7	1	2	2592	7	1

(continues on the following page)

DAYS	CHRGS	PHYS	PAYOR	DAYS	CHRGS	PHYS	PAYOR
1	2081	12	1	2	2804	2	0
1	1791	14	1	2	2801	13	0
2	3729	2	1	2	2638	7	1
2	2117	10	1	2	2263	10	0
3	4510	11	1	2	2449	11	0
2	2017	7	1	2	2473	11	0
2	3118	14	1	3	2864	6	1
3	3392	2	1	2	2534	11	1
2	1854	4	1	2	2550	10	1
14	14898	2	1	2	3768	12	1
3	3819	2	1	2	3194	10	0
3	2644	4	1	1	2153	12	0
3	2331	13	0	1	1840	7	0
3	3492	11	1	2	2522	12	1
2	2666	7	0	1	2145	12	1
3	4248	2	1	2	2403	6	1
3	2500	4	1	3	3287	2	1
2	3112	10	1	2	2979	6	1
3	2955	7	1	2	3182	7	0
2	2761	14	1	2	2954	10	0
2	1905	2	1	2	2158	11	0
3	3394	10	1	2	3455	12	0
2	3663	10	1	2	2367	14	1
3	3059	6	1	3	2798	10	1
2	2528	10	1	2	2408	11	1
2	2823	2	1	2	2468	11	1
6	6710	11	1	3	3081	12	1
2	2543	13	1	2	2218	7	1
3	2785	2	0	3	2133	11	0
2	2589	12	0	4	5499	13	1
2	3183	11	1	2	2251	12	1
1	1806	10	1	2	2310	11	0
2	3115	10	0	1	2220	11	0
2	3204	7	1	2	2176	10	0
2	2921	2	0	1	2048	2	1
3	4933	2	0	2	2566	11	1
2	4722	13	1	2	2450	11	0
3	5633	13	1	2	2582	11	1
2	2780	10	0	2	2612	7	0
2	2798	4	1	2	2683	6	1
3	2066	7	1	2	2827	14	0
3	2494	11	1	2	3202	10	1
3	2724	10	0	3	3779	10	1
2	2352	11	0	2	2833	12	1
1	1793	7	1	2	3458	10	0
2	2629	4	1	1	2789	7	1
2	3094	14	0	2	1590	7	1

(continues on the following page)

DAYS	CHRGS	PHYS	PAYOR	DAYS	CHRGS	PHYS	PAYOR
3	3868	7	1	2	3032	7	0
2	3077	12	1	3	2891	4	0
2	3109	10	0	1	2886	10	1
3	2697	4	0	1	3381	2	0
2	3182	13	1	2	2752	10	0
2	3034	6	1	1	3675	13	1
1	2179	14	1	2	2840	13	1
1	2178	7	1	2	3105	13	0
3	4185	11	1	2	2288	11	1
2	1567	10	1	2	3128	14	0
2	2347	14	1	2	2289	7	0
1	2268	10	1	2	2602	10	0
3	3617	2	0	2	2604	14	1
2	2858	10	1	2	1023	12	0
2	2542	10	1	1	1899	10	0
2	2665	7	1	2	2334	10	1
2	2436	6	1	2	2310	2	1
2	2392	12	1	3	3636	6	1
2	2739	7	0	2	4206	7	0
3	2181	12	0	2	2414	12	1
2	2308	4	0	2	2206	13	1
2	3800	13	0	3	2907	2	1
2	2242	12	1	2	3017	14	1
2	2314	7	1	2	3050	14	1
1	1960	6	0	2	1973	7	1
3	4183	11	1	2	2692	10	1
1	1663	4	1	3	2564	11	0
1	2070	7	0	2	3693	10	1
3	2753	6	1	2	2741	13	1
3	2209	6	0	2	2888	2	1
3	3218	10	1	2	3279	6	0
2	2419	13	0	3	2873	7	1
3	3289	11	0	2	2582	10	1
2	2222	4	1	2	2640	2	0
1	2741	4	0	1	2068	14	1
3	3454	10	1	2	2123	10	0
2	3848	10	1	3	3480	14	1
2	2292	7	0	3	4475	6	0
2	2594	7	1	2	2312	14	1
2	2378	6	1	2	2898	4	1
1	902	7	0	3	3708	10	0
3	2935	7	1	1	2078	10	0
2	3230	6	1	2	2570	11	1
2	2219	2	0	3	3639	14	1
3	2687	12	1	2	4674	12	0
2	1848	14	1	2	2534	14	0
3	2891	4	0	2	2797	7	0
2	2698	12	0	2	3139	12	0

(concludes on the following page)

DAYS	CHRGS	PHYS	PAYOR
3	3826	2	0
2	2067	10	0
2	2840	2	1
2	2465	12	1
1	1947	12	0
1	2167	7	0
2	2840	10	0
2	2711	6	1
2	3137	2	0
1	2955	2	0
2	2563	10	1
1	1924	4	1
1	2062	6	0
2	2663	13	0
1	2184	2	0
2	3064	10	1

CASE **43**

COAL TESTING

Coal is a heterogeneous bulk commodity, typically sold under contracted terms that include bonuses and penalties associated with various quality parameters and impurity levels of the coal. In many cases, predicting and controlling the quality of the coal accurately can mean the difference between economic success and failure for a coal mine.

The heating value is a quality of particular interest. It is the basis of payment in most coal supply contracts. In the United States, the heating value of coal is measured in British thermal units per pound (BTU/lb), by using a national standard test procedure (ASTM D-2015). ASTM standards also indicate that repeated measurements should give values within 100 BTU/lb.

Normally, the only way to determine heating value is to send a carefully prepared sample off to a laboratory and wait for the results. This can lead to disputes, if the customer and the supplier don't agree on which lab to use or if the results from different laboratories give drastically different results.

In an effort to determine the extent of any intra- or interlaboratory variation in heating value measurements, one coal company prepared 50 samples of coal from a common area (so that results were expected to be very close). They shipped the samples in two batches to five reputable laboratories. Each batch consisted of five samples sent to each lab; the two batches were sent at different times. The samples were drawn randomly, packed in water (to minimize oxidation effects) and sent in sealed plastic containers. The labs were not aware that an experiment was being run, and the coal company hoped that the results would thus reflect the labs' normal operating procedures.

This case is based on a real situation, described to us by H. Erik Scherer, P. E., Mining Engineer and Consultant. He also provided the data.

The labs made several measurements, but only those for heating value are given below. What do you think the coal company should conclude about the performance of the laboratories on its list, at least as far as heating value measurements are concerned?

The data are available in file COALLAB. The variables are: LABNO, the laboratory number; BATCH, the batch number (0 or 1); and BTU, the BTUs per pound as reported by the laboratory.

DATA SET

OBS	LABNO	BATCH	BTU
1	1	0	11655
2	1	0	11690
3	1	0	11654
4	1	0	11653
5	1	0	11665
6	1	1	11754
7	1	1	11676
8	1	1	11712
9	1	1	11711
10	1	1	11684
11	2	0	11665
12	2	0	11618
13	2	0	11572
14	2	0	11686
15	2	0	11725
16	2	1	11602
17	2	1	11582
18	2	1	11764
19	2	1	11618
20	2	1	11637
21	3	0	11586
22	3	0	11588
23	3	0	11599
24	3	0	11596
25	3	0	11622
26	3	1	11428
27	3	1	11429
28	3	1	11488
29	3	1	11632
30	3	1	11478
31	4	0	11651
32	4	0	11780
33	4	0	11663
34	4	0	11769
35	4	0	11800
36	4	1	11630
37	4	1	11587

(concludes on the following page)

OBS	LABNO	BATCH	BTU
38	4	1	11567
39	4	1	11531
40	4	1	11630
41	5	0	11591
42	5	0	11605
43	5	0	11549
44	5	0	11590
45	5	0	11609
46	5	1	11739
47	5	1	11747
48	5	1	11768
49	5	1	11768
50	5	1	11794

CASE **44**

PERFORMANCE EVALUATIONS

Ann, an independent contractor who specializes in human resources issues, has been asked by the XYZ company to undertake an independent study of its employee performance evaluation practices. XYZ, a nonprofit organization, is a professional review organization in the health care industry. Most of its revenue comes from government contracts for Medicare and Medicaid utilization reviews.

The employee performance appraisal involves a fairly complex mix of subjective and objective scoring. Each evaluator fills out a form that covers 11 categories of employee performance, such as adaptability, accountability, communication and decision-making skills, sensitivity to cost-effectiveness issues within the organization, attendance, and training and development of other employees (see Exhibit A). Nonsupervisory personnel are evaluated on 8 of the 11 categories; supervisory and management employees are rated on all 11 categories. The same form is used for all employees, although the weight applied to each category varies with the employee's grade and responsibilities, as shown in Exhibit B. Weights are assigned in such a way that any employee can receive a final score anywhere from 0 to 200. Higher scores mean better performance.

Each employee has an evaluation initiated on the 12-month anniversary of his or her hire date, so evaluations go on at XYZ throughout the year. Any employee who is a new hire, has been transferred between departments, has been promoted, or has been demoted is considered to be on probation. These employees are given three-month evaluations in addition to the annual evaluations.

These data are real, although, because they deal with personnel issues, the name of the company has been disguised. Claudia Roberts Salvano collected the data set and provided it to us. Claudia received an M. S. degree in Management in 1993 and was the Manager of Human Resources at the time that she collected the data.

EXHIBIT A EMPLOYEE PERFORMANCE APPRAISAL FORM

ANNUAL PERFORMANCE EVALUATION

Date Evaluation Due:____________ Time on Job:____________ Employment Date:____________

Employee: ____________ Title: ____________

Evaluated by:____________ Title: ____________

Reviewed by:____________ Title: ____________

RATING KEY

RATING	POOR(P)		FAIR(F)		COMPETENT(CP)		COMMENDABLE(C)		DISTINGUISHED	
INDIVIDUAL RATING CATEGORY POINTS	P− 1	P+ 2	F− 3	F+ 4	CP− 5	CP+ 6	C− 7	C+ 8	D− 9	D+ 10
TOTAL POINTS	0-50		51-70	71-90	91-110	111-130	131-150	151-170	171-190	191-200
Raise %	0		0	1%	2%	3%	4%	5 ¼%	5 ¾%	6 ½%

DESCRIPTION OF CATEGORY

Poor	Performance does not meet requirement of position.
Fair	Performance adequate most of the time. Marginal performer in one or more areas.
Competent	Responsibilities are handled competently. Meets all requirements of position.
Commendable	Knowledge and mastery of position performance marked by high quality.
Distinguished	Performance superior in all aspects. Quality leaves nothing to be desired.

____________ ____________ ____________

Employee Manager Approved by

Date: ____________ Date: ____________ Date:____________

Employee Agrees ____________ Employee Does Not Agree____________

EXHIBIT A *(continued)*

Annual ______________
Three-month __________

EMPLOYEE EVALUATION

Employee Name____ Due Date____________________

1. ( ) INITIATIVE/ADAPTABILITY/ACCOUNTABILITY Rating___ Pts___

Demonstrates ability to work with others in organization. Adjusts to changed work assignments or conditions, accepts guidance and special assignments. Demonstrates capabilities for establishing work priorities and begins tasks on own. Fulfills the full range of accountabilities indicated in his/her job description.

COMMENTS: ____

2. ( ) COMMUNICATION Rating___ Pts___

Clearly, concisely, and persuasively expresses ideas and instructions verbally and in writing and understands the ideas and instructions of others.

COMMENTS: ____

3. ( ) DECISION MAKING Rating___ Pts___

Demonstrates the ability to evaluate and make proper decisions within a reasonable time frame. Evaluates data, reaches sound conclusions, makes appropriate recommendations, and follows through with timely decisions.

COMMENTS: ____

4. ( ) COST EFFECTIVENESS Rating___ Pts___

Demonstrates and promotes an awareness of the importance of cost effectiveness and reduces expenses (through unnecessary labor and time, wasted supplies, unneeded equipment, etc.) and reduces costs when and where practicable.

COMMENTS: ____

5. ( ) QUANTITY OF TASKS Rating___ Pts___

Performs day to day assigned activities and special projects, within prescribed standards and in a timely fashion.

COMMENTS: ____

EXHIBIT A *(continued)*

6. ( ) QUALITY OF TASKS Rating___ Pts___

Produces high quality work with minimal error rate.

COMMENTS: ____

7. ( ) ADHERENCE TO CORPORATE POLICY AND DECISIONS Rating___Pts___

Understands the rationale used in decisions, accepts the final decision, and is able to properly convey the decisions to others when appropriate. Adheres to all Corporate Policies outlined in the Personnel Manual and related policy memos.

COMMENTS: ____

8. ( ) ATTENDANCE Rating___ Pts___

Arrives at work on time, returns from breaks and lunch on time, schedules vacation time in advance, schedules Doctor and Dentist appointments so as to use the least amount of work time. The number of hours of sick leave used is a major component for this category.

COMMENTS: ____

9. ( ) PLANNING AND ORGANIZING Rating___ Pts___

Effectively schedules and coordinates the flow of work. Organizes and utilizes his/her people, materials, and equipment. Organizes and utilizes his/her own time, meets established deadlines. Demonstrates ability to prepare annual budgets and to operate within approved budgets.

COMMENTS: ____

10. ( ) MANAGING, CONTROLLING, AND DELEGATING Rating___ Pts___

Leads and motivates people to effective group and individual effort. Directs, reviews, and checks work of subordinates. Establishes and maintains standards of performance. Takes corrective action when necessary. Defines and delegates authority to others.

COMMENTS: ____

EXHIBIT A *(concluded)*

11. ( ) DEVELOPMENT OF OTHERS Rating___ Pts___

Encourages and assists individuals, through training and other methods, to acquire knowledge and skills essential for effective job performance. Selects individuals with potential for promotion and assists them to qualify for advancement with assurance that equal employment opportunity corporate goals are met.

COMMENTS: ____

____

____

____

EXHIBIT B PERFORMANCE EVALUATION CATEGORY WEIGHTS

Category	Director	Manager	RN	Technical	Secretarial	Clerical	Medical Director
Initiative/Adaptability/Accountability	3	3	3	3	2	2	3
Communication	2	2	2	2	2	2	2
Decision Making	2	2	3	2	2	2	2
Cost Effectiveness	2	1	2	1	1	1	1
Quantity of Tasks	1	1	3	4	4	4	2
Quality of Tasks	1	1	3	4	4	4	2
Adherence to Corp. Policy	2	2	2	2	2	2	2
Attendance	1	1	2	2	3	3	0
SUPERVISORY							
Planning and Organizing	2	2	0	0	0	0	2
Managing, Controlling/ and Delegating	2	3	0	0	0	0	2
Development of Others	2	2	0	0	0	0	2
TOTAL	20	20	20	20	20	20	20

Accurate performance appraisals are important. They provide feedback on an employee's contribution to the company, and the annual appraisal score is used to determine merit raises. When this appraisal system was introduced several years ago, evaluators attended extensive training sessions designed to ensure as much consistency and fairness as possible. Evaluators are also provided with detailed guidelines (Exhibit C). In spite of this, employee complaints continue to surface. Ann's job is to undercover any non-work-related biases in the evaluation process.

Concerns expressed by the employees at XYZ include the following.

- Evaluators are occasionally late with their evaluations. Rumors at the company suggest that the later the evaluation, the lower the score, as the manager may be attempting to postpone a controversial or hostile discussion with the employee about poor performance.
- There may be interdepartmental biases. Some managers may be attempting to raise their own department's status and salaries by artificially inflating the evaluation scores.
- Scores may be bumped up a few points to push a favored employee into the next higher raise category.
- Three-month probationary evaluations are considered unnecessary by the employees, because three months are generally not enough time to adjust to a new position or to reach the peak level of efficiency in a new job.
- There are rumors that the company is attempting to force early retirement by giving the long-time employees lower scores than others.

Ann has a sample of 144 employee evaluations conducted over the past calendar year; 10 variables were collected from each employee record as described below. The first 125 observations in the data set refer to employees who were still employed at XYZ at the end of the year; the remaining group left the company at some point during the year. The data available from these evaluations are contained in the file PERFEVAL. What should Ann's recommendations be?

EXHIBIT C DETAILED EVALUATION GUIDELINES

EVALUATION GUIDELINES

This guide has been prepared to help you conduct the performance appraisal. The objectives are twofold:

1. General guidance (ideas to keep in mind when appraising performance).
2. "How-to's" of preparing the evaluation and conducting the session.

1. BE FAIR AND IMPARTIAL

Try to disregard bias about an employee. Concern yourself only with the facts and concrete instances of employee performance. Bias and prejudice may exist, but if you are aware of them, you will be in a good position to control them and their effects. A positive attempt must be made to be as fair as possible to the employee and to XYZ.

2. APPRAISAL PERFORMANCE

This is your role as a manager. Appraise performance against each accountability. It is not fair to base an evaluation, for example, on a person's educational attainments.

3. PERFORMANCE—NOT POTENTIAL

Accurate assessment of individual potential is important; however, the concern here is performance of the past year. Rate only the performance and do not rate what you think his/her potential is.

4. ACHIEVEMENT—NOT PROGRESS

Performance is rated in achieving results during a fixed time period. Success should be measured against previously agreed upon objectives.

5. APPRAISE PERFORMANCE FOR THE WHOLE YEAR

One month of commendable does not offset 11 months of satisfactory.

6. RATE EACH ACCOUNTABILITY INDEPENDENTLY

Do not let your performance rating on one area affect your rating against others. Try to rate each aspect of the evaluation independently.

7. CROSS RATINGS

Usually, evaluations are completed by the employee's manager. In some cases, another manager within XYZ may have important insights regarding the employee's performance; for example, if the employee worked under another manager for a good part of the year, it would be necessary to seek the views of the previous manager. Keep in mind, the goal is to have a complete, objective, and fair appraisal of performance.

8. BE A CONSCIENTIOUS RATER

Give a favorable rating only when the individual has merited it. Forget about giving people a "break." Often, this is difficult for us to accomplish. We want to be fair and often this means giving the benefit of the doubt. If we continue to do this over a period of time, we will find the standards we have set have strongly been lowered.

By lowering standards, XYZ's chances of attaining organization goals are weakened. Additionally, we hurt the individuals being evaluated by asking them to produce less to reach the same level. Be strict and fair to both parties. These two goals do not necessarily conflict.

9. MORE FREQUENT APPRAISALS

If an employee's performance needs improvement, establish a time frame for corrective action and re-evaluation. Discuss performance expectations, actions the employee can take to meet those expectations, and a specific time when you will re-evaluate the employee's performance.

DEFINITIONS

SCORE	Evaluation score in points
LAG	Difference between the due date of the evaluation by the manager and the day on which the results were discussed with the employee, in days
EMP@EOY	Years with the company at the end of the year
POS@EOY	Years in the current position at the end of the year
TYPE	Type of evaluation (0 = 3 month evaluation, 1 = annual evaluation)
GRADE	Employee's grade scale: 1–7 are clerical or technical, 8–9 are skilled employees, 10–12 are supervisory, 13–17 are management
DEPT	Employee's department number
EDUC	Employee's education level: 12 is high school grad, 13 is one year training beyond high school (e.g., secretarial training), 14 is two years beyond high school (i.e., Associate's degree, nursing diploma), 16 is Bachelor's degree, 18 is Master's degree, 20 is PhD
EMP@EVAL	Years with company at time of evaluation
POS@EVAL	Years in the current position at the time of the evaluation

DATA SET

SCORE	LAG	EMP@EOY	POS@EOY	DEPT	GRADE	EDUC	EMP@EVAL	POS@EVAL	TYPE
184	−7	7.29	7.29	40	8	13	7.120	7.12	1
155	9	2.92	2.12	55	13	13	1.820	1.02	1
132	21	0.77	0.77	55	13	16	0.310	0.31	0
118	−7	0.80	0.80	55	4	13	0.230	0.23	0
102	15	13.78	4.70	41	2	12	13.160	4.07	1
178	4	9.65	9.65	64	3	12	9.020	9.01	1
115	2	2.19	2.19	41	9	16	2.150	2.15	1
128	13	0.50	0.50	62	8	16	0.290	0.29	0
99	35	12.76	2.73	62	2	12	12.100	2.07	1
133	−2	4.49	0.51	71	11	16	4.230	0.25	0
158	12	5.14	2.04	55	10	12	4.130	1.03	1
125	0	6.94	6.94	62	8	16	6.000	6.00	1
113	2	1.88	0.39	73	8	14	1.750	0.26	0
121	−12	6.35	0.28	73	17	18	6.290	0.22	0
172	10	3.78	3.78	72	8	16	3.030	3.03	1
159	49	11.70	11.70	64	8	14	11.670	11.67	1
135	14	6.10	3.17	55	4	13	5.830	2.90	1
155	7	7.59	5.29	55	10	18	7.320	5.02	1
128	22	0.56	0.56	41	2	13	0.310	0.31	0
127	62	3.42	3.42	73	10	16	3.170	3.17	1
134	45	0.31	0.31	62	8	16	0.370	0.37	0
126	13	0.77	0.44	55	2	12	0.290	0.05	0
131	16	4.24	4.24	41	10	18	4.050	4.05	1
136	−2	0.96	0.96	55	2	12	0.240	0.24	0
155	8	0.37	0.37	41	11	16	0.270	0.27	0
125	17	0.42	0.42	40	13	16	0.300	0.30	0

(continues on the following page)

SCORE	LAG	EMP@EOY	POS@EOY	DEPT	GRADE	EDUC	EMP@EVAL	POS@EVAL	TYPE
128	21	0.39	0.39	72	8	14	0.310	0.31	0
164	3	4.87	4.87	64	8	14	4.010	4.01	1
90	27	0.74	0.74	72	2	12	0.320	0.32	0
146	−5	4.28	2.79	73	11	14	3.470	1.99	1
159	19	0.92	0.28	55	14	20	0.950	0.30	0
125	12	4.80	4.80	74	8	16	4.040	4.04	1
129	36	3.42	3.42	62	8	14	3.100	3.10	1
149	4	2.61	1.28	63	8	16	2.340	1.01	1
182	−7	7.34	7.34	72	10	16	6.980	6.98	1
172	42	3.76	2.34	62	10	14	3.790	2.38	1
137	−11	6.52	0.28	71	17	16	6.470	0.22	0
171	30	3.16	2.75	62	3	12	2.500	2.08	1
144	34	1.27	0.73	55	3	12	1.090	0.55	1
163	28	1.39	1.39	61	8	14	1.090	1.08	1
165	4	6.89	5.12	74	8	16	6.750	4.98	1
128	11	6.52	5.39	72	3	12	6.170	5.03	1
133	6	1.29	1.29	63	8	14	1.020	1.02	1
162	−5	4.95	4.95	64	8	14	3.990	3.99	1
119	8	0.53	0.53	65	8	16	0.270	0.27	0
131	34	0.37	0.37	74	4	12	0.350	0.35	0
149	62	2.42	0.74	55	9	12	2.100	0.42	0
167	24	0.37	0.37	41	2	12	0.320	0.32	0
136	13	1.31	1.31	63	3	14	1.040	1.04	1
150	21	0.56	0.56	41	2	14	0.310	0.31	0
135	8	0.81	0.81	73	10	14	0.270	0.27	0
176	64	17.24	17.24	62	3	12	17.190	17.19	1
115	19	0.33	0.33	72	8	18	0.300	0.30	0
155	0	5.86	5.86	65	3	12	5.000	5.00	1
127	−12	1.17	1.17	62	3	13	0.970	0.97	1
61	37	4.72	0.28	62	6	14	4.790	0.35	0
115	15	3.16	0.86	55	2	13	2.580	0.28	0
121	23	1.69	0.71	55	2	14	1.290	0.31	0
160	−14	1.42	1.42	72	8	14	0.960	0.96	1
162	−1	12.58	3.99	72	13	14	11.590	3.00	1
155	4	2.86	2.86	72	8	14	2.010	2.01	1
124	−2	5.32	0.51	73	10	16	5.050	0.25	0
141	12	1.42	0.74	73	5	14	0.960	0.28	0
124	13	1.24	1.24	61	8	14	1.040	1.04	1
131	79	4.76	4.76	63	8	14	4.220	4.22	1
111	9	4.35	2.69	73	3	12	3.690	2.02	1
135	10	4.01	0.56	62	10	14	3.700	0.28	0
164	33	7.64	7.64	64	8	14	7.600	7.60	1
127	5	0.46	0.46	62	8	14	0.270	0.27	0
111	3	4.91	0.51	73	10	16	4.670	0.26	0
171	31	8.10	6.50	61	3	12	7.680	6.09	1
123	27	2.73	2.73	74	8	14	2.070	2.07	1

(continues on the following page)

SCORE	LAG	EMP@EOY	POS@EOY	DEPT	GRADE	EDUC	EMP@EVAL	POS@EVAL	TYPE
128	15	0.94	0.94	62	8	16	0.290	0.29	0
128	−6	0.63	0.63	62	8	16	0.240	0.24	0
132	−12	0.88	0.88	70	6	16	0.210	0.21	1
173	34	8.73	8.73	63	8	14	8.100	8.10	1
126	36	0.90	0.54	72	3	12	0.720	0.35	0
120	8	2.96	1.13	62	8	14	2.020	0.20	0
121	30	0.96	0.96	65	3	12	0.330	0.33	0
111	12	0.92	0.92	74	8	14	0.280	0.28	0
129	58	0.96	0.96	61	3	14	0.410	0.41	0
132	9	2.52	2.52	71	10	16	2.070	2.07	1
75	10	10.90	6.50	55	10	16	10.920	6.53	1
157	6	12.76	12.76	73	4	12	12.020	12.02	1
155	20	7.39	3.12	73	13	16	7.330	3.05	1
152	−2	3.28	3.28	62	8	14	2.990	2.99	1
121	5	5.81	0.92	62	3	12	5.150	0.26	0
112	18	0.94	0.94	62	8	14	1.300	1.30	1
170	4	3.42	3.42	62	8	16	3.010	3.01	1
159	13	4.80	4.80	65	8	16	4.040	4.04	1
124	17	0.65	0.65	72	8	14	0.300	0.30	0
174	−7	3.44	3.44	62	8	16	2.980	2.98	1
120	9	2.44	1.13	74	13	16	1.580	0.28	0
139	0	1.96	0.89	55	10	14	1.310	0.24	0
139	0	2.76	2.76	72	2	12	2.000	2.00	1
160	−3	1.69	1.35	41	14	18	1.330	0.99	1
133	−3	5.35	5.35	72	8	14	4.990	4.99	1
147	14	7.56	7.56	64	8	16	7.560	7.56	1
126	15	0.52	0.52	61	8	18	0.290	0.29	0
112	−3	6.78	0.28	64	13	18	6.740	0.24	0
140	16	0.96	0.74	72	3	13	0.520	0.29	0
124	60	1.88	0.39	62	10	14	1.910	0.42	0
104	10	1.99	1.99	72	2	14	1.030	1.03	1
126	30	2.78	0.27	73	8	14	2.840	0.33	0
140	20	5.05	0.91	41	4	12	4.430	0.30	0
106	7	8.15	0.75	71	9	12	7.670	0.27	0
138	7	0.73	0.73	72	4	12	0.270	0.27	0
135	5	2.23	0.81	55	3	12	1.680	0.27	0
105	35	0.28	0.28	72	8	16	0.350	0.35	0
143	2	14.27	14.27	73	11	16	14.030	14.03	1
131	27	2.94	2.69	55	10	18	1.320	1.07	1
108	−10	11.28	0.28	74	17	14	11.220	0.22	0
171	−7	5.33	2.08	72	10	14	4.380	1.13	1
152	22	2.67	2.23	55	5	13	2.500	2.06	1
171	15	5.89	5.22	64	3	12	5.720	5.04	1
152	8	1.96	1.96	62	2	12	1.020	1.02	1
115	69	0.46	0.32	72	3	12	0.580	0.44	0
80	8	0.48	0.48	41	4	12	0.270	0.27	0

(concludes on the following page)

SCORE	LAG	EMP@EOY	POS@EOY	DEPT	GRADE	EDUC	EMP@EVAL	POS@EVAL	TYPE
115	15	2.13	0.36	41	5	16	2.060	0.29	0
134	−8	7.35	0.28	61	13	14	7.300	0.23	0
132	10	1.69	0.74	55	3	13	1.230	0.28	0
137	19	12.70	0.28	55	17	18	12.720	0.30	0
129	12	2.70	0.94	62	3	13	2.047	0.28	0
154	11	16.85	13.51	41	14	14	16.590	13.25	1
113	2	5.90	0.98	71	13	14	5.180	0.25	0
104	5	1.99	1.99	55	2	12	1.010	1.01	1
131	−4	1.10	1.10	72	8	14	0.240	0.24	0
117	34	0.81	0.81	62	8	14	0.350	0.35	0
171	−7	18.15	13.79	65	13	14	17.340	12.99	1
90	9	1.19	1.19	74	8	14	0.280	0.28	0
123	6	1.25	1.25	74	8	14	0.270	0.27	0
101	−3	8.25	4.49	62	13	16	7.760	3.99	1
40	87	0.96	0.96	62	8	18	0.480	0.48	0
117	6	2.80	2.80	74	8	14	2.020	2.02	1
132	9	1.27	1.27	74	8	14	1.020	1.02	1
159	−14	1.84	1.84	72	2	18	0.960	0.96	1
105	26	11.38	10.13	55	4	12	10.820	9.57	1
106	6	2.05	0.64	73	8	14	1.680	0.27	0
116	10	0.58	0.58	41	3	12	0.280	0.28	0
171	21	2.74	2.74	62	8	16	2.060	2.06	1
143	35	7.52	5.32	65	8	16	7.330	5.12	1
90	3	0.65	0.65	74	8	14	0.260	0.26	0
115	24	0.92	0.92	41	3	12	0.310	0.31	0
117	10	0.42	0.42	71	8	16	0.280	0.28	0

CASE **45**

AMCHEM'S TAX CREDIT

As Harold Davies sat down to finish work on Amchem's tax return, he knew most of the details were available in the huge pile of files on his desk. There was one part, though, where he wasn't sure how to approach things—the sales tax on gas production.

Amchem manufactured chemicals in a plant in the Denver metropolitan area. A recent change in the Colorado tax law provided that all nonproduction use of power was subject to state sales tax. Harold knew that much of Amchem's use of natural gas *was* production-related. The problem was: how much? Public Service Company of Colorado bills and records revealed only the total gas consumption, not its eventual purpose. Harold knew that some part of that consumption went to heat the research facilities and other nonproduction and administrative buildings. That part of it, of course, depended on the weather, among other things, and would be subject to sales tax under the new law.

Harold suspected that most of the gas consumption went into the production processes for two groups of chemicals Amchem produced, labeled generically Group X and Group Y. The chemicals in Group X were fired in electric furnaces that use little gas (what little was used was used in other parts of the production process), while those in Group Y were fired in gas furnaces and used (Harold suspected) quite a lot more gas.

Harold felt that as long as he had a rational, defensible basis for his estimate of how much gas consumption was associated with the production process, he could get a credit of some kind for sales taxes paid on the total amount, and he felt it would be a substantial amount. Looking back over some of the files he'd requested, he found he

This case is based on a real business situation, though the names of the company and its products have been changed. The data were provided by David W. Howell.

had information and some quantities of presumed interest, for each month (MONTH) of the last five years (YEAR), as follows.

- Gas consumption (GAS): in hundred cubic feet (CCF), from gas company records.
- Heating degree days (DEGDAY): a measure of cold weather, obtained from the local weather bureau. The amount of gas required to heat buildings is expected to be related to this measure; that's why the weather bureau tracks it and provides it.
- X production: production of chemicals in group X, from Amchem's monthly records, in thousands of pounds.
- Y production: production of chemicals in group Y, from Amchem's monthly records, in thousands of pounds.

The data are available in file AMCHEM.

Harold knew that all these factors somehow contributed to overall gas consumption at Amchem. He figured other factors such as annual plant shutdowns in summer ("turnaround months") might affect things, too, but their effects were presumably represented by lower production during those months. One particularly extended shutdown in July of year 5 might have distorted things a bit, though.

As it got later in the afternoon, Harold wondered if there were some way to organize the information coherently. What he needed was an estimated amount of gas consumption for production use for year 5 that he could defend rationally. It would be even nicer if the method could be made routine, so it could be used in following years. Of course, it would also be possible to simply not claim the credit, but that was unappealing for any number of reasons.

DATA SET

YEAR	MONTH	DEGDAY	GAS	X	Y
1	1	1206	73950	13.6	23.5
1	2	942	70579	13.8	24.0
1	3	657	66408	15.4	23.2
1	4	421	50294	13.9	29.9
1	5	220	50582	13.9	26.9
1	6	55	45252	13.9	31.9
1	7	0	37145	13.8	14.2
1	8	6	43033	12.1	8.7
1	9	246	34696	9.2	8.6
1	10	460	53204	9.6	8.1
1	11	1017	51097	0.0	11.8
1	12	1070	48270	0.0	5.9
2	1	713	56711	4.5	9.7
2	2	780	60922	9.1	13.8
2	3	457	55902	10.7	13.2
2	4	359	44813	9.3	18.9
2	5	226	46704	7.8	19.4
2	6	12	32222	7.6	12.0
2	7	0	38513	6.3	12.7
2	8	2	39746	0.0	15.5

(concludes on the next page)

YEAR	MONTH	DEGDAY	GAS	X	Y
2	9	233	50892	6.1	19.2
2	10	561	48717	11.4	20.5
2	11	792	51416	10.7	16.9
2	12	1104	57815	7.6	18.5
3	1	1038	71074	13.9	21.7
3	2	770	69414	13.7	35.4
3	3	751	79210	13.8	34.6
3	4	338	65732	10.1	43.8
3	5	258	57291	10.7	36.1
3	6	36	53502	13.7	40.6
3	7	10	50987	14.1	46.2
3	8	39	36257	9.1	17.9
3	9	186	47561	13.6	25.9
3	10	493	60697	13.8	8.7
3	11	797	67631	13.1	21.8
3	12	1214	70093	14.0	23.4
4	1	1243	67630	15.0	21.9
4	2	892	68352	15.0	23.5
4	3	815	77922	15.0	31.8
4	4	453	62700	16.7	34.2
4	5	203	63142	17.8	33.8
4	6	15	53836	17.9	36.3
4	7	13	41154	13.4	26.4
4	8	6	54353	18.3	35.8
4	9	154	57456	18.1	37.0
4	10	412	72824	18.3	34.6
4	11	774	73905	19.8	37.9
4	12	963	70732	15.3	23.5
5	1	942	67563	7.8	9.4
5	2	1126	59740	14.0	24.9
5	3	657	52834	13.7	9.0
5	4	453	61719	13.3	8.8
5	5	218	57658	11.7	18.3
5	6	101	47576	13.6	18.9
5	7	0	12732	2.6	8.6
5	8	4	48714	6.6	12.7
5	9	186	51993	10.4	16.7
5	10	498	61752	10.7	10.2
5	11	710	58444	12.6	16.6
5	12	1233	62857	10.1	13.1

CASE **46**

NATURAL GAS PRICES

Your company produces natural gas and needs a statistical model to provide short-term forecasts of natural gas prices. Of course, forecasting the price of natural gas is difficult, because many different factors influence price. The demand for natural gas from the residential, commercial, and industrial sectors of the United States is a major factor. Natural gas prices may be seasonal, because, for instance, home heating requirements are higher in the winter months. Prices will also reflect the costs of producing natural gas. In addition, natural gas prices tend to mirror prices of *related* energy products such as coal and oil. Gas prices are also influenced by embargoes, production levels in other countries, domestic regulations, and other national and international political factors.

You have agreed to do an initial analysis of trends in natural gas prices. To accomplish this, you have been given monthly data on natural gas prices from January of 1984 to December of 1991. The numbers represent average wellhead price measured in dollars per thousand cubic feet. The data are in the file named GASPRICE.

The Energy Information Administration (EIA) maintains a website of the most current data on natural gas prices. This site is constantly updated as earlier published values are replaced with actual prices, and revisions are made to historical data. The latest data may be located by reviewing the EIA home page *www.eia.doe.gov/emeu/mer/contents.html.* The accuracy of your forecasts may depend on having the most current data available.

Construct a model or other statistical representation of natural gas prices and estimate prices for the first three months beyond the last month in your data. Prepare a report on your methodology and forecasts for presentation at a meeting next week.

Earlier data in this case were collected and analyzed by Thomas G. Funk. He found the data in the *Historical Monthly Energy Review 1973–1988* and the *Monthly Energy Review,* both by the Energy Information Administration, Department of Energy. Subsequent updates were located in the website indicated in the case.

DATA SET

Read across the row:

2.67	2.71	2.67	2.64	2.67	2.70	2.68	2.69	2.62	2.63	2.61
2.57	2.64	2.71	2.62	2.64	2.53	2.58	2.51	2.47	2.42	2.37
2.36	2.28	2.28	2.26	2.16	2.10	1.96	1.85	1.80	1.77	1.78
1.73	1.77	1.76	1.74	1.73	1.73	1.69	1.65	1.65	1.66	1.63
1.56	1.57	1.64	1.70	1.96	1.84	1.70	1.59	1.52	1.53	1.56
1.62	1.53	1.68	1.76	1.89	1.99	1.81	1.69	1.56	1.61	1.65
1.65	1.61	1.55	1.58	1.66	1.92	2.23	1.85	1.55	1.49	1.47
1.48	1.49	1.51	1.56	1.76	1.94	2.04	1.96	1.62	1.49	1.50
1.48	1.43	1.34	1.43	1.59	1.82	1.89	2.00	1.74	1.26	1.35
1.42	1.51	1.62	1.55	1.84	1.92	2.38	2.13	2.07	2.03	1.93
2.00	2.06	2.18	1.98	1.99	2.04	2.09	2.02	2.03	2.15	1.86
1.76	1.82	1.90	2.00	1.83	1.81	1.90	1.94	1.85	1.85	1.98
1.62	1.48	1.47	1.52	1.55	1.58	1.43	1.43	1.52	1.54	1.61
1.84	2.05	1.89	1.95	2.08	2.01	2.08	2.25	2.10	1.85	1.94
2.50	3.26	3.66	2.60	1.72	1.82	2.04	2.18	2.10		

CASE **47**

WESTAR BEVERAGE SALES

The Westar hotel, a 250-room business hotel located in a resort area of the Colorado Rockies, has two lounges on its premises: the Pour Nous Lounge, a pub-style lounge that seats about 50, and Gaieté, a 200-seat nightclub.

Jane Bartley is a staff analyst for Westar's corporate parent, charged (among other things) with predicting and analyzing beverage sales (liquor, beer, and wine, primarily) for the chain's business hotels. In fact, her deadline for producing some sort of forecast for Westar beverage sales was the end of the week. After years in the business, Jane had seen a lot of approaches to forecasting beverage sales, but none that struck her as particularly helpful. Most of them used only room occupancy as a basis for forecasting.

Not that it was unreasonable to consider room occupancy: clearly, the number of people staying in the hotel might well influence beverage sales. Further, hotel occupancy was (at least to some extent) predictable, as people made advance reservations, so it could be used for forecasting. In the case of the Westar, though, Jane suspected other things were involved, and the existing corporate forecasting methods (such as they were) made no allowance for them. Specifically Jane wondered if season, promotional food offered during happy hour, and the presence or absence of a piano player in the Pour Nous Lounge might be factors to consider. She was able to gather data for the most recent 30 months on some relevant variables (described on the next page). The data are available in file WESTAR.

The promotional food had been discontinued after month 12 and the piano player after month 13 as the previous management attempted to cut costs. Data from months 14 and 15 were unavailable because a change in ownership about then had led to some changes in record-keeping techniques (and the subsequent unavailability of certain records).

This case is based upon a real situation, described to us by Gail M. Derby, who provided the data. The names of the hotel chain and the lounges have been disguised.

Jane suspected that with these data she could develop a better understanding of the factors that really influenced beverage sales at the Westar and perhaps even influence her department to consider some new and better ways to do the forecasting.

DEFINITIONS

MONTH	Month number (month 1 was a January)
BEVTOTAL	Total beverage sales (in $), pub and nightclub
BEVPUB	Beverage sales (in $) for the Pour Nous Lounge
BEVNITE	Beverage sales (in $) for the Gaieté night club
FOODPUB	Amount spent on food (in $) for the Pour Nous Lounge
FOODNITE	Amount spent on food (in $) for the Gaieté night club
ROOMS	Total room occupancy for the Westar

DATA SET

MONTH	BEVTOTAL	BEVPUB	BEVNITE	FOODTOT	FOODPUB	FOODNITE	ROOMS
1	24443	7254	17189	2317	210	2107	4154
2	22924	7238	15686	2341	200	2141	4457
3	27501	8150	19351	3353	230	3123	5077
4	29282	8081	21201	2796	210	2586	4456
5	21447	7011	14436	2508	220	2288	4733
6	20404	6076	14328	2674	220	2454	5383
7	19051	8001	11050	2381	210	2171	5010
8	19235	8636	10599	2977	230	2747	5030
9	21498	9064	12434	1624	220	1404	5372
10	26619	8760	17859	1852	210	1642	4889
11	25600	7794	17806	1676	220	1456	4295
12	30373	8171	22202	2706	220	2486	3425
13	23483	8186	15297	2497	0	2497	3893
14	*	*	*	*	*	*	*
15	*	*	*	*	*	*	*
16	26004	7703	18301	2465	0	2465	4395
17	19624	6764	12860	2183	0	2183	4524
18	20459	6524	13935	3589	0	3589	5044
19	20956	4209	16747	3010	0	3010	5174
20	29754	7084	22670	3275	0	3275	6331
21	29098	5321	23777	3225	0	3225	4706
22	30559	5879	24680	3448	0	3448	4259
23	31038	5882	25156	3835	0	3835	4284
24	33833	4440	29393	3785	0	3785	3364
25	30095	6935	23160	3198	0	3198	4250
26	28163	4421	23742	3657	0	2657	4342
27	34051	5561	28490	3054	0	3054	5006
28	33139	6033	27106	2930	0	2930	4761
29	29468	5180	24288	1657	0	1657	3860
30	28896	6222	22674	2407	0	2407	5300

CASE **48**

PRICE-FIXING CASES

Price-fixing (when two or more companies conspire to fix the price of a product or service rather than let it be determined by normal market forces) violates a number of federal laws in the United States, in particular, section 2 of the Sherman Act. Findings of guilty in such cases can lead to severe fines and other criminal penalties for the firms involved. One factor that influences the number of such cases in any given year is the actual number of price-fixing conspiracies, of course, but others include the zeal with which the Department of Justice prosecutes such cases and how it interprets both the law and the evidence. In a sense, then, we may think of the number of cases of price-fixing as the output of a complex process, a process by which society spawns such conspiracies and the government investigates and prosecutes them.

The data below are the number of price-fixing cases for each year from 1961 to 1987. This covers the administrations of Kennedy and Johnson (Democratic), Nixon and Ford (Republican), Carter (Democratic), and Reagan (Republican). The data include only those cases that resulted in either guilty findings or pleas of nolo contendere (which can be considered equivalent to guilty pleas for our purposes here). Each case is listed under the year in which an indictment was returned.

Do these data support the claim that the Reagan administration (1981–1988) adopted a more lenient policy with regard to prosecuting price-fixing cases than had been the case in previous administrations?

This case is based on data provided by E. W. Eckard, derived from the reports in the Commerce Clearing House Trade Regulation Reports.

DATA SET

YEAR	CASES	YEAR	CASES	YEAR	CASES	YEAR	CASES
1961	4	1968	8	1975	10	1982	3
1962	9	1969	1	1976	11	1983	9
1963	4	1970	6	1977	12	1984	9
1964	5	1971	8	1978	9	1985	2
1965	4	1972	12	1979	5	1986	5
1966	5	1973	6	1980	14	1987	7
1967	7	1974	10	1981	5		

CASE **49**

PROPERTY CRIMES

You've been asked to undertake a study of the determinants of property crimes in the United States for presentation at an upcoming meeting of business executives and community leaders. Your task is to provide evidence, for or against, common perceptions about property crime. Are crime rates higher in urban than rural areas? Does unemployment or education level contribute to property crime rates? How about public assistance? What other factors relate to property crimes? The file names PCRIMES may help answer some of these questions. Use these data to prepare a report on the characteristics of and determinants of property crimes in the United States.

DEFINITIONS

CRIMES	Property crime rate per hundred thousand inhabitants (property crimes include burglary, larceny, theft, and motor vehicle theft); calculated as # of property crimes committed divided by total population/100,000
PINCOME	Per capita income for each state
DROPOUT	High school drop out rate (%, 1987)
PRECIP	Average precipitation in inches in the major city in each state over 1951–80

The data for this case come from a variety of U.S. government sources, including: the 1988 Uniform Crime Reports, Federal Bureau of Investigation; the Office of Research and Statistics, Social Security Administration; the Commerce Department, Bureau of Economic Analysis; the National Center for Education Statistics, U.S. Department of Education; the Bureau of the Census, Department of Commerce and Geography Division; the Labor Department, Bureau of Labor Statistics; and the National Climatic Data Center, U.S. Department of Commerce. The data set was originally collected by Louis J. Moritz, an operations manager.

PUBAID	Percentage of public aid recipients (1987)
DENSITY	Population/total square miles
KIDS	Public aid for families with children, dollars per family
UNEMPLOY	Percentage of unemployed workers
URBAN	Percentage of the residents living in urban areas
STATE	1 = Alabama; 2 = Alaska; 3 = Arizona; 4 = Arkansas; 5 = California; 6 = Colorado; 7 = Connecticut; 8 = Delaware; 9 = Florida; 10 = Georgia; 11 = Hawaii; 12 = Idaho; 13 = Illinois; 14 = Indiana; 15 = Iowa; 16 = Kansas; 17 = Kentucky; 18 = Louisiana; 19 = Maine; 20 = Maryland; 21 = Massachusetts; 22 = Michigan; 23 = Minnesota; 24 = Mississippi; 25 = Missouri; 26 = Montana; 27 = Nebraska; 28 = Nevada; 29 = New Hampshire; 30 = New Jersey; 31 = New Mexico; 32 = New York; 33 = North Carolina; 34 = North Dakota; 35 = Ohio; 36 = Oklahoma; 37 = Oregon; 38 = Pennsylvania; 39 = Rhode Island; 40 = South Carolina; 41 = South Dakota; 42 = Tennessee; 43 = Texas; 44 = Utah; 45 = Vermont; 46 = Virginia; 47 = Washington; 48 = West Virginia; 49 = Wisconsin; 50 = Wyoming

DATA SET

STATE	CRIMES	PINCOME	DROPOUT	PUBAID	DENSITY	KIDS	PRECIP	UNEMPLOY	URBAN
1	4003.1	12604	30.5	6.5	80.8	114	59.4	7.2	60.0
2	4398.8	19514	26.4	4.3	0.9	593	53.2	9.3	64.3
3	6861.2	14887	30.0	3.5	30.7	268	7.1	6.3	83.8
4	3796.9	12172	21.4	5.9	46.0	190	49.2	7.7	51.5
5	5705.7	18855	31.5	8.8	181.2	581	17.3	5.3	91.3
6	5705.7	16417	24.0	3.7	31.9	317	15.3	6.4	80.6
7	4642.2	22761	21.8	4.3	663.6	486	44.4	3.0	78.8
8	4347.4	17699	29.0	4.3	341.7	267	41.4	3.2	70.6
9	7819.9	16546	36.5	4.1	227.8	240	55.2	5.0	84.6
10	5661.2	14980	35.0	6.4	109.2	252	48.6	5.8	62.4
11	5731.9	16898	15.5	4.9	170.9	481	23.5	3.2	86.5
12	3738.2	12657	20.4	2.7	12.2	250	11.7	5.8	54.0
13	4810.4	17611	22.2	7.5	208.7	309	33.3	6.8	83.3
14	3770.0	14721	24.1	3.6	154.6	263	39.1	5.3	64.2

(concludes on the next page)

STATE	CRIMES	PINCOME	DROPOUT	PUBAID	DENSITY	KIDS	PRECIP	UNEMPLOY	URBAN
15	3819.8	14764	13.4	5.0	50.6	348	38.6	4.5	58.6
16	4514.7	15905	15.9	3.8	30.5	338	28.6	4.8	66.7
17	2804.7	12795	32.1	7.0	93.9	204	43.6	7.9	50.9
18	5043.3	12193	38.4	8.8	99.0	167	59.7	10.9	68.7
19	3420.3	14976	21.2	6.5	38.9	370	43.5	3.8	47.5
20	4897.9	19316	23.5	5.1	469.9	329	41.8	4.5	80.3
21	4371.3	20701	24.0	5.9	752.7	536	43.8	3.3	83.8
22	5342.7	16387	28.6	8.5	162.2	481	31.0	7.6	70.7
23	4024.7	16787	11.3	4.6	54.1	515	26.4	4.0	66.9
24	3267.6	10992	34.4	11.1	55.5	119	52.8	8.4	47.3
25	4292.2	15492	23.9	5.5	74.6	264	33.9	5.7	68.1
26	4144.0	12670	15.5	4.5	5.5	362	11.4	6.8	52.9
27	3866.8	15184	13.3	3.8	20.9	320	30.3	3.6	62.9
28	5672.5	17440	18.7	2.6	9.6	273	4.2	5.2	85.3
29	3186.1	19016	25.4	1.7	120.7	407	36.5	2.4	52.2
30	4712.5	21882	20.3	5.6	1033.9	357	41.9	3.8	89.0
31	5948.1	12481	26.8	5.7	12.4	225	8.9	7.8	72.1
32	5212.0	19299	33.3	8.0	378.0	523	39.3	4.2	84.6
33	4360.4	14128	30.9	5.0	132.9	243	42.5	3.6	42.9
34	2668.9	12720	11.6	3.1	9.6	350	15.4	4.8	48.8
35	4193.2	15485	20.8	7.5	264.7	298	37.8	6.0	73.3
36	5154.6	13269	24.2	4.8	47.2	278	30.9	6.7	67.3
37	6513.0	14982	29.2	4.0	28.8	348	37.4	5.8	67.9
38	2814.4	12168	18.9	6.1	267.4	347	40.0	5.1	69.3
39	4807.7	16793	28.0	6.0	940.9	450	41.9	3.1	87.0
40	4671.2	12764	32.2	6.3	114.9	186	51.6	4.5	54.1
41	2467.3	12475	13.1	3.9	9.4	270	17.5	3.9	46.4
42	3936.7	13659	32.8	6.4	118.9	155	48.5	5.8	60.4
43	7365.1	14640	34.1	4.5	64.3	169	42.0	7.3	79.6
44	5335.5	12013	17.5	3.2	20.6	343	15.3	4.9	84.4
45	4098.2	15382	17.3	5.6	60.1	469	33.7	2.8	33.8
46	3877.5	17640	23.4	4.0	151.5	257	45.2	3.9	66.0
47	6646.6	16569	21.9	5.8	69.9	443	38.6	6.2	73.5
48	2107.4	11658	22.9	8.2	77.8	238	40.7	9.9	36.2
49	3757.6	15444	16.3	7.7	89.2	473	30.9	4.3	64.2
50	3653.1	13718	20.4	3.2	4.9	303	13.3	6.3	62.7

CASE **50**

ROCKVILLE PHYSICIANS' GROUP

In late November, Dr. Eaton sat down to review the results of a study of cesarean sections (C-sections) at the Rockville Physicians' Group (RPG). He knew there were interesting findings in the data gathered over the past year, and he wanted to understand those findings. For example, he had discovered that the C-section rates differed dramatically among the various RPG physicians, from about 14% of all births to about 30%. He also knew, though, that how he summarized and communicated those findings to the obstetrics/gynecology (OBG) staff at a meeting scheduled for mid-January was probably just as important as the results themselves, at least in terms of getting them to "buy in" to the ideas of quality monitoring and measurement that he wanted to establish at RPG.

RPG was founded in the early 1950s as one of the earliest group practices in its geographic region. Located 30 miles from a major city, Rockville was at that time perhaps best known as an attractive suburb and as the home of a state university. Since then, it had grown enormously to about 100,000 residents, plus more than 10,000 students. Other industrial and governmental offices were located in or near Rockville and a major hi-tech firm had its laboratories in a university-sponsored research park near town. Thanks to its attractive scenery, generally liberal attitudes, and a far-seeing (aggressive?) no-growth policy, Rockville had become an affluent (exclusive?) city, and RPG had prospered along with it. It was easily the largest such practice in the Rockville area and had contracts and arrangements with most major insurance and reimbursement agencies, various health maintenance organizations, and others. In all, 50 physicians worked at the center. Total other staff ran about 185.

Dr. Eaton had joined RPG more than 30 years earlier, after leaving the armed services, and practiced internal medicine there until recently, when he retired from active practice

This care is based on a real situation, though the names of the organization and the individuals have been changed. The situation was described for us by John S. Avery, M. D.

and took an (approximately) half-time job as RPG's medical director. As one of his first major forays into administration, he had played a major role, including negotiations with contractors and various regulatory agencies, in the design and construction of a building expansion for the group. He had been involved in contract talks with some of the insurance and other reimbursement agencies, and his interest in computers and data management had grown in the last few years. He had discovered he enjoyed administration, and felt he'd been reasonably successful at it.

There had been a quality assurance committee for some years at RPG, chaired by one of the physicians. Historically, topics for study had been selected in a more-or-less democratic manner by the physicians involved, but those topics had tended to be administrative rather than medical (studies of waiting times, telephone systems, and so on). Dr. Eaton felt the Group needed to be systematically involved in quality monitoring of the medical aspects of their operations, beyond the usual kinds of case review sessions they routinely held. In his words, he was

> hoping to build into our quality assurance efforts . . . these [attitudes and philosophies]:
>
> • Focus on the center of a quality/frequency bell curve, moving the curve to the right, rather than on the detection of instances of poor care, lopping off the left tail of the curve.
>
> • Encouraging self-examination, becoming aware of what our patterns of practice actually were. Stepping back and looking at ourselves. Escaping habits of which we were scarcely conscious.
>
> • Encouraging a research attitude, being quantitative and systematic in looking at ourselves. Creating the possibility of publication.
>
> • Having the courage to be independent; learning from ourselves, not being bound by the received opinions of the day.
>
> I saw this as a way to enhance the fun of practice as well as the quality of care, to lessen the insidious onset of staleness and routine that you see in all professions.

Accordingly, the ground rules of the committee were changed to require that the topics be restricted to the study of the patterns of care and the use of the results in medical decision making. This year, after some discussion with various people involved, the committee decided to focus on the OBG department, and particularly the study of C-sections. Dr. Eaton set up a small team working under his supervision to gather and process data. The report from this study was to be presented at the mid-January meeting.

Cesarean section rates in the United States have been running about 20% to 25% of births in recent years. This rate is a concern to many in the medical community, both because of the costs involved and because it is dramatically higher (double, in many cases) than the rate in the rest of the developed world. There is perhaps some evidence that it is influenced by fear of clinical failure and/or malpractice claims, but Dr. Eaton didn't know specifically whether that or any other identifiable influence was at work in the case of the medical center, whose rates had been around the national averages in recent years.

Working with Mary Spiegel, an assistant, he identified all births, including C-sections and vaginal deliveries, from October 1 to September 30. From the computer system used for administrative and billing operations, they were able to identify the following information for all the cases.

- Chart number.
- Procedure code.
- Doctor number (six different physicians were involved).
- Date of service.

- Charge.
- Financial code (type of insurer, or none).
- Diagnosis code.

In addition, they identified a sample consisting of all C-sections (178 of them) and 176 randomly chosen vaginal deliveries for more detailed analysis. For the sample of 354 patients, charts were pulled and the following further information was recorded.

- Maternal age.
- Parity.
- Gravida status.
- Immediate and underlying reasons for the C-section.
- Potential reasons for a C-section (this refers to cases in which a C-section was not performed, but where some mention was made of a condition potentially leading to a C-section).
- Time in labor.
- Apgar-1 and Apgar-5 Scores (a measure of general condition at age one minute and five minutes for the baby).
- Gestational age.
- Birth weight.
- Maternal death (Y/N).
- Fetal death (Y/N).

While it was a time-consuming exercise, there had been no particular difficulties in obtaining the information, at least none that Mary (with occasional consultation with Dr. Eaton) had been unable to resolve. The process had been greatly facilitated by the existence of fairly standard (nationwide) forms and coding schemes for OBG procedures.

The combined data had been entered into a database package for a personal computer. A sample page of direct printout from the program is attached (Exhibit A). Using the database package, it was possible to select certain subsets of the data based on user-specified criteria. For example, it was possible to ask to look at all those (and only those) cases in which labor time was greater than two hours and the Apgar-1 score was less than 5, if such a combination was considered interesting. It was also possible to extract data from the data base package and "send" it to spreadsheet programs or graphics packages.

The Quality Assurance and Utilization Management (QAUM) committee consisted of the RPG's administrator, nursing supervisor, the head nurse of the area involved (OBG, in this case), and the area physicians. All were expected to have reviewed the agenda and materials for the meeting beforehand and everyone was expected to comment on the materials or the topic generally. At the meeting scheduled for mid-January, the topic would be the results of the C-section study. Dr. Eaton's original goals for the study had been basically to provide information to the staff, help them understand what the data suggested, elicit their suggestions for changes that might be made, and obtain their support for the monitoring and quality activities he hoped to continue at RPG. Now armed with the results of the C-section study, he wondered how best to present them to the OBG staff.

EXHIBIT A SAMPLE PAGE OF PRINTOUT

```
   100% indexed        44632 Records indexed

 . display structure
 Structure for database: C:
 Number of data records:       1017
 Date of last update    : 03/14/91
 Field  Field Name  Type        Width    Dec
     1  CHART       Character       6
     2  PROC_CODE   Character       5
     3  DOCTOR_NUM  Character       4
     4  AGE         Numeric         2
     5  PARITY      Numeric         2
     6  GRAVIDA     Numeric         2
     7  DIAG_CODE   Character       5
     8  CLASS       Character       1
     9  BIRTH_WGT   Numeric         4      1
    10  GEST_AGE    Numeric         2
    11  REASON_1    Numeric         2
    12  REASON_2    Numeric         2
 ** Total **                       38

 .
 Command Line     :<C:>:                          :Rec: 1/1017          :Ins    :Num

                      Enter a dBASE III PLUS command.

 Record#  CHART  PROC_CODE DOCTOR_NUM AGE PARITY GRAVIDA DIAG_CODE CLASS BIRTH_W
 T GEST_AGE REASON_1 REASON_2
       1  313560 59410          103    33      1       3 V222      2           6
 2      40        0          0
       2  351602 59500          103    39      2       4 V222      1           9
 5      39       11         18
       3  222600 59410           18    34      1       2 V222      9           8
 5      50        0          0
       4  307185 59500           42    30      1       3 6612      1           6
 6      40       22          0
       5  307185 59500           94     0      0       0 6612      1           0
 0       0        0          0
       6  349227 59410           42     0      0       0 V222      1           0
 0       0        0          0
       7  349135 59500           42    27      0       1 V222      1           0
 0      40       24          0
       8  349135 59500           80     0      0       0 7684      1           0
 0       0        0          0
       9  350727 59410          103     0      0       0 V222      8           0
 0       0        0          0
      10  349645 59410           80    21      0       1 V222      1           7
 5      42        0          0
      11  335661 59410           18    31      1       2 V222      1           6
 5      40        0          0
      12  340738 59410           18     0      0       0 V222      7           0
 0       0        0          0
      13  342832 59410           80     0      0       0 V222      1           0
 0       0        0          0
      14  255875 59410          103     0      0       0 V222      1           0
 0       0        0          0
      15  342472 59410           18    28      2       4 V222      9           7
 8      37        0          0
      16  336306 59410           80    29      1       2 V222      1           6
 4      40        0          0
      17  293233 59500          103    36      0       3 V222      2           7
 2      39       14         23
```

Obtained by chart-pull, not from main computer

CASE **51**

ABANDONED CALLS—I

Southern Foods, Inc., based in Atlanta, Georgia, sells food products to restaurants throughout the southeastern United States. Southern has over 50 warehouses located in Georgia, Florida, Tennessee, Alabama, South Carolina, and North Carolina. Warehouses stock the full range of grocery items needed by restaurants. When a restaurant manager places an order, delivery is expected within 24 hours. Because many of the products shipped by Southern are perishable, timely delivery is imperative to retaining customers in this business. And because smaller restaurants have limited storage capacity, Southern's customers sometimes make multiple orders per day. It's a business that requires good record keeping and prompt response to maintain good customer service.

Southern relies on a sales force of several hundred people to contact local restaurants, take orders for food products, see that the items are delivered on time, and make periodic follow-up visits with restaurant managers. The salespeople are almost like independent contractors; each has his or her own sales region and is paid on commission. Southern does not supply its salespeople with offices. Most of the sales staff work at home when necessary, but spend the majority of their day on the road.

In past years, fulfilling an order required cumbersome steps. The salesperson would record the order on a ledger, call the warehouse manager, and check on availability of the requested items. The warehouse supervisor would then begin the process of having the items loaded onto trucks and delivered to the appropriate location. Information technology has expedited the process considerably. Now salespeople connect directly to the Electronic Data Ordering System (EDOS) on Southern's mainframe computer. Once connected, the salesperson can check on availability of food products and delivery trucks, do all record keeping about the transaction, and send immediate requests for

These data were collected and studied by David S. Chaney (Manager, Sales Development, Citicorp Diner's Club), Holden U. Holt, and others who wish to remain unnamed. The name of the company and its primary business have been changed.

information to managers or other salespeople if orders cannot be filled. The new system has cut delivery time by about half. Executives at Southern are so impressed that they have directed the information systems group to continually update the software to suit the needs of the sales force.

To use EDOS, salespeople connect their laptops to the mainframe via one of 14 dedicated mainframe modems. Unfortunately, Southern's modems are dated. They will become inoperable if a salesperson sends a bad instruction. Even worse, the modems need to be manually reset by someone in the computer center once the modem has been diagnosed as inoperable. This, and occasionally heavy volumes of calls, can cause salespeople to hang up in frustration. When a call is abandoned, the salesperson relies on the old system of placing the order manually.

Complaints about access to EDOS are increasing. Salespeople say they can't get onto the system quick enough and often enough. Southern is considering three options: purchasing more modems, upgrading the existing modems, or encouraging salespeople to use the system during off-peak hours. Although the last option is appealing to Southern's top executives, it's uncertain whether this approach will solve the problem. Management would like to see abandonments be 3% or less of all incoming calls.

The data named ABANDON1 provide daily (weekdays only) information about the number of calls answered by EDOS (CALLS), the number of abandoned calls (HANGUPS), and the number of modems operable. The first observation is the first Monday in October of last year; the data run through the end of January of this year. Holidays are indicated by an asterisk.

To address the issue of peak and off-peak usage of the system, you've asked the information systems department to provide a data set broken into half-hour readings. That data set, named ABANDON2, includes the number of calls (CALLS) answered by EDOS and the number of abandoned calls (HANGUPS) during that 30-minute interval. Unfortunately, the number of operable modems was not recorded in 30-minute intervals. The best the IS department could do was supply the minimum number of modems available at some point during that day. The first observation in the data set is the 30-minute period beginning at midnight on the first Monday in December of last year. The data set runs through Friday of the same week.

Examine the data and provide any observations or recommendations about the problem that might be suggested by the data.

DATA SETS
ABANDON1

CALLS	HANGUPS	MODEMS
2832	5	13
2703	10	13
2600	8	12
2803	13	12
2790	8	12
3510	28	13
2968	7	13
3418	17	14
2758	7	14
2494	4	14
2964	13	14
2587	4	14
2890	14	14
2797	6	12
2504	4	14
2653	7	14
2764	3	14
2745	9	14
2955	13	14
2751	6	14
2808	6	14
*	*	*
2725	6	14
2796	5	14
2882	7	14
2821	10	14
2757	6	14
2711	6	13
2734	6	14
2836	4	14
2962	4	14
2838	7	13
2847	13	14
2766	22	14
2895	7	14
2758	4	12
2895	2	12
2848	6	12
*	*	*
*	*	*
3888	343	14
2753	6	14
2868	11	12
2766	16	13

CALLS	HANGUPS	MODEMS
2872	9	14
3394	255	12
3288	329	12
3112	339	13
3096	339	13
3373	347	13
3635	431	14
3163	221	14
2747	11	14
2782	6	14
2607	3	14
3148	5	14
2943	5	14
2768	3	14
3047	9	14
2385	7	14
*	*	*
2506	0	12
2747	1	13
2730	3	14
2693	5	14
*	*	*
2594	5	14
2568	1	14
2704	5	14
2927	4	13
2955	4	14
2266	4	14
2713	4	14
2722	8	14
2822	8	14
2713	7	14
2914	6	14
2661	7	14
2817	6	13
2864	8	13
3261	358	12
2758	253	12
2717	253	11
2704	279	12
2671	291	12
2702	279	12
2701	305	11
2748	339	11

(continued on the following page)

ABANDON2

CALLS	HANGUPS	MODEMS	CALLS	HANGUPS	MODEMS
20	1	10	13	3	10
16	2	10	8	1	10
9	1	10	5	0	10
16	1	10	16	1	10
15	1	10	9	1	10
20	2	10	4	1	10
20	2	10	4	0	10
20	2	10	16	0	12
35	3	10	4	1	12
54	4	10	15	1	12
67	8	10	15	0	12
83	10	10	14	0	12
164	23	10	10	1	12
186	29	10	24	2	12
182	32	10	22	2	12
176	31	10	19	1	12
195	25	10	60	5	12
166	20	10	54	5	12
149	17	10	79	5	12
160	20	10	126	11	12
157	14	10	134	15	12
121	14	10	206	27	12
109	15	10	148	36	12
115	14	10	167	34	12
113	14	10	212	24	12
116	13	10	169	18	12
118	14	10	132	15	12
120	12	10	120	12	12
131	13	10	159	1	12
132	13	10	91	1	12
129	13	10	110	0	12
97	13	10	89	1	12
86	9	10	93	0	12
56	8	10	89	0	12
51	6	10	102	0	12
33	5	10	96	0	12
31	3	10	81	0	12
24	3	10	81	0	12
27	3	10	61	0	12
28	2	10	58	0	12
29	2	10	43	0	12
32	0	12	38	0	12

(continues on the following page)

CALLS	HANGUPS	MODEMS	CALLS	HANGUPS	MODEMS
28	0	12	102	0	14
20	0	12	101	0	14
22	0	12	54	0	14
18	0	12	51	0	14
20	0	12	54	0	14
13	0	12	25	1	14
18	0	12	31	0	14
6	0	12	20	0	14
15	0	12	31	0	14
11	0	12	27	1	14
7	0	12	24	0	14
16	0	12	22	0	14
23	0	14	29	0	14
13	0	14	11	0	14
14	0	14	9	0	14
13	0	14	8	0	14
14	0	14	8	0	14
15	0	14	11	0	14
24	0	14	6	0	14
17	0	14	10	0	14
21	0	14	8	0	14
54	0	14	6	0	14
59	0	14	8	0	14
76	0	14	10	0	14
75	0	14	10	0	14
119	0	14	12	0	14
176	1	14	12	0	14
161	1	14	15	0	14
134	0	14	24	0	14
129	0	14	58	0	14
122	0	14	53	0	14
101	1	14	114	0	14
102	0	14	141	1	14
85	1	14	154	1	14
67	0	14	177	4	14
101	0	14	175	1	14
57	1	14	123	1	14
93	0	14	114	0	14
72	0	14	96	2	14
87	0	14	99	1	14
99	1	14	68	1	14
95	0	14	88	0	14

(concludes on the following page)

CALLS	HANGUPS	MODEMS
102	1	14
90	0	14
95	0	14
109	0	14
97	0	14
118	0	14
104	0	14
95	0	14
58	0	14
56	0	14
21	0	14
24	0	14
27	0	14
23	0	14
14	0	14
9	0	14
21	0	14
9	0	14
13	0	14
10	0	14
10	0	14
8	0	14
11	0	14
9	0	14
12	0	14
8	0	14
12	0	14
12	0	14
14	0	14
12	0	14
12	0	14
17	0	14
15	0	14
28	0	14
67	0	14
65	0	14
97	0	14
143	0	14
166	2	14
154	1	14

CALLS	HANGUPS	MODEMS
156	0	14
154	0	14
108	0	14
118	0	14
84	0	14
89	0	14
86	0	14
97	0	14
85	0	14
80	1	14
113	0	14
80	0	14
127	1	14
137	1	14
88	1	14
74	1	14
63	0	14
54	0	14
32	0	14
16	0	14
16	0	14
13	0	14
14	0	14
7	0	14
11	0	14
17	0	14
9	0	14
10	0	14
5	0	14
5	0	14
4	0	14
4	0	14

CASE **52**

ABANDONED CALLS—II

Southern Foods, Inc., based in Atlanta, Georgia, sells food products to restaurants throughout the southeastern United States. Southern has over 50 warehouses located in Georgia, Florida, Tennessee, Alabama, South Carolina, and North Carolina. Warehouses stock the full range of grocery items needed by restaurants. When a restaurant manager places an order, delivery is expected within 24 hours. Because many of the products shipped by Southern are perishable, timely delivery is imperative to retaining customers in this business. And because smaller restaurants have limited storage capacity, Southern's customers sometimes make multiple orders per day. It's a business that requires good record keeping and prompt response to maintain good customer service.

Southern relies on a sales force of several hundred people to contact local restaurants, take orders for food products, see that the items are delivered on time, and make periodic follow-up visits with restaurant managers. The salespeople are almost like independent contractors; each has his or her own sales region and is paid on commission. Southern does not supply its salespeople with offices. Most of the sales staff work at home when necessary, but spend the majority of their day on the road.

To place orders, salespeople connect to the Electronic Data Ordering System (EDOS) on Southern's mainframe computer. Once connected, the salesperson can check on availability of food products and delivery trucks, do all record keeping about the transaction, and send immediate requests for information to managers or other salespeople if orders cannot be filled.

To use EDOS, salespeople connect their laptops to the mainframe via one of 14 dedicated mainframe modems. As a result of a study completed this past summer, the old EDOS modems were slowly replaced with new models that are operable most of the

These data were collected and studied by David S. Chaney (Manager, Sales Development, Citicorp Diner's Club), Holden U. Holt, and others who wish to remain unnamed. The name of the company and its primary business have been changed.

time. The salespeople have remarked on the improvement the new modems have made in their ability to connect to EDOS. Hangup calls now occur only during peak times of the day.

Of the 14 modems currently in use, only one of the old models remains. These days, it is out of service about 50% of the time. The information systems (IS) department would like to disconnect the old modem, but is unsure whether it needs to be replaced or not. The manager of the IS department would like to know whether that 14th modem, when operable, is really needed. Would 13 new modems provide the salespeople with the same level of access to the EDOS as 14?

You've been asked to look into this issue by the IS manager. The file, named ABANDON3, includes data about the number of calls (CALLS) answered by the EDOS and the number of abandoned calls (HANGUPS) during two-hour intervals. The data begins at midnight of the second Monday in October of this year and runs for five days. The IS manager seems to think that this week is fairly typical of use of the system these days. During Monday and Tuesday of the recorded week, the old modem was out of commission. All 14 modems were operable for the remainder of the week.

What do you recommend to the IS manager about replacing the old modem? The issue is complicated somewhat by a new directive from Southern's CEO: abandoned calls are not to exceed 1% of all attempted calls to the EDOS system.

DATA SET

CALLS	HANGUPS	CALLS	HANGUPS	CALLS	HANGUPS	CALLS	HANGUPS
46	0	606	0	505	5	77	0
57	0	594	2	417	0	59	0
251	0	394	2	178	0	43	0
582	13	392	2	73	0	46	0
681	9	348	0	58	0	56	0
632	3	174	0	44	0	148	0
513	2	81	1	44	0	525	2
366	1	47	0	53	0	487	1
186	0	36	0	170	0	275	1
102	0	27	0	520	0	360	0
51	0	42	0	494	2	351	0
43	0	239	0	367	0	127	0
30	0	715	5	346	1	48	0
55	0	716	5	395	4	32	0
202	0	394	2	190	0	39	0

CASE **53**

EVERYBODY DOES IT

"Why *can't* they examine all the files? They manage to absorb 30% per year increases in their budgets. Let them show what they can do with all that money! They call themselves 'Information Services,' don't they? Well, I need some information, and they're the ones who have it."

The president's tone of voice did not seem to allow for much disagreement, so Fred let the issue drop, but on the way back to his office, he had a "why me?" feeling about the whole affair. Somehow, what started as a simple case of an employee's using a company computer for personal business had gone out of control, and now the president was irritated enough to decree that the Information Services (IS) division would examine every file in every company computer to see if it represented appropriate use of company resources. As the vice president to whom the IS division reported, it fell to Fred to implement this idea. His only hope at the moment was that perhaps the president was just having a bad day and would relent if an alternative approach could be developed.

BACKGROUND

Fred's company, MegaMicro, had 40,000 employees in 10 sites in the United States. They manufactured a variety of software, micro-code, and hardware products sold to many different kinds of customers and industries.

MegaMicro's 10 sites used approximately 100 mainframe computers, linked by various networks, as near as Fred could remember offhand, and no doubt thousands of personal computers, many of *them* linked to servers, mainframes, and each other. "That must add up to millions of files," he mused, "and there would be no way of automating the examination of them. We simply have to talk him out of the idea."

This case was inspired by an actual situation with which we are familiar, but all specific characters and the company name are fictitious.

THE INCIDENT

About a month ago, a supervisor at a Louisiana MegaMicro site had passed the desk of an employee who reported to her. While the employee was not at his desk, the supervisor did notice a printer actively churning out paper, which to all appearances could be nothing but the up-to-date standings of the local bowling association, and apparently about 50 copies. The individual involved had been warned about similar activities before, and the supervisor felt she had no choice but to invoke disciplinary action, which (in view of the employee's past record) had led to a notice of dismissal.

The employee in turn appealed the dismissal to higher levels of the company, using the novel defense of "everybody does it." While at first no one took this defense particularly seriously, the employee persevered, and when the company attorneys were consulted, they advised that if, in fact, everybody *did* do it, the employee might have a case. To be sure, there were stated policies against use of company resources (and computers in particular) for nonbusiness purposes, but there were similar policies against using company telephones, too, and, as the employee pointed out, no one was ever fired for calling home to ask if the family needed milk, to say he or she would be 10 minutes late, or to check to see if the kids were OK, even though such acts did violate stated company policies. The *unwritten* rule permitted them.

In short, if everybody did it, it *might* actually be "OK." Thus the higher levels of the firm decided they needed to know the extent to which people actually did use company computers for nonbusiness purposes.

Fred presumed that the president hadn't really thought about the details when he suggested that the IS division should examine every file. Perhaps he could be persuaded to settle for an examination of a sample of files, but Fred wasn't sure. Even if only a sample was used, the problem loomed large: how could IS determine if a file was or was not appropriate for company business? Some, of course, would clearly be inappropriate, but many would be hard or impossible to label "OK" or "BAD." He had no ready alternative, though: it was unlikely that employees would respond to a survey saying, in effect, "yes, I cheat." For present purposes, Fred had no interest in knowing *who* might cheat. (Indeed, he had no *interest* at all in the matter, but apparently it wasn't going to go away on its own.) He just needed a defensible estimate of, say, what *fraction* of the employees did so, or what percentage of the system resources were so used, or something similar.

Devise a practical procedure, survey, or other process that Fred and/or the IS division could use to obtain a reasonable estimate of the extent to which MegaMicro's employees *do* use company computers for personal purposes.

CASE **54**

MARRIAGE PREP

According to a recent report[1] a 12-hour course entitled "Prevention and Relationship Enhancement Program" (PREP, for short) may offer a kind of "Geneva convention for marriage." The article continues:

> In lectures and coaching sessions, couples learn and practice ground rules for airing complaints so they do not get stuck in the negativity that is so corrosive to love.
>
> In a study in Germany, 64 couples who took PREP are being tracked along with 32 members of a comparison group, 18 who received standard premarital counseling and 14 who received no counseling at all. Five years after the course, given in six weekly sessions of two hours, the PREP couples have a divorce rate of 4 percent. In comparison, 24 percent of the couples in the comparison group have divorced. Those who took PREP also reported half the incidence of physical aggression.

Do these results convince you that PREP makes a difference? Is the difference statistically "significant"? How would you rate the article as an example of statistical reporting?

[1] See "Rescuing Marriages Before They Begin," by Hara Estroff Marano, *The New York Times,* page C9, May 28, 1997. Copyright © 1997 by The New York Times. Reprinted by permission.

CASE **55**

PREDICTING SNOWFALL

Ski resorts depend critically on snow, of course. Even with snow-making machinery, snow from nature is still the most important part of providing good skiing. Machinery can make only a small part of the snow needed (and the water to make the snow with can be expensive).

For planning purposes, it would be nice to be able to predict snowfall, at least approximately. That information would help estimate traffic to the slopes, required snow-making, inventories to keep in local shops, and help in many other ways.

Is it possible to predict snowfall from past data? The data set in file SNOW (and listed on the next pages) provides recorded snowfall for the most recent 60 years for Aspen, Colorado, a resort city in the Rocky Mountains. The variables are

WINTER	An index (1 through 60). For these purposes, winter begins in September and ends in June.
SEPT	Recorded snowfall (in inches) for September
OCT	Recorded snowfall (in inches) for October
⋮	
JUNE	Recorded snowfall (in inches) for June

Snowfall for July and August was not recorded. "T" stands for a "trace." Missing values are recorded as "*".

Summarize the relevant patterns you observe in the data. Develop a method for predicting snowfall for this city, and use your method to predict snowfall for winters 61, 62, and 63.

The data used here were compiled and provided by Jim Markalunas, and are used here with permission. Thanks also to Lance Luckett.

DATA SET

WINTER	SEPT	OCT	NOV	DEC	JAN	FEB	MAR	APR	MAY	JUNE
1.00	3.00	0.00	27.90	24.60	23.50	10.00	22.00	16.40	0.00	0.00
2.00	5.00	8.20	19.30	20.60	39.60	79.00	17.60	5.20	12.80	0.00
3.00	4.50	2.80	15.00	28.50	21.20	46.00	24.30	24.40	0.00	3.00
4.00	0.00	1.00	18.50	28.10	18.80	23.90	47.00	15.00	21.00	0.00
5.00	0.00	1.50	31.50	52.00	39.50	34.50	22.00	4.20	0.00	0.00
6.00	0.00	3.00	2.50	19.50	46.50	29.80	8.30	9.80	0.00	0.00
7.00	0.00	3.30	19.70	14.40	23.80	18.20	40.60	14.40	T	0.00
8.00	0.00	4.90	7.20	26.80	34.10	33.60	30.00	19.60	0.00	0.00
9.00	1.50	3.50	33.20	11.90	29.20	28.00	36.70	1.50	0.00	1.00
10.00	0.00	6.50	8.00	14.30	28.40	17.00	48.50	14.00	T	0.00
11.00	0.00	0.00	14.00	13.80	15.00	25.50	21.50	22.00	1.00	0.00
12.00	0.00	16.50	19.00	31.50	26.00	11.00	13.50	15.00	2.00	0.00
13.00	1.00	12.30	33.70	11.30	24.50	29.20	19.90	19.10	T	11.20
14.00	0.00	4.10	15.90	16.70	20.80	25.30	41.60	31.40	0.00	0.00
15.00	0.00	1.10	24.60	25.20	20.50	8.60	17.70	3.10	1.00	0.00
16.00	0.00	*	*	12.60	44.10	22.50	27.50	2.00	0.00	0.00
17.00	0.00	T	9.80	29.40	39.90	23.00	20.40	15.00	1.90	0.00
18.00	0.00	10.50	17.20	58.30	27.80	18.90	28.30	6.70	0.00	0.00
19.00	0.00	*	18.90	11.80	22.90	*	22.50	10.80	3.00	0.00
20.00	0.00	2.00	18.10	22.00	13.50	2.00	*	0.00	1.50	0.00
21.00	0.00	0.00	14.00	6.50	14.00	36.00	22.00	10.00	0.00	0.00
22.00	0.00	0.00	31.00	*	30.00	30.50	8.50	6.00	0.00	0.00
23.00	0.00	4.00	26.50	26.00	71.50	5.00	15.50	25.00	8.00	0.00
24.00	0.00	0.00	0.00	20.00	21.00	31.00	19.00	11.00	0.00	0.00
25.00	0.00	T	16.00	19.00	33.50	20.00	*	*	*	0.00
26.00	10.00	23.00	13.00	10.20	21.20	35.50	24.00	1.00	4.00	0.00
27.00	0.00	8.00	13.00	21.00	5.00	26.00	23.00	21.00	14.00	0.00
28.00	27.00	16.00	26.50	29.50	25.00	39.00	16.00	19.50	2.50	T
29.00	0.00	T	11.00	10.50	22.50	16.00	27.00	4.00	0.00	0.00
30.00	0.00	5.00	12.50	14.50	22.50	18.00	38.00	10.50	5.00	1.00
31.00	0.00	5.00	37.50	47.00	21.50	19.50	76.50	10.00	2.00	0.00
32.00	7.00	T	22.50	15.70	33.00	15.50	7.20	23.00	5.50	0.00
33.00	0.00	11.50	16.50	30.10	11.00	31.20	9.00	11.00	2.30	0.00
34.00	0.00	6.00	19.50	34.20	14.80	26.10	14.50	31.10	8.30	0.00
35.00	0.00	5.60	24.90	18.00	19.10	10.40	13.20	6.50	0.00	2.30
36.00	0.00	35.60	9.80	23.70	43.00	11.80	28.10	36.00	2.50	0.00
37.00	8.10	20.10	15.30	22.70	18.20	28.60	30.90	17.00	3.40	0.00
38.00	6.80	10.40	23.70	38.10	10.20	14.00	19.30	10.00	0.00	0.00
39.00	8.70	14.50	36.30	19.40	12.00	26.70	24.60	4.20	3.50	0.00
40.00	0.00	8.50	20.20	32.00	6.50	21.80	26.30	15.00	1.50	0.00
41.00	T	7.50	17.80	27.70	35.80	21.85	36.40	14.90	7.00	1.30
42.00	0.00	13.00	12.00	20.50	15.20	24.50	24.70	7.50	0.00	4.50
43.00	0.00	T	9.00	6.00	13.00	19.50	24.00	12.20	2.00	0.00
44.00	0.00	T	26.00	45.00	34.00	25.50	34.00	6.50	15.00	0.00
45.00	1.00	T	14.00	53.00	41.50	26.00	19.50	7.50	2.00	14.50
46.00	0.00	1.50	26.00	10.00	41.60	31.00	26.00	10.00	2.00	0.00

(concludes on the next page)

WINTER	SEPT	OCT	NOV	DEC	JAN	FEB	MAR	APR	MAY	JUNE
47.00	0.00	11.80	9.50	12.50	9.90	9.00	36.50	16.90	6.00	0.00
48.00	0.00	14.60	19.50	57.50	48.50	14.50	30.50	12.50	8.50	0.00
49.00	2.50	12.60	31.20	11.50	12.80	31.50	45.00	28.00	41.00	0.00
50.00	0.00	8.00	55.00	72.00	10.50	28.00	47.50	28.50	13.00	15.50
51.00	2.50	41.45	15.00	36.00	21.00	18.50	44.50	25.00	7.00	0.50
52.00	4.00	7.50	48.75	12.40	9.25	37.90	14.30	18.80	8.00	0.00
53.00	4.50	16.00	22.50	10.50	18.50	33.50	23.00	6.00	0.00	0.00
54.00	0.00	0.00	20.50	35.00	40.50	13.50	26.50	6.00	9.00	0.00
55.00	0.00	0.00	31.00	22.00	21.00	40.00	35.00	28.00	3.00	0.00
56.00	0.00	6.50	24.00	18.00	13.50	20.00	19.75	24.00	4.00	1.00
57.00	0.00	18.00	31.50	26.25	20.00	17.25	42.00	34.00	3.00	0.00
58.00	0.00	24.00	26.10	13.25	19.50	14.85	34.00	10.00	0.00	0.00
59.00	0.00	6.00	56.00	24.00	21.00	53.00	32.00	32.00	3.50	0.50
60.00	0.00	9.70	25.00	22.00	18.00	38.50	20.00	33.25	0.25	0.00

CASE **56**

CIGARETTE RANKINGS

In marketing research studies it is common to have different panels of judges rank competing brands or products. The various panels are chosen to represent different kinds of customers or potential customers, and the rankings of the products presumably reflect their reactions to the products, based on whatever criteria the panel members feel are important. Sometimes the main point of interest is to see where a particular product of interest ranks in comparison to its competitors, or to discover the top-ranked (or bottom-ranked) product. It may also be of interest to see if certain products and/or panels of judges are similar to each other. Such groups may reflect natural classes of customers or, alternatively, groups of products that consumers view as being similar.

The data that follow (and in file CIGRANK) reflect how 20 panels of judges rated 8 cigarettes.[1] The rankings given are numbers from 1 to 8 (1 = best cigarette, 8 = worst). For example, panel 1 rated cigarette 8 the best (rank 1) and cigarette 3 the worst.

Which cigarette(s) seem to be perceived as best? Worst? Are there natural groupings among the cigarettes? Among the panels of judges? Describe any potentially meaningful patterns you find.

[1]The data were published in Y. Baba, "Graphical Analysis of Rank Data," *Behaviormetrika 19 (1986),* pp. 1–15, and are used here with the kind permission of Dr. Baba, Dr. Akagi (who originally provided the data), and Dr. Kazuo Shigemasu, editor of *Behaviormetrika.*

DATA SET

	RANKING OF CIGARETTE							
Panel	1	2	3	4	5	6	7	8
1	2	5	8	6	3	4	7	1
2	1	5	8	4	6	3	7	2
3	1	5	7	3	6	4	8	2
4	4	2	5	6	1	7	8	3
5	1	4	2	6	3	7	8	5
6	1	7	5	4	2	6	8	3
7	4	1	2	3	6	5	7	8
8	4	1	2	3	5	7	6	8
9	3	2	1	4	6	5	8	7
10	4	1	7	3	2	5	8	6
11	6	7	1	5	4	2	8	3
12	3	1	2	4	6	5	8	7
13	8	1	7	6	5	3	4	2
14	3	1	6	2	4	5	8	7
15	3	1	6	4	2	5	8	7
16	3	1	6	2	5	4	8	7
17	8	5	4	2	1	3	6	7
18	1	7	2	8	4	5	6	3
19	3	8	2	5	4	7	6	1
20	1	6	2	5	4	3	8	7

CASE **57**

UTILITY BILLS

Terry is the owner and sole employee of Flowers To Go (FTG), a florist shop in Denver, Colorado. Having raised three children and sent them off to lives of their own, Terry is able to pursue a lifelong wish: to own a small business that would generate income and complement her love of horticulture.

Flowers To Go is heated with natural gas; electricity runs lights, coolers for the flowers, and other incidentals. Since most of her business is local, FTG's long-distance phone charges never run very high. Access to e-mail helps keep phone bills down, too. Terry would like to be able to forecast FTG's monthly utility bills for budgeting purposes.

The data set named UTILITY contains monthly information on gas, electric, water, and phone bills since FTG opened in June of 1995. The numbers come from a compilation of past billing statements from Public Service (the electric and natural gas company serving the metropolitan Denver area), the City of Denver, and US West. In this data set, the column labeled GAS is the consumption of natural gas at FTG in hundreds of cubic feet (CCF), and ELEC is electricity consumption in kilowatts (kwh). The dollar amounts of gas and electric consumption are shown in the columns GAS($) and ELEC($). Water consumption appears in thousands of gallons (WATER) and dollar amount of the bill (WATER($)). FTG's monthly phone bill in dollars is labeled PHONE($). Unfortunately, Terry has misplaced the statement for 1/97; the asterisk in PHONE($) column indicates this missing value. Finally, Terry noticed that Public Service just recently began including temperature information on the monthly statements; TEMP is the average temperature in Denver that month, measured in degrees Fahrenheit (°F).

You've been hired by Terry to perform two tasks. First, provide a description of historical patterns in her utility bills. Second, develop a forecasting methodology that Terry might use to forecast FTG's utility bills. Describe to Terry how forecasts are to

These data are real, though the name of the company is not.

be derived from your chosen method; use enough detail so that she can prepare her own forecasts in the future. As an example, derive forecasts for each of the first three months immediately following the data set and explain how you arrived at these numbers.

DATA SET

GAS	ELEC	TEMP	GAS($)	ELEC($)	WATER	WATER($)	PHONE($)
22	194	*	10.99	16.56	4	8.76	58.10
24	395	*	14.85	34.09	1	5.28	60.84
24	346	*	14.85	30.58	6	11.08	45.99
89	385	*	38.04	33.29	4	8.76	50.83
97	439	*	38.09	37.00	4	8.76	62.88
105	505	*	40.60	41.65	4	8.76	44.54
179	447	*	64.04	37.58	4	9.60	55.78
192	475	*	68.27	39.61	3	8.40	38.60
146	430	*	53.69	36.43	3	8.40	28.90
116	358	*	43.44	31.35	3	8.40	14.87
59	284	*	24.33	26.30	4	9.60	35.04
36	453	*	17.75	39.46	5	10.80	35.71
32	464	75	16.61	40.25	4	9.60	38.43
22	391	76	13.75	35.07	2	7.20	35.08
25	354	73	14.61	32.49	5	10.80	29.28
64	294	60	25.52	28.02	4	9.60	28.91
91	408	49	32.68	35.41	3	8.40	44.11
149	446	37	47.48	37.85	3	8.40	39.74
205	557	27	67.67	45.41	4	10.41	86.29
165	515	30	62.15	42.42	4	10.41	*
168	467	37	64.44	38.87	3	9.11	31.08
105	363	43	43.20	31.84	2	7.81	28.71
81	397	49	35.05	34.13	4	10.41	38.65
29	438	62	17.50	36.87	3	9.11	38.33
24	480	73	15.88	39.77	3	9.11	50.56
25	550	74	16.28	44.56	4	10.41	46.11
22	368	72	15.07	32.27	2	7.81	47.74
41	405	64	21.99	34.87	3	9.11	28.75
111	451	45	48.44	38.19	3	9.11	28.75
139	611	31	69.26	49.13	3	9.11	36.84
113	412	31	57.75	35.55	5	12.88	50.51
158	565	35	78.62	46.03	3	10.20	31.93
143	505	32	71.13	41.93	3	10.20	29.04
94	439	45	49.39	37.43	3	10.20	36.37

CASE **58**

JOB HOPPING AT CHIPS, INC.

Chips, Inc. is a major software/applications computing technology firm with about 15,000 employees around the country. The firm produces data processing and programming applications, and delivers specialized on-site training courses about management of business information technology. Recent courses include software maintenance and development, client-server system development, database management, and object-oriented systems development. Consulting services now account for about 25% of the firm's revenues. Chips consultants provide in-house evaluations of information technology needs, and oversee purchasing, installation, and training of new applications. Chips' clientele includes a number of Fortune 500 companies. Business is booming.

Like many of the technology-oriented companies, Chips is hard hit by job hopping, a phrase used to describe the tendency of employees to continually change jobs in search of better positions or attractive geographic locations. The days of company loyalty and long-term employment are long gone.

Job hopping is costly to Chips for two reasons. The hiring process of advertising, screening, and interviewing takes time and money. More importantly, new recruits undergo extensive training during their first few months. The Human Resources (HR) department has noticed a recent trend of employees leaving shortly after their training period. Chips executives are anxious to ensure their employees have the latest technological skills, but balk at training their competitors' future employees!

As head of the HR department, you've been asked to study this problem. You've estimated that a new recruit needs to stay with Chips for 12 months if the company is to break even on its training investment. You've collected information from 71 randomly chosen personnel records about two groups of new employees: those who stayed for at

These data come from actual employee records. Since they relate to personnel issues, the company has been disguised. The data were collected and supplied by Hugh E. Cunningham, consultant and Project Manager at CIBER, Inc., and Vanessa D. Eastland, Advertising Coordinator, AWWA.

least one year with Chips and those who did not. From these data you hope to more fully understand what characteristics of a new recruit indicate a tendency to job hop.

In the data set, the variable labeled KEEPER indicates whether that employee left the firm within the first year (1 = "keeper": someone who stayed at least 12 months at Chips; 0 = "hopper": someone who left within 12 months of his or her initial employment date). You suspect that a variety of factors might determine whether an employee will be a keeper or a hopper:

- A history of job hopping. The variables JOBS_DP, JOBS_PRG, and JOBS_CON measure the number of jobs in data processing, programming, and consulting before coming to Chips.
- Experience and educational background. YRS_DP and YRS_PRG refer to the number of years experience in data processing or programming. COLLEGE is the number of years of undergraduate coursework taken; GRAD is years in graduate school.
- Employment status. Chips retains both part-time and full-time employees. Part-time employees are less costly to Chips because no or few benefits are paid. FULLTIME is 1 for a full-time employee and 0 for a part-time employee.
- Portfolio of marketable technical skills. ENDUSER indicates whether an employee has end-user computing experience (1 = yes, 0 = no). GENER3 and GENER4 are the number of third- and fourth-generation programming languages known by the employee. DATABASE is the number of database management systems in which the employee is proficient.

Do the data have anything to say about the difference between hoppers and keepers? If so, how does this information help Chips with its problem? The data file is named CHIPS.

DATA SET

KEEPER	YRS_DP	YRS_PRG	JOBS_DP	JOBS_PRG	JOBS_CON	COLLEGE	GRAD	ENDUSER	GENER3	GENER4	DATABASE	FULLTIME
1	11	11	4	4	1	4	3	1	2	0	1	1
1	8	8	5	5	3	4	0	1	3	0	1	1
1	27	27	5	5	2	3	0	1	3	2	2	1
0	11	11	4	4	1	4	0	0	4	1	1	0
0	20	20	13	13	0	4	1	1	2	0	1	0
1	11	11	4	4	0	2	0	1	4	1	1	1
0	5	4	6	4	0	4	0	1	3	1	1	0
0	7	7	4	4	0	2	0	1	2	2	0	1
1	16	16	11	11	3	4	0	1	3	0	0	1
0	5	5	3	3	0	4	0	1	0	0	0	0
0	11	10	5	4	1	4	3	0	4	3	2	1
1	8	4	2	2	0	4	0	0	3	0	0	1
0	9	3	1	1	0	4	0	1	1	2	1	0
0	8	7	3	2	1	2	0	1	2	0	0	0
1	3	3	1	1	0	4	0	1	2	2	1	0
1	4	4	1	4	1	4	0	0	1	1	1	1
1	11	11	6	5	0	4	3	1	3	1	1	1
0	4	4	1	1	0	4	0	0	3	1	1	0

(continues on the following page)

KEEPER	YRS_DP	YRS_PRG	JOBS_DP	JOBS_PRG	JOBS_CON	COLLEGE	GRAD	ENDUSER	GENER3	GENER4	DATABASE	FULLTIME
1	26	26	1	1	0	4	0	0	2	1	1	0
1	9	9	4	4	0	0	0	0	2	1	2	1
1	5	5	2	2	0	4	0	1	2	0	2	1
1	12	12	2	2	1	4	0	1	3	1	2	0
1	14	14	9	9	0	4	0	1	3	0	0	1
1	7	7	3	3	1	4	0	1	4	2	4	1
1	16	16	4	4	0	4	0	1	3	1	2	0
0	2	2	1	1	0	4	0	1	2	1	0	1
1	11	11	4	4	1	0	0	1	2	2	2	0
0	14	6	3	3	0	0	0	0	3	2	2	0
1	8	2	1	1	0	3	0	0	1	0	1	0
0	12	11	4	3	0	4	0	1	4	1	1	0
1	6	6	4	4	0	4	3	1	2	1	2	0
0	22	22	8	8	1	0	0	0	3	2	2	0
0	9	9	4	4	0	4	3	0	2	0	0	0
0	15	11	6	6	0	4	0	1	1	1	1	0
1	8	8	4	4	0	4	0	0	2	0	2	0
1	19	19	7	7	1	4	3	0	4	1	3	0
0	5	5	4	4	1	4	0	1	2	0	1	0
1	25	25	9	9	3	0	0	1	3	1	0	0
1	7	7	4	4	0	2	0	0	3	1	0	1
0	5	5	5	5	0	4	0	1	2	0	0	0
0	6	5	2	1	0	4	0	0	2	0	0	1
0	23	23	3	3	0	0	0	1	2	0	0	0
0	3	1	2	1	0	4	0	1	2	0	1	1
0	23	23	8	7	3	2	0	0	2	0	1	0
1	15	15	6	6	1	4	3	1	3	0	1	1
0	13	7	5	5	0	2	0	0	2	0	0	0
0	6	6	1	1	0	4	0	0	2	0	0	0
1	11	11	5	5	0	4	0	0	3	0	1	1
1	4	4	1	1	0	4	3	0	1	0	1	1
1	20	20	22	22	9	2	0	1	4	2	1	0
0	6	6	6	6	1	0	0	1	2	0	1	0
1	18	17	8	8	3	1	0	1	3	2	2	0
0	13	4	4	2	0	4	0	1	1	0	1	0
1	3	3	1	1	0	4	0	1	2	1	2	0
0	6	6	1	1	0	4	3	1	4	0	1	0
0	8	7	3	2	0	4	0	0	1	1	0	0
0	11	11	1	1	0	0	0	1	1	0	1	0
1	8	7	7	6	4	2	0	1	2	0	0	0
1	8	8	3	3	1	2	0	0	2	1	0	0
0	19	16	4	4	0	0	0	1	2	2	1	0
0	3	3	3	3	2	4	3	0	2	1	1	0
1	13	13	5	5	0	4	0	0	1	0	1	1
0	5	5	4	4	0	4	0	1	2	0	1	0
1	13	13	8	8	2	4	3	1	3	0	0	1
0	12	12	4	4	0	4	0	1	3	2	1	1

(concludes on the following page)

KEEPER	YRS_DP	YRS_PRG	JOBS_DP	JOBS_PRG	JOBS_CON	COLLEGE	GRAD	ENDUSER	GENER3	GENER4	DATABASE	FULLTIME
0	11	11	4	4	1	4	0	0	1	1	2	0
0	16	16	6	6	3	4	3	0	3	2	2	0
1	25	25	4	4	0	2	0	0	3	2	0	1
0	16	16	7	7	4	4	1	1	4	1	0	0
0	3	3	3	3	1	4	0	1	5	2	3	0
0	4	4	3	3	0	0	0	0	2	0	0	1

CASE **59**

MEDICATION ERRORS

When a patient receives the wrong medicine, receives the right medicine, but at the wrong time or in the wrong dosage, or doesn't receive a scheduled medication, the results can be disastrous. Most hospitals, clinics, and similar medical organizations have some sort of more or less formal mechanism to be sure that such medication errors are reported and tracked. Appropriate study of the resulting records can presumably provide insight into possible changes to procedures or policies so as to improve future performance. In many cases, they also help satisfy accreditation or other regulatory requirements.

One hospital chain provided the data that follow, covering the first three quarters of 1995. Tables 1, 2, and 3 provide breakdowns of the department reporting the error, and the cause of the errors, indicated by code numbers, for quarters 1, 2, and 3, respectively. A few of the codes and their meanings are:

Code	Type error
101	Prescription not available
104	Medication omitted
105	Duplicate or extra medication
106	Medication not given close to correct time
108	Wrong dose given
109	Wrong medication given
110	Prescription filled wrong
Others	Various other causes

Table 4 is a summary table derived from Tables 1, 2, and 3.

The situation was described and the data were provided by Jeffrey S. Mason, M.D., VPMA, Santé Health System, Fresno California.

Administrators and medical staff are interested in understanding whether performance with respect to medication errors has been getting better or worse, and in learning if particular departments or causes of error dominate recent experience.

Provide an appropriate written (text and/or graphics) summary of the available data, suitable to be used as the basis for a quick (perhaps three to five minutes) presentation to the board of advisors.

TABLE 1 RESULTS FOR FIRST QUARTER, 1995

Department	101	104	105	106	107	108	109	110	112	113	114	115	116	117
CANCER	0	0	0	1	0	0	1	0	0	0	0	0	0	0
CARDIOLOGY	2	3	1	1	0	2	1	0	1	0	0	0	1	3
SUBACUTE CENTER	0	3	0	0	0	1	0	0	0	0	0	0	1	0
EMERGENCY	0	0	0	0	0	0	1	0	0	0	1	0	0	0
MEDICAL	0	14	6	9	1	6	10	5	2	1	1	3	3	1
OB	0	7	2	11	1	3	3	0	0	0	0	1	0	0
PSYCH	0	4	1	1	0	2	1	2	3	0	0	0	1	0
REHAB	0	1	0	1	1	1	0	0	0	0	0	0	1	0
SURGERY	0	0	0	2	0	1	2	2	0	0	1	0	2	0
TCU	0	0	1	1	0	0	3	0	0	0	0	0	1	0
HOME HEALTH	0	0	0	0	0	0	0	0	0	0	0	0	0	0
OTHER SERVICES	0	0	0	0	0	0	0	2	0	0	0	0	0	0
RESOURCE DISTRIBUTION	0	0	0	1	0	0	0	0	0	0	0	0	0	0
Grand Total	2	32	11	28	3	16	22	11	6	1	3	4	10	4

TABLE 2 RESULTS FOR SECOND QUARTER, 1995

Department	101	103	104	105	106	107	108	109	110	111	112	113	114	115	116	117
CANCER	0	0	0	1	1	0	0	0	0	0	0	0	0	0	0	0
CARDIOLOGY	0	0	1	1	2	0	1	3	1	0	0	0	1	0	2	0
SUBACUTE CENTER	0	0	0	0	0	0	0	0	0	0	0	0	0	0	0	0
EMERGENCY	0	0	0	0	0	0	0	0	0	0	0	0	0	0	2	0
MEDICAL	2	0	9	4	4	2	7	8	3	0	2	1	0	4	1	1
OB	0	1	4	3	3	1	4	5	0	0	3	0	0	0	2	0
PSYCH	0	0	0	1	0	0	5	1	1	0	0	0	0	0	0	0
REHAB	0	0	3	0	2	0	0	0	0	0	0	0	0	0	1	0
SURGERY	0	0	3	1	0	0	0	2	0	0	0	0	0	0	1	0
TCU	1	0	1	1	0	0	1	1	0	0	0	0	0	0	0	0
HOME HEALTH	0	0	0	0	0	0	1	0	0	1	0	0	0	0	2	0
OTHER SERVICES	0	0	0	0	0	0	1	0	0	0	0	0	0	0	1	0
RESOURCE DISTRIBUTION	0	0	0	0	0	0	0	0	2	0	0	0	0	0	0	0
Grand Total	3	1	21	12	12	3	20	20	7	1	5	1	1	4	12	1

TABLE 3 RESULTS FOR THIRD QUARTER, 1995

Department	101	104	105	106	107	108	109	110	111	112	113	114	115	116	117
CANCER	0	1	0	2	0	1	1	0	0	0	0	0	0	1	0
CARDIOLOGY	0	5	0	1	0	2	4	2	0	1	0	0	1	1	2
SUBACUTE CENTER	0	0	0	0	0	0	0	0	0	0	0	0	0	0	0
EMERGENCY	0	1	0	0	0	1	0	0	0	0	0	0	1	1	0
MEDICAL	0	6	1	4	1	10	1	3	0	3	1	0	0	5	1
OB	0	4	2	3	0	0	2	1	0	1	0	0	1	1	0
PSYCH	0	4	1	0	0	2	0	0	0	0	0	0	0	0	0
REHAB	0	0	0	2	0	0	0	0	0	0	0	0	0	1	0
SURGERY	0	1	1	1	0	1	2	0	0	1	1	1	0	1	0
TCU	0	1	0	0	0	0	1	0	0	0	0	0	0	0	0
HOME HEALTH	0	0	1	0	0	0	0	0	1	0	0	0	0	0	0
OTHER SERVICES	0	0	0	0	0	0	0	0	0	0	0	0	0	0	0
RESOURCE DISTRIBUTION	1	0	0	0	0	0	0	2	0	0	0	0	0	1	0
Grand Total	1	23	6	13	1	17	11	8	1	6	2	1	3	12	3

TABLE 4 SUMMARY OF INCIDENTS FOR FIRST THREE QUARTERS, 1995

Department	1st Quarter	2nd Quarter	3rd Quarter	Total
MEDICAL	62	48	36	146
CARDIOLOGY	15	12	19	46
OB	28	26	15	69
SURGERY	10	7	10	27
PSYCH	15	8	7	30
CANCER	2	2	6	10
EMERGENCY	2	2	4	8
RESOURCE DISTRIBUTION	1	2	4	7
REHAB	5	6	3	14
TCU	6	5	2	13
HOME HEALTH	0	4	2	6
SUBACUTE CENTER	5	0	0	5
OTHER SERVICES	2	2	0	4

CASE **60**

CLINICAL TREATMENT OF DEPRESSION

New laws aimed at creating parity between treatment of physical illnesses and mental illnesses have been passed in many jurisdictions. Prior to such legislation, insurance companies often restricted reimbursement for mental health treatment by limiting inpatient hospital stays for mental illness. New laws typically forbid restrictions on hospital stays for certain mental illnesses.

Depression is a widespread, often treatable affliction, but some events (e.g., early childhood trauma or death of a loved one) can trigger serious depression requiring hospitalization. Depression symptoms include low self-esteem, inability to concentrate, feelings of hopelessness, or thoughts of suicide. Recent scientific studies point to another cause of depression: biochemical imbalances in the brain. Regardless of the source, depression is expensive. Friends and family take on added caretaking responsibilities and are deprived of healthy companionship. Depression also takes its toll in the workplace through lost productivity.

Costs for care of depressed patients are estimated to run as high as $145,000 per patient per year, and this worries Max, a local hospital administrator. He wonders if there is some way to control these projected cost increases *and* provide needed care.

Max acquired a data set and a written report prepared for the hospital's board. He's asked you to examine the data and report back to him.

These data were collected and studied by Juan Carlos Bravo, Timothy Woodworth (Kaiser-Permanente), and others. The names of physicians and administrators have been concealed to ensure confidentiality.

BACKGROUND

The data set includes information on two medical diagnoses about depression: major depressive disorder and bipolar disorder. Major depressive episodes are those in which the patient becomes incapacitated with feelings of sadness, incompetence, or worthlessness. A depressed patient may experience notable loss of interest in activities or people, changes in body weight and/or loss of appetite, insomnia or other sleep disruptions, inability to concentrate, or thoughts of death. Bipolar disorder is the term used to describe patients whose moods alternate between depressive bouts and periods of unrealistic elation. During the latter phase, a patient may form unrealistic self-assessments, engage in risky physical or social behavior, or find little need for sleep. Major depressive episodes and bipolar disorders account for about 91% of all hospitalizations at Max's institution.

Hospitalizations can be of two types. Inpatient treatment means the patient stays in the hospital around the clock. With partial hospitalization, the patient receives care through a daytime program at the hospital and returns home at night. Partial hospitalization is reserved for the less serious cases of depression.

Depression can be treated with psychotherapy, medication, or both. During psychotherapy, patients and the therapist explore issues that contribute to depressive episodes and uncover possible routes for recovery. Medication such as selective serotonin reuptake inhibitors (SSRIs, such as Prozac™, Zoloft™, or Paxil™) influence chemical reactions in the brain. Drug intervention can hasten recovery and prevent future depressive bouts for some patients.

Finally, the report to the board describes two attending physicians' treatment modalities, medical backgrounds, and personalities:

- Dr. Jones has eight years of service and has practiced solely in the hospital. Dr. Jones often uses a combination of psychotherapy and medications. At times it appears Dr. Jones is unsure of the diagnosis and orders additional testing (e.g., psychological testing) in an effort to clarify the diagnosis before beginning medication. Dr. Jones often refuses to see the patients' families, deferring them to the social work staff, and has an antagonistic relationship with the nursing staff.
- Dr. Smith, an experienced MD, has practiced for 30 years in another medical subspeciality before coming to inpatient psychiatry. Dr. Smith brings a broad range of experience from his prior position. Even though he is a more experienced physician, he is relatively new to inpatient psychiatry. This recent training may present an advantage because he practices the latest medical techniques. His primary method for treating patients is medication intervention, which he learned during his residency training. Dr. Smith has a tremendous customer service philosophy, demonstrating exceptional bedside manner with patients and concern for family members. Although an experienced physician, he is often compared to a young doctor fresh out of residency. He is eager to take care of the most difficult patients.

Do certain types of depression respond more quickly to certain treatments? Can a patient's need for hospitalization be predicted for budgeting purposes? Is there *anything* you can recommend to Max that might help keep costs under control? The data set in the file DEPRESS includes information about all patients admitted to the hospital over the last calendar year with a diagnosis of either major depressive disorder or bipolar disorder.

DEFINITIONS

Age	Age of the patient in years at time of admission
Gender	Gender of patient: Male = 0, Female = 1
Phy	Treating physician: Dr. Jones = 0, Dr. Smith = 1
I/P	Inpatient or partial hospitalization: 0 = Inpatient (patient stays in the hospital 24 hours a day), 1 = Partial (patient attends a daytime hospital program)
Los	Number of days patient is associated with the hospital (either inpatient or partial)
Md/Bi	Medical diagnosis: 0 = Major depressive disorder, 1 = Bipolar disorder

DATA SET

Age	Gender	Phy	I/P	Los	Md/Bi
25	1	0	0	1	0
25	1	0	0	1	0
27	1	0	0	2	0
29	1	0	0	4	0
30	1	0	0	1	0
32	1	0	0	4	0
32	1	0	0	3	0
33	1	0	0	10	0
33	1	0	0	1	0
34	1	0	0	1	0
35	1	0	0	2	0
36	1	0	0	2	0
37	1	0	0	3	0
37	1	0	0	3	0
38	1	0	0	8	0
38	1	0	0	2	0
39	1	0	0	11	0
39	1	0	0	3	0
40	1	0	0	5	0
41	1	0	0	13	0
41	1	0	0	5	0
41	1	0	0	4	0
41	1	0	0	2	0
42	1	0	0	2	0
42	1	0	0	4	0
44	1	0	0	14	0
44	1	0	0	2	0
44	1	0	0	2	0
44	1	0	0	1	0
45	1	0	0	19.5	0
45	1	0	0	3	0
46	1	0	0	2	0
47	1	0	0	3	0

Age	Gender	Phy	I/P	Los	Md/Bi
47	1	0	0	5	0
48	1	0	0	9	0
48	1	0	0	4	0
49	1	0	0	4	0
51	1	0	0	4	0
51	1	0	0	12	0
63	1	0	0	14	0
63	1	0	0	19	0
67	1	0	0	5	0
68	1	0	0	3	0
68	1	0	0	17	0
70	1	0	0	11	0
71	1	0	0	11	0
76	1	0	0	35	0
79	1	0	0	12	0
83	1	0	0	5	0
18	1	0	0	1	0
19	1	0	0	1	0
23	1	0	0	5	0
27	1	0	0	1	0
39	1	0	0	1	0
39	1	0	0	5	0
44	1	0	0	6	0
44	1	0	0	2	0
45	1	0	0	6	0
45	1	0	0	2	0
45	1	0	0	3	0
49	0	0	0	4	0
67	0	0	0	3	0
68	0	0	0	30	0
70	0	0	0	6	0
70	0	0	0	5	0
74	0	0	0	19	0

(continues on the following page)

Age	Gender	Phy	I/P	Los	Md/Bi
23	1	1	0	1	0
24	1	1	0	3	0
26	1	1	0	4	0
27	1	1	0	1	0
27	1	1	0	7	0
28	1	1	0	2	0
31	1	1	0	1	0
32	1	1	0	3	0
32	1	1	0	3	0
33	1	1	0	4	0
34	1	1	0	3	0
34	1	1	0	6	0
34	1	1	0	3	0
37	1	1	0	7	0
37	1	1	0	2	0
37	1	1	0	3	0
38	1	1	0	7	0
39	1	1	0	10	0
41	1	1	0	7	0
41	1	1	0	4	0
41	1	1	0	7	0
41	1	1	0	2	0
42	1	1	0	3	0
42	1	1	0	1	0
42	1	1	0	7	0
42	1	1	0	1	0
42	1	1	0	1	0
42	1	1	0	7	0
42	1	1	0	1	0
42	1	1	0	9	0
42	1	1	0	9	0
42	1	1	0	5	0
43	1	1	0	4	0
43	1	1	0	6	0
43	1	1	0	1	0
44	1	1	0	6	0
45	1	1	0	6	0
45	1	1	0	8	0
45	1	1	0	2	0
47	1	1	0	3	0
48	1	1	0	3	0
49	1	1	0	4	0
50	1	1	0	2	0
52	1	1	0	2	0
54	1	1	0	6	0
54	1	1	0	2	0
55	1	1	0	4	0

Age	Gender	Phy	I/P	Los	Md/Bi
55	1	1	0	4	0
56	1	1	0	2	0
58	1	1	0	9	0
59	1	1	0	7	0
68	1	1	0	7	0
72	1	1	0	2	0
73	1	1	0	4	0
75	1	1	0	1	0
18	0	1	0	1	0
19	0	1	0	1	0
23	0	1	0	9	0
29	0	1	0	4	0
29	0	1	0	0	0
41	0	1	0	4	0
42	0	1	0	1	0
44	0	1	0	3	0
48	0	1	0	2	0
49	0	1	0	3	0
50	0	1	0	4	0
51	0	1	0	4	0
53	0	1	0	2	0
56	0	1	0	3	0
56	0	1	0	21	0
70	0	1	0	3	0
70	0	1	0	10	0
73	0	1	0	2	0
75	0	1	0	18	0
85	0	1	0	5	0
25	1	0	1	0	0
26	1	0	1	0	0
28	1	0	1	9	0
33	1	0	1	3	0
38	1	0	1	4	0
41	1	0	1	1	0
45	1	0	1	0	0
51	1	0	1	6	0
76	1	0	1	0	0
27	1	0	1	4	0
40	1	0	1	3	0
43	1	0	1	10	0
26	1	1	1	1	0
26	1	1	1	17	0
31	1	1	1	1	0
34	1	1	1	0	0
37	1	1	1	7	0
39	1	1	1	7	0
41	1	1	1	4	0

(continues on the following page)

Age	Gender	Phy	I/P	Los	Md/Bi
52	1	1	1	4	0
52	1	1	1	4	0
18	0	1	1	1	0
29	0	1	1	2	0
33	0	1	1	5	0
41	0	1	1	7	0
44	0	1	1	4	0
48	0	1	1	4	0
56	0	1	1	0	0
70	0	1	1	5	0
19	1	0	0	3	1
19	1	0	0	1	1
21	1	0	0	4	1
21	1	0	0	2	1
22	1	0	0	6	1
22	1	0	0	10	1
24	1	0	0	5	1
26	1	0	0	4	1
28	1	0	0	2	1
28	1	0	0	1	1
31	1	0	0	5	1
31	1	0	0	20	1
32	1	0	0	8	1
33	1	0	0	3	1
33	1	0	0	2	1
34	1	0	0	2	1
35	1	0	0	4	1
35	1	0	0	7	1
36	1	0	0	42	1
36	1	0	0	8	1
36	1	0	0	5	1
36	1	0	0	5	1
37	1	0	0	6	1
37	1	0	0	3	1
38	1	0	0	10	1
40	1	0	0	2	1
40	1	0	0	2	1
42	1	0	0	3	1
44	1	0	0	12	1
46	1	0	0	3	1
46	1	0	0	7	1
47	1	0	0	3	1
47	1	0	0	8	1
50	1	0	0	18	1
51	1	0	0	3	1
52	1	0	0	3	1
52	1	0	0	6	1

Age	Gender	Phy	I/P	Los	Md/Bi
59	1	0	0	12	1
65	1	0	0	15	1
73	1	0	0	9	1
76	1	0	0	51	1
19	1	0	0	29	1
19	1	0	0	8	1
19	1	0	0	12	1
21	1	0	0	15	1
21	1	0	0	5	1
25	1	0	0	6	1
26	1	0	0	2	1
31	1	0	0	8	1
31	1	0	0	3	1
32	1	0	0	8	1
44	1	0	0	3	1
45	1	0	0	1	1
59	0	0	0	2	1
60	0	0	0	7	1
69	0	0	0	2	1
18	1	1	0	6	1
18	1	1	0	3	1
18	1	1	0	5	1
21	1	1	0	9	1
22	1	1	0	3	1
22	1	1	0	3	1
22	1	1	0	19	1
22	1	1	0	11	1
23	1	1	0	4	1
23	1	1	0	10	1
27	1	1	0	1	1
31	1	1	0	5	1
32	1	1	0	3	1
32	1	1	0	13	1
32	1	1	0	12	1
33	1	1	0	4	1
33	1	1	0	2	1
34	1	1	0	6	1
38	1	1	0	7	1
40	1	1	0	14	1
40	1	1	0	4	1
41	1	1	0	35	1
41	1	1	0	7	1
41	1	1	0	20	1
41	1	1	0	7	1
42	1	1	0	12	1
43	1	1	0	3	1
43	1	1	0	5	1

(concludes on the following page)

Age	Gender	Phy	I/P	Los	Md/Bi
44	1	1	0	2	1
45	1	1	0	2	1
45	1	1	0	15	1
45	1	1	0	5	1
45	1	1	0	12	1
46	1	1	0	15	1
47	1	1	0	8	1
47	1	1	0	6	1
49	1	1	0	7	1
50	1	1	0	17	1
50	1	1	0	17	1
51	1	1	0	14	1
51	1	1	0	15	1
52	1	1	0	2	1
60	1	1	0	1	1
65	1	1	0	10	1
65	1	1	0	28	1
69	1	1	0	12	1
71	1	1	0	7	1
77	1	1	0	20	1
89	1	1	0	3	1
19	0	1	0	7	1
19	0	1	0	6	1
19	0	1	0	7	1
19	0	1	0	2	1
21	0	1	0	8	1
21	0	1	0	5	1
21	0	1	0	3	1
22	0	1	0	4	1
23	0	1	0	6	1
23	0	1	0	11	1
27	0	1	0	2	1
28	0	1	0	4	1
31	0	1	0	7	1
31	0	1	0	4	1
33	0	1	0	5	1
33	0	1	0	5	1
33	0	1	0	20	1
34	0	1	0	7	1
38	0	1	0	6	1
42	0	1	0	10	1
42	0	1	0	9	1
42	0	1	0	5	1
49	0	1	0	5	1
49	0	1	0	7	1

Age	Gender	Phy	I/P	Los	Md/Bi
50	0	1	0	6	1
51	0	1	0	4	1
53	0	1	0	3	1
53	0	1	0	1	1
54	0	1	0	3	1
54	0	1	0	1	1
56	0	1	0	1	1
60	0	1	0	3	1
68	0	1	0	7	1
68	0	1	0	12	1
81	0	1	0	1	1
22	1	0	1	6	1
24	1	0	1	3	1
26	1	0	1	4	1
26	1	0	1	5	1
32	1	0	1	4	1
35	1	0	1	9	1
36	1	0	1	2	1
36	1	0	1	5	1
46	1	0	1	1	1
47	1	0	1	6	1
59	1	0	1	3	1
65	1	0	1	3	1
19	1	0	1	0	1
25	1	0	1	1	1
34	1	0	1	5	1
18	1	1	1	1	1
21	1	1	1	0	1
22	1	1	1	14	1
22	1	1	1	2	1
32	1	1	1	3	1
32	1	1	1	4	1
33	1	1	1	3	1
38	1	1	1	7	1
41	1	1	1	2	1
43	1	1	1	1	1
45	1	1	1	5	1
49	1	1	1	4	1
51	1	1	1	5	1
19	0	1	1	2	1
19	0	1	1	6	1
19	0	1	1	0	1
29	0	1	1	9	1
41	0	1	1	3	1

CASE **61**

DEER-TRAIN COLLISIONS IN EASTERN HOKKAIDO

The Kushiro Branch of the Hokkaido Railway Company is concerned because collisions between Sika deer *(Cervus nippon yesoensis)* and the company's trains increased substantially between 1987 to 1995. Hokkaido is the large island at the northern end of today's Japan. The largest city on the island is Sapporo. The railway line involved runs about 300 kilometers (km) eastward from Sapporo to Kushiro and then along the coast another 125 km to Nemuro. Most of the deer-train collisions occurred along the stretch of track between Kushiro and Nemuro (see Figure 1).

Relatively little is known about deer-train collisions, though deer-automobile collisions (and measures to prevent them) have been studied to some extent. Aside from injuries and deaths among the deer, which cause justifiable public concern, the collisions often damage train equipment and the delays required to deal with the collisions interfere with railway service.

Conceivable measures to deal with the problem could involve training, schedule (or train speed) adjustments, physical barriers, and the like, but the first order of business is to understand the extent of the problem and its nature. In particular, both the railway management and consultants from the Obihiro University of Agriculture and Veterinary Medicine would like to know such things as:

- Have collisions been increasing? If so, are there simply more deer in the vicinity, or are other factors at work?

An analysis of the data given here was published in Keiichi Onoyama, Noboru Ohsumi, Naoko Mitsumochi, and Tsuyoshi Kishihara, "Data Analysis of Deer-Train Collisions in Eastern Hokkaido, Japan," in C. Hayashi, N. Ohsumi, K. Yajima, Y. Tanaka, H.-H. Bock, and Y. Baba (Eds.), *Data Science, Classification, and Related Methods,* Tokyo, Springer-Verlag, 1998, pp. 746–751, from which some of the description is used here with permission. Dr. Keeichi Onoyama, Professor of Obihiro University of Agriculture and Veterinary Medicine, and Dr. Noboru Ohsumi, Professor of the Institute of Statistical Mathematics, kindly provided the data.

FIGURE 1 JAPAN AND HOKKAIDO

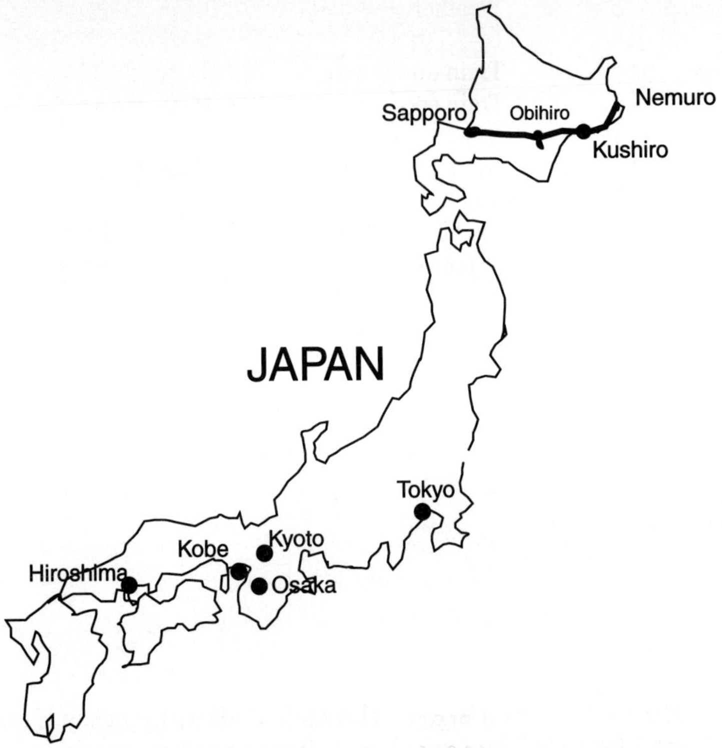

- Where do the collisions occur? When?
- What other patterns might suggest effective means of dealing with the problem?

The data available are extracted from train driver reports from April 1987 to March 1995, consisting of 696 collision reports. The number of train runs per year along this stretch of track was approximately the same over this period.

Provide an appropriate summary report addressing the questions above, at least to the extent that the available data permit.

The data are listed below and are available in file DEER. The (*) indicates missing data. The variables in the data set are:

ID	Record identification number
YR	Year (1987, 1988, . . .)
MO	Month (January = 1, February = 2, etc.)
DA	Day of month
HR	Hour of collision (0 = Midnight, 12 = Noon, 23 = 11 P.M., etc.)
MI	Minute of collision (0 to 59)
PLACE	Location of collision, measured in kilometers from Sapporo (Kushiro = About 310, Nemuro = About 435
TRN	Identifying number of the train. Normally, a train with a particular number runs at the same time every day. For example, train numbers 5638, 5639, 5640, 5641, 5642, 5645, 5647, and 3644 provided hourly service between Kushiro and Nemuro stations during late afternoon and evening hours essentially every day during the period).

DW	Day of week (1 = Sunday, . . ., 7 = Saturday)
WE	Weather (1 = Fine, 2 = Cloudy, 3 = Rainy, 4 = Misty, 5 = Snowy, 6 = Sleety)
DI	Train direction (1 = Westwards, 2 = Eastwards)
SP	Train speed (km/h) when driver noticed the deer
NF	Number of deer noticed by driver
DS	Distance (m) between the point driver noticed the deer and the location of the deer
NH	Number of deer that collided with the train
DH	Distance (m) between the point where the train hit the deer and the point where the train stopped.

DS + DH gives the total distance for the train to stop, beginning at the moment the driver sees the deer, including distance after the collision, if any.

Notes to specific entries:

- ID numbers 326 and 665 refer to special trains such as maintenance trains
- For ID number 371, the train was actually an extra section of train 9647, though this does not appear in the data as listed.
- ID numbers 92, 100, 131, 142, 333, 349, 357, 360, 367, 390, 391, 411, and 415 refer to special runs of their respective train numbers.

DATA SET

ID	YR	MO	DA	HR	MI	PLACE	TRN	DW	WE	DI	SP	NF	DS	NH	DH
1	1987	4	3	6	03	419.500	524	6	1	1	70	5	50	0	200
2	1987	4	6	19	05	273.500	232	2	1	1	70	1	20	1	220
3	1987	7	19	9	46	230.300	224	1	1	1	70	1	100	1	110
4	1987	7	25	21	51	336.100	3540	7	2	1	65	2	100	1	100
5	1987	8	24	9	07	381.700	526	2	2	1	70	2	50	1	160
6	1987	8	27	18	55	273.200	37	5	2	2	65	3	250	0	-50
7	1987	9	20	5	35	123.366	1455	1	2	2	80	6	100	1	350
8	1987	9	22	18	05	364.500	533	3	2	2	80	4	100	1	160
9	1987	9	26	22	03	420.440	537	7	3	2	65	1	10	1	210
10	1987	10	19	20	59	386.900	3540	2	4	1	80	2	0	1	280
11	1987	10	21	19	19	261.650	232	4	1	1	60	4	80	1	100
12	1987	10	27	22	00	339.830	539	3	1	2	70	1	50	1	170
13	1987	10	28	5	47	433.700	524	4	1	1	65	1	20	0	190
14	1987	11	23	20	53	364.100	537	2	*	2	*	*	*	*	*
15	1987	12	1	19	30	364.500	538	3	1	1	75	4	50	2	200
16	1987	12	3	20	35	351.800	537	5	1	2	75	10	80	1	180
17	1987	12	10	17	15	420.100	531	5	1	2	65	1	10	1	180
18	1988	1	13	18	27	425.800	538	4	1	1	70	1	0	1	210
19	1988	1	14	22	15	420.100	537	5	2	2	68	3	0	1	250
20	1988	1	16	17	25	373.750	536	7	1	1	70	2	30	1	200
21	1988	1	31	15	40	376.440	534	1	1	1	65	1	0	1	170
22	1988	2	12	20	23	426.570	3540	6	1	1	50	1	50	1	80
23	1988	2	29	16	28	420.000	536	2	2	1	60	8	120	1	30
24	1988	3	7	19	15	424.100	533	2	5	2	60	3	20	1	150

(continues on the following page)

ID	YR	MO	DA	HR	MI	PLACE	TRN	DW	WE	DI	SP	NF	DS	NH	DH
25	1988	3	18	19	30	429.500	533	6	1	2	70	1	30	1	190
26	1988	3	19	20	31	422.000	3540	7	5	1	60	1	100	1	60
27	1988	4	2	18	33	424.900	538	7	1	1	60	1	50	1	130
28	1988	4	4	8	59	418.750	525	2	2	2	70	8	100	1	120
29	1988	4	6	15	00	126.300	402	4	2	1	75	10	60	3	180
30	1988	4	9	20	57	365.000	537	7	1	2	75	1	30	1	220
31	1988	4	15	7	23	434.300	523	6	4	2	60	6	50	1	100
32	1988	4	29	18	33	426.400	538	6	4	1	60	6	30	1	150
33	1988	7	14	19	39	213.400	39	5	2	2	75	1	50	1	220
34	1988	7	22	17	41	273.300	44	6	1	1	80	2	200	1	100
35	1988	8	7	19	29	428.750	533	1	2	2	55	2	30	1	100
36	1988	8	14	18	55	404.680	538	1	2	1	80	1	100	1	200
37	1988	8	22	22	22	299.850	9416	2	4	1	80	4	40	1	260
38	1988	9	12	18	08	261.500	232	2	2	1	65	1	30	1	150
39	1988	9	20	18	34	425.700	538	3	3	1	70	1	20	1	200
40	1988	9	28	17	20	387.060	536	4	2	1	75	1	30	1	220
41	1988	10	12	18	06	350.300	536	4	1	1	70	1	50	1	190
42	1988	10	21	15	46	373.800	534	6	1	1	65	2	110	1	100
43	1988	10	24	17	45	348.000	533	2	1	2	70	1	0	1	230
44	1988	10	24	21	40	350.500	3540	2	1	1	70	1	0	1	220
45	1988	10	30	22	32	419.800	537	1	3	2	70	2	0	1	230
46	1988	10	31	11	44	433.600	527	2	2	2	60	1	0	1	150
47	1988	11	2	19	19	336.750	535	4	1	2	60	1	30	1	140
48	1988	11	9	5	48	433.500	524	4	1	1	60	6	20	1	150
49	1988	11	12	20	23	334.850	540	7	1	1	50	1	0	1	110
50	1988	11	12	17	00	409.000	538	7	2	1	65	1	20	1	170
51	1988	11	18	19	23	420.250	535	6	1	2	70	1	20	1	200
52	1988	11	21	16	02	364.500	536	2	6	1	75	1	20	1	230
53	1988	11	21	16	50	422.000	538	2	5	1	65	5	20	1	160
54	1988	11	22	16	28	240.120	42	3	1	1	80	1	200	1	140
55	1988	11	22	20	58	365.450	539	3	2	2	80	1	100	1	180
56	1988	11	23	20	33	424.200	3542	4	1	1	70	1	50	1	180
57	1988	11	24	22	10	417.700	539	5	3	2	65	1	0	1	200
58	1988	11	25	22	38	327.700	541	6	1	2	55	1	30	1	150
59	1988	12	2	17	42	364.200	540	6	1	1	80	1	50	1	240
60	1988	12	13	8	12	422.300	526	3	1	1	75	1	150	1	100
61	1988	12	14	6	00	424.730	524	4	*	1	60	4	20	1	150
62	1988	12	18	20	12	345.900	540	1	1	1	75	3	50	1	200
63	1989	1	11	17	28	418.660	533	4	2	2	70	5	50	1	190
64	1989	1	13	16	51	412.550	538	6	1	1	75	5	100	1	160
65	1989	1	21	21	42	327.850	541	7	5	2	60	1	20	1	100
66	1989	2	6	17	40	370.000	538	2	5	1	80	2	50	1	250
67	1989	2	6	22	15	423.100	539	2	5	2	70	2	50	1	200
68	1989	3	1	16	39	429.650	538	4	4	1	60	1	50	1	105
69	1989	3	4	22	55	287.150	231	7	2	2	75	5	50	1	230
70	1989	3	6	19	37	368.800	540	2	1	1	80	2	100	1	100
71	1989	3	17	17	45	433.550	533	6	2	2	60	2	30	2	150
72	1989	3	24	23	28	284.700	231	6	1	2	50	2	50	1	70

(continues on the following page)

ID	YR	MO	DA	HR	MI	PLACE	TRN	DW	WE	DI	SP	NF	DS	NH	DH
73	1989	4	9	21	03	262.600	285	1	2	2	60	6	70	1	80
74	1989	4	29	18	35	275.550	37	7	1	2	80	3	100	1	200
75	1989	5	3	18	46	410.000	542	4	1	1	75	5	50	1	200
76	1989	6	18	19	38	441.500	537	1	2	2	60	1	40	1	120
77	1989	7	1	10	07	315.400	526	7	2	1	70	3	100	1	120
78	1989	7	4	22	54	285.900	231	3	1	2	70	3	10	1	200
79	1989	7	11	18	38	317.900	540	3	3	1	60	4	50	1	100
80	1989	7	24	22	02	316.600	3544	2	1	1	65	1	30	1	160
81	1989	7	30	19	28	433.650	537	1	2	2	65	1	50	1	120
82	1989	8	14	17	26	421.880	535	2	4	2	70	3	0	1	230
83	1989	9	5	20	24	424.100	3544	3	2	1	65	1	50	1	130
84	1989	9	26	20	41	404.700	3544	3	1	1	75	5	180	1	90
85	1989	10	13	21	51	424.000	541	6	2	2	70	3	70	2	130
86	1989	10	14	21	37	344.200	3544	7	*	1	80	2	80	1	200
87	1989	10	19	16	56	407.050	540	5	2	1	70	2	80	1	150
88	1989	10	20	23	58	242.600	412	6	1	1	52	4	20	0	100
89	1989	10	21	18	31	424.800	542	7	1	1	70	2	20	1	210
90	1989	10	25	17	24	329.500	537	4	1	2	60	2	50	1	90
91	1989	11	2	22	03	278.600	280	5	1	1	65	2	40	1	130
92	1989	11	7	22	25	347.250	f546	3	2	1	70	2	50	1	170
93	1989	11	8	22	08	254.750	231	4	2	2	70	8	100	1	120
94	1989	11	8	17	16	415.060	535	4	1	2	80	1	50	1	250
95	1989	11	8	18	32	424.300	542	4	1	1	70	1	100	1	130
96	1989	11	9	20	11	245.100	39	5	3	2	60	1	50	1	110
97	1989	11	9	20	10	337.750	542	5	3	1	65	2	50	1	130
98	1989	11	10	20	11	245.100	39	6	2	2	60	2	100	1	65
99	1989	11	10	21	50	425.900	541	6	1	2	65	6	50	1	130
100	1989	11	13	22	33	338.800	f546	2	3	1	65	1	40	1	140
101	1989	11	17	21	49	254.000	41	6	1	2	75	4	70	1	210
102	1989	11	17	20	49	364.400	541	6	1	2	75	3	50	1	200
103	1989	11	23	16	34	428.300	540	5	1	1	70	5	70	1	140
104	1989	12	1	17	39	365.800	540	6	1	1	70	1	0	1	210
105	1989	12	7	21	57	321.800	3544	5	2	1	60	2	50	1	110
106	1989	12	13	17	39	364.600	540	4	1	1	80	1	90	1	200
107	1989	12	25	16	26	345.010	538	2	2	1	80	2	30	1	250
108	1989	12	31	20	29	419.350	3544	1	1	1	70	3	70	1	140
109	1990	1	6	8	23	392.100	525	7	1	2	80	1	80	1	200
110	1990	1	6	20	24	425.000	3544	7	5	1	65	2	50	1	140
111	1990	1	9	16	50	414.750	540	3	1	1	75	3	50	1	200
112	1990	1	10	10	21	350.250	528	4	2	1	73	10	80	2	150
113	1990	1	14	17	28	424.640	535	1	1	2	60	1	0	1	160
114	1990	1	28	19	52	238.800	234	1	1	1	60	4	100	1	70
115	1990	2	3	18	30	425.510	542	7	1	1	60	5	10	2	150
116	1990	2	6	17	27	426.150	535	3	1	2	70	3	60	2	150
117	1990	2	11	17	31	427.800	535	1	1	2	60	3	60	1	100
118	1990	2	15	20	24	424.350	3544	5	1	1	60	3	50	1	120
119	1990	2	18	18	29	248.100	229	1	1	2	60	2	30	1	120
120	1990	2	23	16	32	255.500	42	6	2	1	65	12	220	0	-10

(continues on the following page)

ID	YR	MO	DA	HR	MI	PLACE	TRN	DW	WE	DI	SP	NF	DS	NH	DH
121	1990	2	25	10	33	378.100	529	1	*	2	*	*	*	*	*
122	1990	3	1	20	13	334.870	542	5	1	1	60	4	50	2	100
123	1990	3	9	18	25	429.050	542	6	1	1	70	3	100	1	120
124	1990	3	15	8	08	433.100	526	5	2	1	60	4	50	1	100
125	1990	3	19	16	02	364.050	538	2	1	1	75	4	120	1	130
126	1990	3	29	18	37	418.600	542	5	1	1	60	2	60	1	100
127	1990	4	3	20	16	247.300	39	3	1	2	80	5	210	1	100
128	1990	4	4	16	42	420.100	540	4	3	1	70	2	20	1	200
129	1990	4	8	18	37	348.300	539	1	2	2	60	5	20	1	130
130	1990	4	10	17	22	420.500	535	3	1	2	70	5	40	1	180
131	1990	4	12	6	27	340.300	f521	5	1	2	60	10	30	1	130
132	1990	4	15	12	13	369.250	3530	1	2	1	75	6	100	1	150
133	1990	4	22	16	32	338.200	538	1	2	1	65	15	50	1	130
134	1990	5	2	21	29	284.800	285	4	1	2	50	1	30	1	71
135	1990	5	6	20	12	336.450	542	1	1	1	55	6	80	1	60
136	1990	5	6	21	41	338.950	3544	1	1	1	58	5	55	1	160
137	1990	5	6	20	27	423.620	3544	1	1	1	65	1	50	1	125
138	1990	5	12	3	39	273.960	5480	7	3	1	60	1	40	1	260
139	1990	5	20	22	20	286.150	41	1	2	2	80	2	10	1	300
140	1990	5	27	22	40	276.400	231	1	1	2	60	1	100	1	50
141	1990	6	23	20	24	222.850	234	7	1	1	80	2	70	*	230
142	1990	7	1	22	44	329.760	f546	1	1	1	60	1	50	1	120
143	1990	7	7	9	36	278.800	36	7	1	1	74	2	100	0	120
144	1990	7	8	4	45	223.300	1457	1	2	2	65	3	50	1	250
145	1990	7	9	22	20	285.500	41	2	1	2	70	3	30	2	190
146	1990	8	5	22	17	280.900	41	1	2	2	70	1	20	1	220
147	1990	8	28	9	46	337.850	526	3	1	1	70	2	20	1	200
148	1990	9	4	18	43	392.500	537	3	1	2	80	1	0	1	300
149	1990	9	20	20	08	238.050	540	5	2	1	75	3	50	1	250
150	1990	9	23	15	53	421.500	9552	1	1	1	60	2	50	1	100
151	1990	10	1	21	05	374.400	3544	2	3	1	65	*	0	*	200
152	1990	10	8	18	55	132.500	40	2	2	1	78	2	80	1	150
153	1990	10	9	20	36	269.000	3058	3	3	1	65	1	50	1	210
154	1990	10	15	18	46	347.080	539	2	1	2	72	2	100	1	120
155	1990	10	24	17	19	386.650	540	4	1	1	80	1	0	1	280
156	1990	10	28	20	10	237.350	234	1	1	1	50	1	0	1	100
157	1990	10	30	16	55	390.300	535	3	1	2	80	1	0	1	300
158	1990	10	31	18	34	422.880	542	4	1	1	55	2	30	1	90
159	1990	11	5	4	52	240.930	411	2	3	2	45	1	0	1	95
160	1990	11	5	19	03	317.850	540	2	3	1	60	1	0	1	150
161	1990	11	5	19	49	420.100	542	2	3	1	70	1	40	1	150
162	1990	11	11	21	46	242.780	231	1	1	2	60	2	50	1	120
163	1990	11	12	16	45	331.650	538	2	1	1	60	1	0	1	140
164	1990	11	13	9	53	350.450	528	3	1	1	80	3	150	1	140
165	1990	11	16	6	32	420.650	524	6	1	1	70	6	150	1	50
166	1990	11	26	22	12	238.200	4011	2	2	2	60	2	50	1	90
167	1990	11	26	21	20	274.900	285	2	2	2	80	1	0	1	300
168	1990	11	26	19	39	288.000	278	2	2	1	60	6	80	1	80

(continues on the following page)

ID	YR	MO	DA	HR	MI	PLACE	TRN	DW	WE	DI	SP	NF	DS	NH	DH
169	1990	11	26	21	46	339.750	543	2	2	2	70	1	40	1	170
170	1990	11	27	19	00	287.700	234	3	1	1	80	1	0	1	290
171	1990	11	27	21	42	421.150	541	3	1	2	65	1	30	1	170
172	1990	11	29	22	31	251.200	4011	5	1	2	75	1	50	1	230
173	1990	11	29	21	45	423.400	541	5	1	2	70	3	50	1	140
174	1990	12	4	23	51	251.200	412	3	1	1	60	1	50	1	110
175	1990	12	9	16	42	418.500	540	1	1	1	80	5	100	1	180
176	1990	12	16	6	32	418.700	524	1	1	1	75	4	30	1	200
177	1990	12	17	17	45	363.450	540	2	1	1	80	1	100	1	170
178	1990	12	17	16	56	406.450	540	2	1	1	70	2	100	1	120
179	1990	12	25	21	43	285.450	280	3	1	1	75	2	100	1	160
180	1990	12	25	16	31	351.800	538	3	1	1	80	2	20	1	250
181	1990	12	31	21	43	424.000	541	2	1	2	70	5	30	1	190
182	1991	1	1	7	23	420.000	523	3	1	2	70	8	30	1	200
183	1991	1	4	6	33	419.200	524	6	1	1	70	1	0	1	200
184	1991	1	10	6	23	428.900	524	5	1	1	80	1	50	1	240
185	1991	1	14	21	29	405.650	541	2	1	2	80	2	50	1	230
186	1991	1	18	20	10	433.350	3544	6	2	1	60	1	20	1	150
187	1991	1	22	17	31	279.400	232	3	1	1	80	3	20	2	260
188	1991	1	24	18	12	279.300	4012	5	1	1	75	3	150	1	100
189	1991	1	26	11	45	367.200	3530	7	1	1	80	3	100	1	190
190	1991	1	26	21	55	432.565	541	7	1	2	70	3	0	2	230
191	1991	2	3	17	15	411.700	535	1	1	2	80	2	0	1	270
192	1991	2	5	12	04	143.500	4003	3	1	2	80	5	180	1	130
193	1991	2	11	14	26	255.200	228	2	2	1	65	15	100	0	100
194	1991	2	15	23	00	287.100	412	6	2	1	70	16	0	1	250
195	1991	2	20	18	21	432.000	542	4	1	1	50	1	0	1	95
196	1991	2	23	13	55	144.250	227	7	2	2	75	3	70	1	200
197	1991	2	26	19	20	422.500	537	3	1	2	70	2	40	1	170
198	1991	2	26	18	26	429.900	542	3	1	1	65	1	80	1	100
199	1991	2	26	14	22	430.500	3533	3	1	2	65	26	120	2	100
200	1991	2	27	18	26	430.200	542	4	1	1	80	1	0	1	280
201	1991	3	1	18	26	428.700	542	6	1	1	60	4	0	1	150
202	1991	3	16	10	23	143.700	34	7	2	1	75	16	300	0	-30
203	1991	3	16	18	23	431.800	542	7	2	1	60	2	80	1	80
204	1991	3	17	20	35	287.050	4009	1	2	2	75	4	60	1	200
205	1991	3	17	19	15	423.000	537	1	1	2	60	2	50	1	120
206	1991	3	23	13	40	255.850	4003	7	2	2	40	7	70	0	-10
207	1991	3	26	18	19	367.250	537	3	1	2	80	3	40	1	250
208	1991	3	28	23	45	255.750	412	5	2	1	60	1	60	1	160
209	1991	3	29	19	30	366.000	542	6	1	1	80	2	50	1	240
210	1991	3	29	9	19	433.550	525	6	1	2	60	6	0	1	130
211	1991	3	31	19	32	364.010	542	1	1	1	60	3	140	1	10
212	1991	4	1	19	25	369.600	542	2	2	1	75	4	0	2	250
213	1991	4	6	17	45	364.000	540	7	2	1	80	12	100	1	180
214	1991	4	18	4	29	243.700	411	5	2	2	60	1	30	0	170
215	1991	4	21	19	02	279.100	4007	1	1	2	70	7	100	1	140
216	1991	4	27	19	57	243.000	4009	7	1	2	65	3	120	0	100

(continues on the following page)

ID	YR	MO	DA	HR	MI	PLACE	TRN	DW	WE	DI	SP	NF	DS	NH	DH
217	1991	5	7	21	54	275.100	280	3	2	1	85	1	50	1	280
218	1991	5	8	18	12	244.250	232	4	2	1	65	8	20	1	150
219	1991	5	22	19	20	273.200	234	4	1	1	80	6	50	1	250
220	1991	6	12	19	23	372.900	542	4	2	1	70	1	50	1	180
221	1991	6	14	20	09	236.700	234	6	2	1	60	2	20	1	150
222	1991	7	1	20	35	423.000	3544	2	1	1	70	1	0	1	200
223	1991	7	2	18	28	136.160	4009	3	1	2	75	2	100	2	350
224	1991	7	3	18	27	338.500	540	4	1	1	65	2	100	1	70
225	1991	7	11	18	32	249.600	4007	5	3	2	90	1	50	1	190
226	1991	7	18	19	32	364.500	542	5	2	1	75	3	50	1	200
227	1991	8	12	20	00	347.150	542	2	2	1	70	2	40	1	180
228	1991	8	26	22	56	287.100	4011	2	1	2	75	2	100	1	160
229	1991	8	26	21	14	361.550	3544	2	1	1	75	1	150	1	100
230	1991	8	30	19	19	425.900	537	6	1	2	70	2	10	1	200
231	1991	9	22	22	59	287.600	4011	1	2	2	80	15	50	2	250
232	1991	9	22	20	08	421.100	9549	1	4	2	70	2	15	1	240
233	1991	9	22	20	23	425.400	3544	1	4	1	70	1	10	1	210
234	1991	9	23	22	37	330.750	546	2	3	1	60	3	20	1	130
235	1991	9	24	21	35	286.950	285	3	1	2	70	4	50	1	170
236	1991	9	25	23	01	285.500	231	4	1	2	65	2	20	1	170
237	1991	9	30	17	45	363.900	540	2	1	1	80	4	40	1	250
238	1991	10	8	19	48	287.160	229	3	1	2	75	6	100	1	150
239	1991	10	22	22	40	328.850	546	3	3	1	60	2	50	1	115
240	1991	10	26	22	22	244.200	4011	7	2	2	65	4	50	1	160
241	1991	10	27	16	24	255.900	4005	1	2	2	65	1	30	1	140
242	1991	11	1	4	24	258.370	411	6	1	2	60	1	0	1	180
243	1991	11	1	18	29	336.400	540	6	1	1	60	3	20	3	120
244	1991	11	7	17	27	317.850	537	5	1	2	60	1	50	1	100
245	1991	11	9	20	08	236.850	234	7	1	1	50	1	0	1	110
246	1991	11	14	17	59	346.250	537	5	2	2	80	2	20	1	270
247	1991	11	14	16	29	431.850	540	5	2	1	60	6	100	1	40
248	1991	11	20	20	20	423.200	3544	4	1	1	65	1	20	1	130
249	1991	11	24	21	36	338.465	3544	1	1	1	65	1	20	1	160
250	1991	11	27	16	43	419.652	540	4	2	1	70	3	70	1	130
251	1991	12	7	16	41	338.630	538	7	1	1	65	5	150	1	30
252	1991	12	15	20	25	345.100	541	1	*	2	*	6	100	1	200
253	1991	12	19	17	35	323.700	537	4	*	2	*	1	30	1	120
254	1991	12	22	21	30	408.400	541	1	*	2	*	1	50	1	200
255	1991	12	23	11	37	262.800	4001	2	2	2	65	4	100	1	120
256	1992	1	2	18	45	260.200	4007	5	1	2	45	3	0	1	90
257	1992	1	3	22	57	286.700	4011	6	1	2	70	1	0	1	200
258	1992	1	6	18	43	414.700	542	2	1	1	80	7	100	1	180
259	1992	1	11	17	23	280.050	227	7	1	2	82	5	100	1	200
260	1992	1	13	22	57	285.900	4011	2	5	2	75	8	170	3	100
261	1992	1	30	17	39	370.000	540	5	1	1	75	5	70	2	180
262	1992	2	7	22	39	276.200	231	6	2	2	60	7	50	1	90
263	1992	2	14	17	34	430.300	535	6	*	2	*	2	50	1	110
264	1992	2	14	22	56	286.800	4011	6	2	2	70	1	0	1	220

(continues on the following page)

ID	YR	MO	DA	HR	MI	PLACE	TRN	DW	WE	DI	SP	NF	DS	NH	DH
265	1992	2	27	8	00	248.000	4002	5	1	1	60	6	150	0	10
266	1992	2	27	18	15	366.500	537	5	*	2	*	5	100	1	120
267	1992	3	1	18	28	428.400	542	1	*	1	*	1	50	1	170
268	1992	3	8	18	37	419.900	542	1	*	1	*	1	10	1	200
269	1992	3	9	19	23	429.300	537	2	1	2	75	3	70	1	180
270	1992	3	12	17	22	426.300	535	5	1	2	60	5	30	1	110
271	1992	3	15	17	41	333.100	537	1	3	2	60	3	130	1	35
272	1992	3	23	18	21	327.000	539	2	1	2	60	2	80	1	70
273	1992	3	26	19	15	421.750	537	5	*	1	*	2	20	1	140
274	1992	4	16	22	43	273.800	4011	5	1	2	80	1	100	0	190
275	1992	4	16	21	49	279.100	280	5	1	1	70	6	80	1	120
276	1992	5	8	4	41	126.700	1753	6	*	2	63	1	*	1	250
277	1992	5	8	18	56	273.050	4007	6	2	2	65	2	50	1	140
278	1992	5	8	21	35	287.710	285	6	4	2	65	3	70	1	110
279	1992	5	8	21	44	423.800	541	6	4	2	70	1	20	1	200
280	1992	6	20	22	58	287.700	4011	7	2	2	80	2	0	1	300
281	1992	7	9	22	45	242.310	1458	5	2	1	60	2	20	*	150
282	1992	7	9	20	28	279.100	2578	5	2	1	75	3	50	2	190
283	1992	7	15	18	36	422.350	5642	4	1	1	70	1	20	1	180
284	1992	7	18	5	51	275.900	2522	7	2	1	60	2	20	1	160
285	1992	7	19	18	53	236.790	2529	1	3	2	70	1	0	1	200
286	1992	7	20	21	56	245.200	2531	2	2	2	60	1	0	1	160
287	1992	7	25	22	08	214.000	4011	7	3	2	75	1	0	1	250
288	1992	8	4	4	51	277.900	t9372	3	2	1	50	6	180	1	25
289	1992	8	4	18	17	360.080	5641	3	*	2	80	1	50	1	260
290	1992	8	5	8	29	327.600	5624	4	2	1	60	1	20	1	160
291	1992	8	9	21	18	261.600	2585	1	2	2	70	1	100	1	100
292	1992	8	12	20	52	376.000	5645	4	*	2	70	1	70	1	180
293	1992	8	13	22	18	228.700	4011	5	*	2	70	1	30	1	200
294	1992	8	19	17	30	422.450	5639	4	*	2	70	4	50	1	180
295	1992	8	23	18	56	243.570	4012	1	2	1	60	1	50	1	100
296	1992	8	23	18	51	346.600	5643	1	*	2	70	1	50	1	165
297	1992	8	26	18	19	436.500	5642	4	*	1	70	3	100	1	110
298	1992	9	19	22	13	133.400	39	7	2	2	90	1	0	1	400
299	1992	9	27	4	59	258.100	411	1	3	2	60	2	200	0	10
300	1992	10	3	18	21	273.200	4012	7	1	1	85	5	100	1	260
301	1992	10	4	0	00	242.800	412	1	1	1	55	2	10	1	120
302	1992	10	6	21	45	425.700	5645	3	1	2	70	2	50	1	160
303	1992	10	7	23	13	287.600	4011	4	2	2	80	2	50	1	230
304	1992	10	7	20	05	288.300	2529	4	2	2	70	2	50	1	170
305	1992	10	8	3	45	242.600	1455	5	2	2	50	1	0	1	120
306	1992	10	8	17	20	256.100	2527	5	2	2	60	3	0	1	150
307	1992	10	9	21	41	421.000	5645	6	3	2	60	1	50	1	110
308	1992	10	12	22	30	244.300	4011	2	4	2	60	3	0	0	180
309	1992	10	14	6	22	420.600	5624	4	1	1	70	1	20	1	180
310	1992	10	15	21	54	431.850	5645	5	1	2	65	1	20	1	160
311	1992	10	16	22	52	261.800	4011	6	1	2	70	5	100	1	100
312	1992	10	23	21	42	331.150	3644	6	1	1	60	1	20	1	120

(continues on the following page)

ID	YR	MO	DA	HR	MI	PLACE	TRN	DW	WE	DI	SP	NF	DS	NH	DH
313	1992	10	26	21	44	338.900	5647	2	1	2	65	1	30	1	160
314	1992	10	28	17	26	420.200	5639	4	1	2	60	4	50	*	120
315	1992	10	30	17	06	286.700	2530	6	2	1	80	1	30	1	270
316	1992	10	30	16	40	424.700	5640	6	2	1	70	2	50	1	190
317	1992	10	31	17	27	423.900	5639	7	2	2	60	1	50	1	120
318	1992	11	1	21	10	363.900	3644	1	1	1	75	2	50	1	200
319	1992	11	2	22	34	245.100	4011	2	1	2	65	1	50	1	160
320	1992	11	2	18	28	329.700	5643	2	1	2	75	1	50	1	200
321	1992	11	8	20	40	365.900	5645	1	1	2	75	8	200	1	50
322	1992	11	9	21	36	420.500	5645	2	1	2	60	6	130	1	30
323	1992	11	10	21	21	287.400	1458	3	1	1	70	2	0	0	250
324	1992	11	10	21	38	273.200	2578	3	3	1	80	3	100	1	200
325	1992	11	10	21	38	332.100	3644	3	3	1	65	5	50	*	120
326	1992	11	12	5	48	143.500	*	5	1	*	30	7	50	1	*
327	1992	11	12	2	32	144.400	411	5	1	2	60	2	50	*	150
328	1992	11	12	3	38	236.900	1455	5	1	2	60	7	0	*	190
329	1992	11	13	17	13	325.250	5638	6	1	1	65	5	50	*	150
330	1992	11	13	21	53	435.800	5645	6	1	2	75	2	50	1	200
331	1992	11	15	21	25	403.850	5645	1	2	2	75	1	50	1	180
332	1992	11	19	17	27	418.750	5639	5	2	2	80	1	80	1	220
333	1992	11	26	22	17	339.700	f5646	5	1	1	75	7	60	1	190
334	1992	11	26	21	38	420.100	5645	5	1	2	60	5	50	1	110
335	1992	11	27	16	51	415.450	5640	6	1	1	80	1	100	1	160
336	1992	11	28	22	19	262.980	2531	7	1	2	70	4	100	1	140
337	1992	11	28	17	48	365.300	5640	7	1	1	80	7	100	1	190
338	1992	11	29	21	32	338.700	3644	1	1	1	70	1	100	1	110
339	1992	11	29	19	09	421.700	5641	1	1	2	65	1	30	1	160
340	1992	11	29	21	42	424.350	5645	1	1	2	70	3	50	1	160
341	1992	11	30	16	26	243.850	2528	2	1	1	65	1	80	1	80
342	1992	11	30	21	51	432.250	5645	2	1	2	70	1	80	1	160
343	1992	12	4	16	20	364.100	5638	6	1	1	80	1	50	1	230
344	1992	12	4	21	10	365.600	3644	6	1	1	80	6	80	1	200
345	1992	12	6	16	22	361.500	5638	1	2	1	80	1	80	1	210
346	1992	12	6	21	50	431.400	5645	1	2	2	65	1	50	1	140
347	1992	12	10	6	16	425.550	5624	5	1	1	70	2	50	1	170
348	1992	12	11	7	11	337.600	5622	6	2	1	70	5	50	1	150
349	1992	12	12	6	20	351.400	f5621	7	1	2	80	2	80	1	200
350	1992	12	15	17	39	332.400	5641	3	1	2	60	6	0	1	160
351	1992	12	15	20	22	419.900	3644	3	1	1	70	3	20	1	180
352	1992	12	16	19	22	434.300	5641	4	1	2	70	8	50	2	160
353	1992	12	17	20	37	364.500	5645	5	1	2	70	1	100	1	110
354	1992	12	18	19	52	347.550	5642	6	1	1	70	3	50	1	160
355	1992	12	18	21	29	410.100	5645	6	1	2	80	1	50	1	240
356	1992	12	18	16	40	425.100	5640	6	1	1	60	2	30	1	130
357	1992	12	20	6	18	350.350	f5621	1	1	2	85	4	50	1	280
358	1992	12	20	16	16	351.500	5639	1	5	2	85	15	100	1	230
359	1992	12	21	20	02	284.000	2529	2	1	2	50	1	0	1	100
360	1992	12	21	6	19	350.350	f5621	2	1	2	80	2	50	1	250

(continues on the following page)

ID	YR	MO	DA	HR	MI	PLACE	TRN	DW	WE	DI	SP	NF	DS	NH	DH
361	1992	12	22	20	11	336.700	5645	3	2	2	65	2	50	1	130
362	1992	12	25	16	55	335.650	5638	6	1	1	65	4	70	1	100
363	1992	12	26	19	10	422.400	5641	7	1	2	65	3	50	1	140
364	1992	12	27	16	13	372.500	5638	1	1	1	75	6	50	1	220
365	1992	12	27	16	52	413.750	5640	1	1	1	80	3	70	1	220
366	1992	12	30	21	47	429.400	5645	4	1	2	75	3	100	1	150
367	1993	1	1	22	19	337.900	f5646	6	1	1	70	5	70	1	150
368	1993	1	1	16	19	364.300	5638	6	1	1	80	8	100	2	190
369	1993	1	2	17	12	143.500	4007	5	2	2	80	2	70	1	250
370	1993	1	3	17	30	425.650	5639	1	1	2	70	6	50	1	160
371	1993	1	18	22	56	420.100	9647	2	2	2	70	3	80	1	160
372	1993	1	19	16	51	341.400	5638	3	5	1	80	4	180	1	100
373	1993	1	23	17	01	287.400	2581	7	1	2	65	1	20	1	180
374	1993	1	23	17	43	434.050	5639	7	1	2	60	4	50	1	100
375	1993	1	24	16	06	128.400	6028	1	2	1	40	1	300	1	1
376	1993	1	24	16	50	339.500	5638	1	1	1	85	4	150	1	180
377	1993	1	25	20	14	339.500	5645	2	1	2	75	1	50	1	200
378	1993	1	28	6	41	344.700	3625	5	3	2	85	2	100	1	250
379	1993	1	28	6	21	420.200	5624	5	3	1	65	3	30	1	140
380	1993	2	6	16	23	352.300	5639	7	5	2	60	2	30	1	130
381	1993	2	9	16	35	430.500	5639	3	1	2	75	8	30	1	230
382	1993	2	10	12	25	323.500	5630	4	1	1	60	1	50	1	100
383	1993	2	11	23	24	285.000	4011	5	1	2	55	7	100	0	30
384	1993	2	13	12	14	271.290	2524	7	1	1	80	3	50	1	240
385	1993	2	14	17	35	430.050	5639	1	1	2	70	5	100	1	125
386	1993	2	16	12	35	143.500	36	3	1	1	80	1	50	1	200
387	1993	2	17	17	13	287.300	4005	4	1	2	75	1	50	1	210
388	1993	2	18	17	59	350.900	5641	5	1	2	85	7	200	1	130
389	1993	2	21	19	53	143.500	1454	1	1	1	65	2	0	2	200
390	1993	2	25	17	48	409.500	f8356	5	1	1	80	16	100	1	180
391	1993	2	26	0	06	287.050	f8666	6	1	1	75	1	150	1	100
392	1993	2	28	23	23	286.000	4011	1	5	2	65	3	70	0	120
393	1993	3	1	22	56	286.700	412	2	2	1	60	8	100	1	60
394	1993	3	5	22	56	287.100	412	6	2	1	65	1	200	1	*
395	1993	3	6	4	38	287.000	1455	7	1	2	*	*	*	*	*
396	1993	3	7	7	45	143.650	32	1	1	1	75	1	200	1	25
397	1993	3	7	7	25	144.600	2545	1	1	2	70	12	0	1	200
398	1993	3	7	16	39	273.550	4005	1	1	2	80	8	100	*	160
399	1993	3	9	14	55	144.400	38	3	1	1	80	5	200	0	50
400	1993	3	9	17	41	370.500	5640	3	1	1	80	10	100	1	200
401	1993	3	11	16	29	432.400	5640	5	2	1	65	2	50	*	130
402	1993	3	16	6	27	426.400	5624	3	1	1	70	4	0	1	220
403	1993	3	17	18	31	420.000	5642	4	1	1	70	4	50	1	140
404	1993	4	4	4	53	255.800	4013	1	5	2	60	4	50	1	90
405	1993	4	4	23	23	285.800	4011	1	2	2	75	4	200	1	70
406	1993	4	8	17	43	436.600	5639	5	1	2	55	7	30	1	90
407	1993	4	10	18	17	279.500	4012	7	1	1	80	3	0	2	290
408	1993	4	24	17	37	273.800	2530	7	3	1	75	7	100	1	150

(continues on the following page)

ID	YR	MO	DA	HR	MI	PLACE	TRN	DW	WE	DI	SP	NF	DS	NH	DH
409	1993	4	24	20	16	237.370	2532	7	3	1	50	1	0	1	90
410	1993	5	1	22	06	238.880	2531	7	1	2	60	2	0	1	140
411	1993	5	2	22	44	339.550	f5646	1	1	1	75	3	100	1	140
412	1993	5	3	18	57	242.450	4012	2	3	1	65	2	100	1	50
413	1993	5	3	20	42	125.700	4012	2	5	1	80	3	100	1	150
414	1993	5	5	22	28	332.800	3644	4	1	1	70	2	50	1	140
415	1993	5	6	5	21	332.950	t9371	5	2	2	50	1	80	1	40
416	1993	5	6	18	56	407.900	5641	5	1	2	70	2	50	1	160
417	1993	5	13	20	04	243.100	4009	5	1	2	70	1	200	1	-5
418	1993	5	15	18	26	426.200	5642	7	1	1	70	7	100	1	130
419	1993	5	19	21	45	285.900	2585	4	2	2	75	4	50	1	200
420	1993	5	23	7	00	367.650	3625	1	3	2	75	1	0	1	250
421	1993	5	26	19	15	373.800	5642	4	2	1	70	1	50	1	170
422	1993	6	7	10	08	243.500	3523	2	1	2	70	2	0	1	250
423	1993	6	9	19	34	258.200	2529	4	2	2	65	7	100	1	100
424	1993	6	11	18	31	420.450	5642	6	4	1	70	3	60	1	130
425	1993	6	28	19	28	361.100	5642	2	2	1	70	2	50	1	170
426	1993	6	29	22	58	270.400	4011	3	1	2	70	2	100	1	120
427	1993	7	11	19	24	364.000	5642	1	4	1	80	4	100	1	190
428	1993	7	11	20	44	441.500	3644	1	4	1	75	1	50	1	200
429	1993	7	12	10	14	327.800	5628	2	1	1	60	1	50	1	100
430	1993	7	14	17	48	124.250	4010	4	1	1	80	1	100	1	150
431	1993	7	21	8	44	418.720	5628	4	2	1	75	3	35	1	200
432	1993	7	21	19	05	274.700	4007	4	2	2	60	1	100	0	60
433	1993	7	23	16	49	382.220	5639	6	2	2	75	2	50	1	170
434	1993	7	25	18	33	386.650	5641	1	2	2	80	2	100	1	180
435	1993	7	27	18	47	352.320	5643	3	2	2	60	2	50	1	150
436	1993	8	2	6	35	337.200	3625	2	2	2	70	4	80	1	140
437	1993	8	8	4	55	132.000	1753	1	2	2	65	*	0	*	200
438	1993	8	9	19	25	364.450	5642	2	2	1	80	2	50	1	250
439	1993	8	17	6	20	420.500	5624	3	1	1	65	1	150	1	50
440	1993	9	2	20	06	329.770	5645	5	1	2	70	1	20	1	190
441	1993	9	3	18	32	332.600	5641	6	1	2	65	2	80	1	120
442	1993	9	5	21	45	371.000	3644	1	3	1	65	2	50	1	130
443	1993	9	15	18	09	440.900	5642	4	1	1	70	1	50	1	160
444	1993	9	15	19	12	278.600	4007	4	1	2	75	1	0	1	290
445	1993	9	16	20	21	347.700	5645	5	1	2	85	2	10	1	320
446	1993	9	18	18	44	321.100	5640	7	4	1	60	2	50	1	110
447	1993	9	23	17	39	331.100	5641	5	1	2	65	1	30	1	170
448	1993	9	28	19	51	124.750	37	3	1	2	95	2	25	1	330
449	1993	9	29	18	24	326.250	5643	4	1	2	65	4	100	1	70
450	1993	9	29	19	29	361.500	5642	4	1	1	80	1	20	1	280
451	1993	9	30	20	39	285.060	4009	5	3	2	65	2	30	1	140
452	1993	10	2	17	40	373.200	5640	7	1	1	70	3	60	1	180
453	1993	10	2	21	51	242.200	2531	7	1	2	40	1	20	1	50
454	1993	10	9	15	52	285.300	2580	7	1	1	70	1	20	1	200
455	1993	10	10	19	47	125.100	37	1	2	2	95	1	150	0	200
456	1993	10	12	18	25	336.900	5640	3	1	1	65	5	50	1	120

(continues on the following page)

ID	YR	MO	DA	HR	MI	PLACE	TRN	DW	WE	DI	SP	NF	DS	NH	DH
457	1993	10	13	17	23	316.500	5641	4	1	2	75	1	80	1	170
458	1993	10	14	18	50	403.200	5642	5	1	1	80	1	50	1	250
459	1993	10	15	17	39	331.000	5641	6	1	2	45	5	50	1	40
460	1993	10	16	16	59	331.300	5638	7	1	1	65	2	70	1	100
461	1993	10	17	21	40	424.000	5645	1	2	2	60	2	40	1	110
462	1993	10	19	19	34	244.300	4012	3	1	1	68	4	20	1	50
463	1993	10	20	21	39	338.800	5647	4	1	2	75	3	70	1	190
464	1993	10	21	16	22	120.800	4008	5	1	1	80	1	150	0	100
465	1993	10	21	20	26	317.850	5642	5	1	1	60	1	50	1	120
466	1993	10	26	19	49	317.400	5645	3	1	2	60	1	30	1	140
467	1993	10	27	4	00	126.600	9751	4	1	2	65	1	40	1	230
468	1993	10	28	18	32	420.000	5642	5	1	1	70	2	50	1	150
469	1993	10	28	23	20	285.700	2531	5	1	2	75	2	50	1	210
470	1993	10	30	16	05	330.600	5636	7	3	1	65	1	50	1	120
471	1993	10	31	21	13	253.800	2585	1	1	2	70	2	70	1	150
472	1993	11	2	0	30	236.700	4014	3	1	1	50	1	20	1	80
473	1993	11	2	16	57	388.500	5639	3	1	2	80	4	150	1	140
474	1993	11	4	16	43	421.550	5640	5	1	1	60	3	50	1	100
475	1993	11	4	22	05	351.800	3644	5	1	1	85	2	70	1	250
476	1993	11	7	18	29	423.300	5642	1	2	1	65	3	50	1	140
477	1993	11	10	18	11	350.600	5640	4	1	1	85	1	50	1	280
478	1993	11	11	21	25	406.400	5645	5	1	2	70	1	0	1	220
479	1993	11	13	21	01	422.800	3644	7	1	1	70	1	80	1	140
480	1993	11	15	21	45	429.400	5645	2	1	2	80	2	30	1	260
481	1993	11	16	20	38	364.000	5646	3	1	1	80	2	50	1	240
482	1993	11	17	17	18	411.740	5639	4	2	2	80	3	30	1	260
483	1993	11	18	16	22	362.580	5638	5	3	1	80	2	50	1	240
484	1993	11	21	21	03	419.500	3644	1	2	1	70	3	30	1	190
485	1993	11	30	18	55	404.500	5641	3	1	2	80	4	80	1	200
486	1993	12	2	16	34	366.500	5639	5	2	2	80	2	100	1	190
487	1993	12	2	19	02	145.000	2567	5	2	2	75	1	0	1	250
488	1993	12	6	16	21	364.300	5638	2	1	1	80	4	100	*	190
489	1993	12	6	18	12	365.200	5641	2	1	2	80	1	20	1	270
490	1993	12	12	23	12	285.500	4011	1	1	2	60	8	0	1	140
491	1993	12	16	21	35	419.200	5645	5	1	2	65	3	100	1	100
492	1993	12	19	16	17	143.500	35	1	2	2	80	1	50	1	200
493	1993	12	21	16	01	144.300	4008	3	5	1	80	3	150	1	150
494	1993	12	21	21	04	419.700	3644	3	1	1	70	1	130	1	90
495	1993	12	25	23	48	273.230	4014	7	1	1	70	1	50	1	150
496	1993	12	26	21	48	432.650	5645	1	1	2	70	*	100	1	120
497	1993	12	28	22	16	371.400	9647	3	1	2	80	10	50	1	240
498	1993	12	29	17	42	432.500	5639	4	1	2	70	3	50	1	180
499	1993	12	29	22	55	421.300	9647	4	1	2	70	1	50	1	180
500	1993	12	31	22	07	364.350	9647	6	1	2	80	4	50	1	240
501	1994	1	2	19	11	423.800	5641	1	1	2	70	2	50	1	180
502	1994	1	3	22	54	420.150	9647	2	1	2	60	16	80	1	85
503	1994	1	6	17	25	418.000	5639	5	1	2	70	2	70	1	150
504	1994	1	8	2	25	144.500	4014	7	1	1	70	5	100	1	90

(continues on the following page)

ID	YR	MO	DA	HR	MI	PLACE	TRN	DW	WE	DI	SP	NF	DS	NH	DH
505	1994	1	8	22	45	407.400	9647	7	1	2	70	4	100	1	120
506	1994	1	11	16	27	433.400	5640	3	1	1	60	5	50	1	120
507	1994	1	15	18	21	429.900	5642	7	1	1	75	5	60	1	200
508	1994	1	16	22	24	144.300	39	1	1	2	39	4	100	1	260
509	1994	1	20	16	38	426.200	5640	5	1	1	65	20	70	1	125
510	1994	1	23	17	34	429.900	5639	1	1	2	60	4	60	1	100
511	1994	1	24	17	34	428.800	5639	2	1	2	75	4	40	1	210
512	1994	1	31	17	13	405.620	5639	2	1	2	75	5	50	1	200
513	1994	2	2	23	28	285.200	4011	4	2	2	65	3	0	1	170
514	1994	2	5	14	00	278.100	4008	7	1	1	70	12	200	0	-5
515	1994	2	6	17	33	274.200	2530	1	1	1	82	3	25	1	280
516	1994	2	6	23	34	285.100	4014	1	1	1	50	2	50	1	55
517	1994	2	9	23	17	284.430	4011	4	5	2	57	2	30	1	130
518	1994	2	15	19	18	429.600	5641	3	1	2	75	8	150	2	100
519	1994	2	17	18	25	337.000	5640	5	1	1	65	2	100	1	90
520	1994	2	17	21	02	419.300	3644	5	1	1	70	3	50	1	170
521	1994	2	19	18	21	429.700	5642	7	1	1	80	1	50	1	240
522	1994	2	20	19	17	429.300	5641	1	1	2	80	1	40	1	250
523	1994	2	27	17	24	418.900	5639	7	1	2	80	7	80	1	210
524	1994	2	28	8	05	241.200	4002	2	1	1	70	4	120	1	80
525	1994	3	3	21	56	359.500	3644	5	1	1	60	1	80	1	85
526	1994	3	9	21	48	287.200	2582	4	1	1	70	15	100	1	120
527	1994	3	10	17	39	429.400	5639	5	2	2	70	6	100	1	120
528	1994	3	19	18	31	144.500	3440	7	1	1	70	6	100	1	120
529	1994	3	19	22	39	247.000	4011	7	1	2	60	3	70	0	80
530	1994	3	20	22	07	247.100	2531	1	1	2	55	3	50	2	90
531	1994	3	20	23	14	287.020	4011	1	1	2	70	16	50	1	150
532	1994	3	24	18	10	287.000	4012	5	2	1	70	3	100	3	100
533	1994	3	24	20	20	346.800	5645	5	3	2	75	2	100	1	150
534	1994	3	24	22	20	333.150	3644	5	1	1	60	2	50	1	110
535	1994	3	25	17	39	332.800	5641	6	1	2	60	3	80	1	90
536	1994	3	26	17	57	274.400	2527	7	1	2	85	7	50	1	280
537	1994	3	26	21	54	362.300	3644	7	1	1	80	1	150	1	140
538	1994	3	29	18	39	246.408	4007	3	1	2	65	3	130	1	60
539	1994	3	30	15	00	143.600	38	4	1	1	80	5	200	0	*
540	1994	3	31	23	14	287.500	4011	5	1	2	70	4	100	1	100
541	1994	4	9	18	23	428.900	5642	7	5	1	65	3	90	1	100
542	1994	4	9	23	13	286.800	4011	7	1	2	75	5	0	3	200
543	1994	4	25	20	09	242.500	4009	2	1	2	62	3	100	1	50
544	1994	5	10	23	13	287.300	4011	3	1	2	85	5	80	1	210
545	1994	5	12	4	36	126.000	1753	5	1	2	60	4	200	0	150
546	1994	5	16	20	29	273.100	4009	2	2	2	63	3	100	0	70
547	1994	5	18	18	26	126.000	4009	4	1	2	94	2	20	1	370
548	1994	5	19	19	47	226.200	4009	5	1	2	85	2	150	1	140
549	1994	5	24	21	09	414.500	3644	3	1	1	80	1	80	1	210
550	1994	6	12	19	47	274.000	2529	1	1	2	50	12	50	1	60
551	1994	6	14	3	55	249.800	1455	3	2	2	60	4	50	4	150
552	1994	6	16	6	35	241.100	2522	5	1	1	60	1	0	1	160

(continues on the following page)

ID	YR	MO	DA	HR	MI	PLACE	TRN	DW	WE	DI	SP	NF	DS	NH	DH
553	1994	6	22	20	11	336.500	5645	4	1	2	65	1	50	1	140
554	1994	7	7	18	24	329.000	5643	5	2	2	65	2	100	1	90
555	1994	7	7	19	05	275.500	4007	5	2	2	85	2	130	1	150
556	1994	7	13	6	22	325.360	3625	4	1	2	70	3	100	1	120
557	1994	7	25	3	54	249.250	1455	2	2	2	60	3	100	3	90
558	1994	7	28	18	34	132.050	4009	5	2	2	80	1	30	1	260
559	1994	7	28	21	30	266.810	2585	5	2	2	50	2	30	1	80
560	1994	7	31	21	25	326.050	5649	1	1	2	65	1	50	1	140
561	1994	8	3	20	20	131.500	2437	4	1	2	90	2	50	1	330
562	1994	8	10	22	32	318.500	3644	4	1	1	65	3	40	1	150
563	1994	8	14	18	54	398.700	5642	1	4	1	75	1	50	1	200
564	1994	8	14	19	14	285.500	4007	1	1	2	70	2	50	1	150
565	1994	8	21	21	48	285.500	2582	1	2	1	75	1	50	1	200
566	1994	8	22	7	16	327.950	5622	2	1	1	65	1	100	1	100
567	1994	8	24	7	47	273.300	2521	4	1	2	70	2	100	1	120
568	1994	8	26	9	21	383.760	5628	6	1	1	80	2	70	1	220
569	1994	8	26	20	04	329.000	5645	6	1	2	60	2	50	1	110
570	1994	8	27	18	32	424.100	5642	7	2	1	65	6	40	1	150
571	1994	8	29	19	26	363.700	5642	2	4	1	80	2	120	1	170
572	1994	9	5	18	35	388.600	5641	2	1	2	80	1	70	1	220
573	1994	9	6	18	56	406.700	5641	3	4	2	70	2	70	1	150
574	1994	9	6	23	11	286.300	4011	3	1	2	78	1	50	0	200
575	1994	9	13	3	50	245.800	1455	3	2	2	55	*	0	1	160
576	1994	9	13	23	14	288.100	4011	3	2	2	65	2	100	1	70
577	1994	9	15	17	15	410.400	5639	5	2	2	80	2	50	1	240
578	1994	9	16	19	00	404.200	5641	6	2	2	70	6	30	1	190
579	1994	9	18	22	35	244.380	4011	1	3	2	65	3	0	1	160
580	1994	9	19	17	41	369.250	5640	2	3	1	75	1	30	1	230
581	1994	9	20	18	59	248.200	4007	3	2	2	65	2	130	1	40
582	1994	9	20	20	03	336.700	5642	3	3	1	65	1	30	1	150
583	1994	9	20	22	09	276.000	2582	3	2	1	75	8	60	1	200
584	1994	9	21	18	28	328.900	5643	4	1	2	65	2	30	2	160
585	1994	9	22	2	48	133.400	4014	5	3	1	60	6	50	1	100
586	1994	9	22	5	00	239.750	4013	5	2	2	60	4	40	1	35
587	1994	9	23	17	41	261.700	2527	6	3	2	60	1	100	0	60
588	1994	9	23	20	19	263.700	4009	6	3	2	60	2	50	0	90
589	1994	9	24	22	28	321.100	3644	7	4	1	80	1	80	1	210
590	1994	9	26	19	09	225.956	4012	2	1	1	85	2	80	1	200
591	1994	10	2	17	33	327.000	5641	1	1	2	65	1	50	1	140
592	1994	10	2	21	49	285.400	2582	1	1	1	70	3	50	1	180
593	1994	10	3	17	30	322.000	5641	2	1	2	75	1	100	1	150
594	1994	10	10	22	34	243.400	4011	2	1	2	70	5	150	1	50
595	1994	10	11	18	53	321.850	5640	3	2	1	75	2	150	1	110
596	1994	10	21	18	02	350.600	5641	6	1	2	85	2	100	1	230
597	1994	10	22	21	48	350.630	5647	7	1	2	85	2	100	1	230
598	1994	10	23	18	04	248.900	2530	1	2	1	60	1	30	1	130
599	1994	10	26	18	42	346.200	5643	4	1	2	85	3	0	3	330
600	1994	10	28	22	30	280.680	5480	6	1	1	65	1	0	1	250

(continues on the following page)

ID	YR	MO	DA	HR	MI	PLACE	TRN	DW	WE	DI	SP	NF	DS	NH	DH
601	1994	10	31	20	40	285.975	4009	2	1	2	75	2	50	1	175
602	1994	11	3	21	08	420.700	3644	5	1	1	65	5	50	1	140
603	1994	11	4	17	46	337.970	5641	6	1	2	65	2	50	1	140
604	1994	11	6	22	21	333.100	3644	1	1	1	60	8	80	1	80
605	1994	11	8	17	35	328.700	5641	3	1	2	70	1	50	1	170
606	1994	11	10	16	53	327.800	5638	5	1	1	60	1	50	1	115
607	1994	11	12	20	46	439.050	3644	7	1	1	80	2	100	1	190
608	1994	11	12	21	27	325.400	5647	7	1	2	70	1	50	1	160
609	1994	11	13	16	58	407.100	5640	1	1	1	60	7	50	1	110
610	1994	11	13	23	19	241.200	5480	1	1	1	60	1	0	1	210
611	1994	11	14	18	36	420.500	5642	2	1	1	70	2	100	1	120
612	1994	11	15	16	08	375.200	5638	3	1	1	65	4	90	1	100
613	1994	11	15	18	13	366.000	5641	3	1	2	75	3	80	1	180
614	1994	11	20	16	32	367.100	5639	1	2	2	75	2	100	1	150
615	1994	11	20	18	59	133.700	40	1	1	1	90	1	0	1	300
616	1994	11	20	22	01	241.000	2531	1	1	2	45	2	50	0	30
617	1994	11	23	17	15	276.500	1453	4	1	2	55	1	0	1	130
618	1994	11	24	17	32	425.800	5639	5	1	2	70	2	50	1	170
619	1994	11	28	17	37	327.800	5641	2	1	2	65	2	50	1	140
620	1994	11	30	22	20	332.250	3644	4	1	1	65	2	50	1	140
621	1994	12	7	6	30	332.700	3625	4	1	2	70	1	100	1	120
622	1994	12	7	16	53	410.700	5640	4	1	1	70	1	100	1	120
623	1994	12	7	21	18	286.500	1458	4	1	1	65	2	50	1	200
624	1994	12	17	16	56	387.250	5639	7	1	2	80	2	80	1	210
625	1994	12	17	18	13	364.070	5641	7	1	2	75	4	200	1	55
626	1994	12	20	6	16	431.650	5624	3	1	1	55	8	80	1	60
627	1994	12	20	21	07	418.700	3644	3	1	1	80	2	50	1	240
628	1994	12	24	16	49	250.320	1453	7	1	2	60	3	0	1	200
629	1994	12	26	17	10	406.900	5639	2	1	2	70	2	180	1	40
630	1994	12	29	18	20	373.000	5641	5	1	2	70	4	30	2	190
631	1994	12	30	17	42	368.600	5640	6	1	1	50	1	10	1	100
632	1994	12	31	17	54	346.960	5641	7	1	2	80	5	70	2	220
633	1995	1	2	0	25	240.900	4014	2	1	1	55	2	50	1	80
634	1995	1	3	16	32	365.500	5639	3	2	2	70	1	30	1	190
635	1995	1	3	21	47	430.800	5645	3	2	2	45	1	5	1	85
636	1995	1	4	15	40	407.150	5638	4	1	1	75	2	50	1	220
637	1995	1	5	0	14	251.700	4014	5	2	1	45	4	0	1	80
638	1995	1	7	15	43	350.300	5636	7	1	1	75	2	200	1	200
639	1995	1	9	6	20	431.100	5624	2	1	1	45	7	50	1	40
640	1995	1	18	20	11	252.450	4009	4	1	2	57	5	40	1	80
641	1995	1	19	21	47	433.000	5645	5	2	2	65	4	40	1	150
642	1995	1	25	16	56	407.500	5640	4	1	1	75	2	30	1	240
643	1995	1	26	7	54	335.600	5624	5	2	1	80	2	50	1	240
644	1995	1	26	20	55	428.500	3644	5	1	1	75	1	80	2	180
645	1995	1	27	6	18	429.050	5624	6	2	1	80	10	100	1	200
646	1995	1	27	17	25	418.700	5639	6	2	2	80	10	100	1	185
647	1995	1	27	18	26	429.000	5642	6	2	1	75	1	50	1	200
648	1995	1	29	7	52	426.100	3625	1	2	2	40	12	50	1	20

(concludes on the following page)

ID	YR	MO	DA	HR	MI	PLACE	TRN	DW	WE	DI	SP	NF	DS	NH	DH
649	1995	1	29	11	15	409.100	3630	1	5	1	80	10	100	1	190
650	1995	2	4	18	15	369.200	5641	7	1	2	80	1	50	1	240
651	1995	2	10	9	23	371.150	5629	6	1	2	75	5	100	1	160
652	1995	2	12	10	10	420.300	5629	1	1	2	70	7	100	1	130
653	1995	2	12	17	40	432.500	5639	1	1	2	65	3	50	1	145
654	1995	2	13	7	46	420.100	3625	2	1	2	65	2	150	1	45
655	1995	2	14	17	33	274.300	2530	3	2	1	85	20	150	4	180
656	1995	2	14	17	38	329.500	5641	3	2	2	65	2	50	1	140
657	1995	2	15	17	24	422.400	5639	4	1	2	65	20	100	2	90
658	1995	2	15	21	41	425.950	5645	4	1	2	70	2	50	1	170
659	1995	2	16	18	17	370.200	5641	5	2	2	75	3	50	1	200
660	1995	2	16	20	43	369.400	5645	5	2	2	75	1	150	1	100
661	1995	2	19	22	21	331.250	3644	1	1	1	65	1	40	1	150
662	1995	2	22	17	49	341.300	5641	4	1	2	85	3	50	1	280
663	1995	2	23	6	18	428.700	5624	5	1	1	70	10	100	1	120
664	1995	2	23	15	14	121.900	38	5	1	1	85	1	100	1	200
665	1995	2	24	12	43	275.800	*	6	2	1	60	3	50	0	70
666	1995	2	24	17	42	368.950	5640	6	2	1	70	5	50	1	170
667	1995	2	25	18	11	364.000	5641	7	1	2	80	5	50	1	240
668	1995	2	28	6	21	424.300	5624	3	2	1	70	3	150	1	80
669	1995	2	28	18	28	425.700	5642	3	1	1	60	1	50	1	110
670	1995	2	28	19	08	287.300	2532	3	1	1	75	2	0	1	250
671	1995	2	28	21	51	366.400	3644	3	1	1	75	1	30	1	220
672	1995	3	1	21	03	421.900	3644	4	5	1	70	2	100	1	120
673	1995	3	5	10	20	430.850	5629	1	1	2	70	1	50	1	170
674	1995	3	8	23	12	285.400	4011	4	1	2	70	3	115	1	85
675	1995	3	9	6	15	431.100	5624	5	1	1	70	10	70	2	150
676	1995	3	9	17	57	271.500	2527	5	1	2	70	3	50	1	180
677	1995	3	11	19	09	421.600	5641	7	2	2	65	2	100	1	100
678	1995	3	11	21	37	421.700	5645	7	2	2	60	2	50	1	110
679	1995	3	12	21	48	369.700	3644	1	1	1	80	4	150	1	140
680	1995	3	13	17	26	421.500	5639	2	1	2	60	6	80	1	85
681	1995	3	13	19	13	421.300	5641	2	1	2	70	5	50	1	175
682	1995	3	14	8	40	331.400	5629	3	1	2	60	8	50	1	110
683	1995	3	15	17	53	346.200	5641	4	1	2	80	3	100	1	190
684	1995	3	15	18	34	274.500	4096	4	1	1	60	7	0	1	200
685	1995	3	17	19	45	258.500	2532	6	2	1	65	3	0	1	185
686	1995	3	17	22	22	248.200	1458	6	2	1	55	2	10	1	150
687	1995	3	17	22	52	255.750	4011	6	2	2	60	6	50	1	95
688	1995	3	20	17	45	365.350	5640	2	1	1	65	2	0	1	190
689	1995	3	21	18	55	244.300	4012	3	3	1	65	3	50	1	120
690	1995	3	22	17	38	433.100	5637	4	1	2	60	2	70	1	90
691	1995	3	23	18	36	418.700	5642	5	1	1	75	3	100	1	150
692	1995	3	24	20	54	429.350	3644	6	1	1	80	1	*	1	240
693	1995	3	25	19	58	247.100	2532	7	1	1	50	10	*	1	100
694	1995	3	27	23	12	285.500	4011	2	2	2	80	4	*	2	150
695	1995	3	30	17	07	249.500	4096	5	2	1	55	5	0	*	*
696	1995	3	30	17	34	429.600	5637	5	2	2	80	10	*	1	190

CASE **62**

PUBLIC DEBT IN GERMANY

When Germany was reunited after the Berlin wall fell, two dramatically different social, economic, and political systems were merged. Public debt had been treated very differently in the two Germanys, and this became increasingly important as the adoption of the European common currency (the Euro) loomed (various treaty clauses required countries to manage their debt carefully to be allowed to be part of the first group of countries to adopt the new currency). In the data below (also found in file PUBDEBT), the public debt (including states and localities within them) is given for each of the German states in 1991 and in 1996. The variables are:

STATE	State name (note that Bremen, Hamburg, and Berlin are states as well as being cities)
EAST	1 if formerly a state in the German Democratic Republic (East Germany), 0 if formerly part of the Federal Republic of Germany (West Germany); note that Berlin was part eastern and part western; its value is "missing" for this variable
DEBT 1991	Public debt in 1991 in Deutschmarks per inhabitant
DEBT 1996	Public debt in 1996 in Deutschmarks per inhabitant

Prepare a brief summary outlining any important trends you find.

Based on data reported in *Die Zeit,* November 14, 1997, Nr. 47/97. Reprinted by permission of GLOBUS Infographik GmbH.

DATA SET

STATE	EAST	DEBT1991	DEBT1996
Bremen	0	22334	24547
Hamburg	0	11425	16385
Saar	0	13745	14516
Berlin	*	4543	14043
Schleswig-Holstein	0	8707	11193
North Rhine-Westfalia	0	8284	9974
Lower Saxony	0	7780	9905
Rhineland-Palatinate	0	7615	9487
Hessen	0	7591	9120
Brandenburg	1	815	8869
Saxony-Anhalt	1	733	8803
Thuringia	1	861	7837
Mecklenburg-Pomerania	1	684	7317
Baden-Wuerttemberg	0	5440	6517
Saxony	1	748	5834
Bavaria	0	3900	4584

CASE **63**

JOBS IN DENVER

The mayor of Denver has just called you into his office. A press conference is scheduled for tomorrow at noon. The purpose of the meeting is to report to the media information about the economic health of the city of Denver. You've been asked to examine employment trends in the Denver Metropolitan Statistical Area (MSA). The Denver MSA includes Adams, Arapahoe, Denver, Douglas, and Jefferson counties.

The data set below (and in the file JOBS) contains monthly data beginning January, 1990. Seven series are available: services; trade; manufacturing; construction; finance, insurance, and real estate; transportation, communications, and utilities; and the total for the Denver MSA. The numbers in the data set are estimates of full-time and part-time civilian employees (in thousands) who worked during or received pay for any part of the pay period including the 12th of the month. Self-employed and unpaid family and domestic workers are not included. None of the data series are seasonally adjusted.

The mayor has asked you to do two things to help him prepare for tomorrow's news conference. First, describe the historical trends of Denver employment found in the data. Secondly, prepare numerical forecasts for each of the upcoming three months (November and December, 1997 and January, 1998) for all six of the series. In your report to the mayor, briefly describe how you obtained these forecasts.

A portion of these data were originally collected by David R. Myers (Senior Analyst, Mortgage Opportunity Fund, LLC), Chris Stites, and others as part of a student project. The data come from the Colorado Department of Labor and Employment, Labor Market Information Section, Non-agricultural Wage and Salary Data Reports.

DEFINITIONS

BUILD	Construction jobs; includes general building contractors, heavy construction, and special trade contractors
TRADE	Jobs in the trade industries, including wholesale (durable and nondurable goods) and retail (general merchandise and apparel; food stores; automotive dealers and service stations; and eating and drinking places)
MAKE	Manufacturing jobs in durable goods industries (lumber and wood products; stone, clay, and glass products; primary, fabricated metal products, industrial machinery) and nondurable goods industries (food products; textiles; apparel and leather products, paper products, printing and publishing; chemical products; and petroleum, coal, and plastics)
SERVICES	Jobs in service-providing sectors, including hotels and other lodging places; personal services; business services; amusements and motion pictures; health services and hospitals; legal services; educational and social services; engineering and management services; and agricultural services
TCU	Jobs in transportation (railroad, trucking, air, and warehousing), communications, and electric, gas, and sanitary public utilities
FIRE	Finance, insurance, and real estate employees
TOTAL	Total jobs in the Denver MSA; this is the sum of the previous categories plus: mining; federal, state, and local government jobs; state and local educational employees; and miscellaneous other employees

DATA SET

ROW	BUILD	TRADE	MAKE	SERVICES	TCU	TOTAL	FIRE
1	29.0	201.4	94.1	221.4	69.6	813.9	63.2
2	28.4	199.1	93.9	223.9	69.9	819.3	63.7
3	29.0	199.9	94.5	226.5	69.7	824.7	64.2
4	30.0	200.6	94.3	229.1	68.6	830.3	64.6
5	31.8	207.1	94.6	232.1	68.8	844.8	65.0
6	33.7	207.2	95.9	236.8	69.6	853.4	65.6
7	34.6	206.6	95.6	236.3	69.8	841.6	65.3
8	36.0	209.1	95.4	239.2	70.1	848.1	65.3
9	35.6	207.9	94.8	237.2	70.0	852.4	64.9
10	35.6	206.6	94.1	234.8	70.3	849.3	64.5
11	35.1	210.2	93.9	234.7	70.7	854.1	64.4
12	34.1	213.5	93.9	234.4	71.4	855.9	64.5
13	31.2	204.3	92.7	228.7	70.4	830.9	63.6
14	31.3	202.5	92.3	231.2	70.3	834.6	63.7
15	31.8	203.2	92.5	233.5	70.5	839.2	64.1
16	33.4	202.3	91.8	236.8	70.1	842.8	64.2
17	34.7	205.1	92.2	239.3	70.5	851.7	64.7
18	36.3	207.1	93.2	243.2	70.8	861.5	65.6
19	37.2	206.6	92.9	242.6	70.0	851.7	65.9
20	37.9	208.1	92.5	244.1	70.7	854.0	65.7
21	37.6	207.6	92.3	242.1	70.6	858.7	65.2

(continues on the following page)

ROW	BUILD	TRADE	MAKE	SERVICES	TCU	TOTAL	FIRE
22	38.3	207.3	91.7	239.4	70.8	859.2	65.3
23	37.5	211.1	91.5	239.7	71.3	862.9	65.2
24	36.4	214.2	91.3	240.9	71.9	866.2	65.9
25	34.9	206.8	91.0	235.4	70.5	845.9	65.0
26	36.6	205.1	90.8	237.7	70.5	851.4	65.5
27	36.4	205.0	91.0	240.0	70.9	856.2	66.0
28	38.6	207.1	91.2	243.8	70.9	864.8	66.5
29	40.5	210.5	91.1	246.0	71.0	873.8	67.0
30	42.3	211.6	91.3	249.5	71.4	880.0	67.6
31	43.6	212.2	92.0	250.7	71.8	876.7	67.5
32	43.9	213.0	91.7	251.5	71.9	879.5	67.7
33	43.8	212.0	91.2	250.2	72.0	883.4	67.5
34	44.3	213.1	91.1	249.4	71.9	884.4	67.5
35	43.3	216.7	90.8	248.5	72.1	886.8	67.8
36	41.3	220.2	90.9	250.1	73.1	890.5	68.0
37	38.3	211.9	90.0	246.1	71.9	871.5	68.1
38	39.7	211.6	90.0	248.3	71.9	879.2	68.5
39	42.0	212.9	90.3	251.2	72.0	886.9	69.0
40	44.0	216.8	90.5	255.0	72.2	898.8	69.9
41	46.2	218.6	90.1	256.2	72.8	904.6	70.2
42	48.2	221.2	91.2	260.8	73.7	916.0	71.1
43	50.4	220.2	91.0	262.3	74.8	910.9	71.1
44	51.3	221.6	90.4	263.9	74.7	914.1	71.4
45	51.1	220.5	90.0	262.8	74.4	916.8	71.5
46	51.7	221.6	90.1	261.8	74.3	921.3	71.9
47	50.0	226.1	89.2	261.3	74.3	922.7	72.1
48	48.9	231.1	88.9	263.5	75.7	930.4	72.9
49	47.4	221.6	88.0	260.5	74.1	908.5	72.0
50	46.7	221.0	88.1	263.7	74.6	915.1	72.5
51	47.7	221.8	88.2	267.3	74.6	921.4	73.0
52	49.0	225.6	88.7	271.4	73.9	931.7	73.2
53	49.8	228.0	89.0	274.9	74.3	938.9	73.2
54	51.6	229.9	90.0	279.7	74.9	949.1	73.7
55	52.7	213.1	90.3	281.4	75.1	945.5	73.8
56	52.8	231.2	90.0	281.4	74.9	945.4	73.6
57	52.2	231.0	89.8	280.4	74.3	949.5	73.3
58	52.1	234.2	89.7	280.0	74.2	954.2	73.1
59	51.9	238.6	90.5	280.9	74.3	960.3	73.1
60	50.9	240.5	90.4	281.8	75.0	962.4	73.3
61	49.7	236.5	85.4	277.6	81.3	947.5	73.3
62	49.9	235.5	85.8	281.6	81.7	956.4	73.6
63	51.1	237.1	86.1	285.8	82.6	965.6	74.2
64	51.3	238.9	86.2	285.9	83.3	968.2	74.5
65	52.5	241.7	86.8	290.4	83.9	978.7	75.2
66	54.1	245.4	87.6	295.7	84.3	990.8	75.9
67	55.6	245.0	87.9	296.6	83.8	985.6	75.9
68	55.7	245.6	88.0	299.2	84.0	989.8	76.1

(concludes on the following page)

ROW	BUILD	TRADE	MAKE	SERVICES	TCU	TOTAL	FIRE
69	55.7	244.1	88.1	299.0	84.1	994.8	75.9
70	55.5	245.1	88.3	297.6	84.4	996.5	76.6
71	54.7	249.7	88.3	297.8	85.2	1003.0	77.4
72	54.9	252.9	88.1	298.4	86.0	1007.9	77.6
73	51.6	241.9	88.5	291.5	84.3	977.6	76.1
74	52.4	240.1	88.6	293.2	84.6	983.7	76.8
75	54.0	241.5	89.2	297.3	84.7	993.5	77.4
76	55.3	242.5	88.8	300.2	84.7	998.3	77.7
77	57.4	246.7	89.2	304.2	84.9	1010.6	78.3
78	59.4	249.1	89.6	310.1	85.0	1021.5	79.2
79	59.7	243.2	89.6	309.4	84.9	1006.6	79.0
80	60.2	250.0	89.7	310.2	85.0	1014.6	78.9
81	59.5	247.4	89.6	308.0	84.2	1015.6	78.5
82	59.5	248.0	89.5	308.3	84.4	1017.2	77.5
83	59.9	251.3	89.8	307.9	83.6	1022.1	77.8
84	58.6	254.7	89.9	308.7	83.2	1024.7	78.5
85	54.5	247.7	89.2	302.1	81.7	1001.3	78.7
86	56.2	246.7	89.6	305.6	81.5	1008.4	78.8
87	57.3	247.1	89.6	309.0	81.2	1013.7	79.0
88	58.5	248.7	90.0	309.6	81.2	1017.3	79.2
89	59.9	249.9	89.4	312.7	80.2	1022.3	79.7
90	61.4	252.9	90.2	317.2	81.0	1033.2	80.9
91	62.0	252.4	90.6	317.7	80.7	1026.9	81.6
92	62.1	253.5	91.1	318.0	78.8	1026.9	82.3
93	62.2	252.7	90.9	317.9	80.4	1036.0	82.2
94	62.3	252.8	91.5	316.6	81.0	1038.2	82.3

CASE **64**

THE ADVERTISING EXPERIMENT

"I need help with this one," said the vice president. "Let me explain. About six months ago, the marketing manager proposed that we try a massive advertising campaign to assess its effectiveness. After some pretty strenuous debate, we approved it. Basically, we took 5 of our 10 regions and just about doubled the advertising budget in those regions, while leaving it pretty much unchanged in the other regions. The agreement was that after a trial period, which ended last month, we would evaluate the performances of the various regions and see whether the advertising was working and, if so, to what extent. Some people were pretty upset at the time, mostly those in the regions where the advertising budget was not raised. I guess in retrospect, we should have adjusted quotas for them or done something to get their participation, but we didn't. Anyhow, it's time to decide whether the program worked or not, and no one seems to agree on it. We collected data from various sources, as shown below.

"SALESPER is the number of sales representatives in the region, VISITS is how many sales visits they made during the interval in question, ADEXP is total advertising expenses for the region (in thousands), SALES is total sales for the region (in dollars) and SPECIAL is a code (0 if the region did not have its ad budget increased, 1 if it did).

"It looks simple enough to me, but I have reports recommending all sorts of different conclusions. One says the program was a complete failure—it brought in about $500 for each $1000 we spent. Another says the data show that every ad dollar spent returned more than $1.30 in sales. One says that on a per salesperson basis, sales were actually lower in regions with the higher ad budgets; another says every dollar spent in advertising resulted in average sales anywhere from $3 to more than $6. What should I conclude? How can so many different opinions sprout from one set of data like that?

"I have to report to the board next week on how this whole project went. I want you to give the pitch to the board and answer questions about the data. Various other people will be there, so you should be prepared to defend your interpretation of the data, and explain why you think the others are wrong (if you do think they're wrong). I'll back you up, but I'm over my head on the technical part of things."

The data on the next page are available in file ADVERT.

DATA SET

OBS	SALESPER	VISITS	ADVEXP	SALES	SPECIAL
1	5	1510	100	733000	0
2	3	885	100	484050	0
3	7	2106	120	997341	0
4	4	1193	110	622235	0
5	5	1518	140	781154	0
6	5	1114	200	710976	1
7	7	1582	230	930748	1
8	3	671	210	551403	1
9	4	903	180	580632	1
10	5	1122	250	786250	1

CASE **65**

STAFFING THE PHONE DESK

TIAA-CREF (Teachers Insurance and Annuity Association and the College Retirement Equities Fund) based out of New York, New York, provides financial retirement annuities to college and university educators. TIAA-CREF manages retirement accounts of varying objectives: the TIAA traditional annuity (which guarantees a principal and specified interest rate for low risk investors), and real estate, equity, fixed-income, and balanced funds for those willing to accept additional risk. TIAA-CREF prides itself on low investment, administrative, and distribution expenses and is widely recognized as a premier retirement annuity provider.

TIAA-CREF does more than maintain annuity accounts. Specially trained financial counselors conduct on-site seminars about retirement planning. Informational booklets on various topics are available to TIAA-CREF customers. Videos and specialized publications are distributed to institutional administrators.

Investors make inquiries and maintain their individual accounts through a variety of media. Using an automated telephone service (ATS), policyholders can obtain current information and perform account transactions 24 hours a day, seven days a week. ATS, available either via a toll free phone number or the TIAA-CREF Internet home page, is designed for customers comfortable with electronic communication. Staff at the TIAA-CREF phone counseling centers provide the full range of counseling and account maintenance services for customers who need further information or prefer to speak with a real person about their business needs.

As an electronic medium, the ATS is a less costly method of serving customers than personal phone counseling. Even so, phone counseling is an important component of customer service at TIAA-CREF. With over a million phone calls answered at the

These data were collected by Oswaldo A. Grenardo and others.

phone service centers in 1994 alone, staffing the phone desks is a nontrivial matter for TIAA-CREF management.

TIAA/CREF's Western Phone Counseling Center located in Denver, Colorado employs about 50 phone consultants to handle inbound calls from 8:00 A.M. to 9:00 P.M. Mondays through Fridays. A counselor works a single four-hour shift each day on the phones. The rest of the day is filled with other projects, meetings, and special assignments. Management wants to ensure that representatives are available to answer incoming calls while minimizing idle time at the phone desk.

The data set below, and available in the file named PHONES, contains two types of information about calls to the Western Phone Counseling Center: hourly and weekly. DAY, HR, and CALLS represent day of the week, hour of the day, and total number of calls over that hour during the month of March 1996. MONTH, WEEK, and AVGCALLS describe the average number of weekly calls from March (MONTH = 3) of 1995 to February 1996. Data of this nature are collected on an ongoing basis at the Western Phone Counseling Center.

You've been hired on a one-time basis to study and interpret the data. Your task is to enlighten management about future staffing needs at the phone desk. Several scenarios come into play. If the 1995–96 data are fairly typical and expected to be representative of future calling patterns, they provide a basis for understanding future staffing needs. If not, management needs to know how to adjust desk assignments in light of new information that will be available in the future. The challenge for you is to prepare management for either outcome. To this end, prepare a report that addresses the following.

- Describe patterns in the 1995–96 data. Provide specific recommendations about how this information might be used to dictate staffing needs.
- Describe how to use subsequent data to determine if there has been an important change in calling patterns. Be specific.

Indicate in your report further information needed to translate incoming phone call patterns into staffing needs. Again, be specific.

DATA SET

DAY	HR	CALLS	MONTH	WEEK	AVGCALLS
4	8	17	3	1	1604
4	9	39	3	2	2321
4	10	65	3	3	1722
4	11	100	3	4	1542
4	12	97	3	5	1428
4	13	90	4	1	1536
4	14	115	4	2	2009
4	15	111	4	3	1561
4	16	84	4	4	1706
4	17	63	5	1	1419
4	18	71	5	2	1347
4	19	77	5	3	1263
4	20	57	5	4	1247
5	8	23	5	5	1712
5	9	58	6	1	1741
5	10	80	6	2	1245

DAY	HR	CALLS	MONTH	WEEK	AVGCALLS
5	11	77	6	3	1206
5	12	74	6	4	1199
5	13	86	7	1	1475
5	14	126	7	2	1753
5	15	103	7	3	2171
5	16	68	7	4	1599
5	17	92	8	1	1642
5	18	78	8	2	1407
5	19	64	8	3	1303
5	20	48	8	4	1279
6	8	23	8	5	1384
6	9	49	9	1	1866
6	10	55	9	2	1402
6	11	98	9	3	1305
6	12	92	9	4	1205
6	13	92	10	1	1240

(continues on the following page)

DAY	HR	CALLS	MONTH	WEEK	AVGCALLS
6	14	108	10	2	1945
6	15	95	10	3	1859
6	16	90	10	4	1626
6	17	89	10	5	1525
6	18	86	11	1	1526
6	19	59	11	2	1452
6	20	53	11	3	1458
7	8	25	11	4	1616
7	9	65	12	1	1749
7	10	59	12	2	1450
7	11	82	12	3	839
7	12	79	12	4	1276
7	13	109	1	1	2238
7	14	100	1	2	1713
7	15	136	1	3	2895
7	16	94	1	4	2176
7	17	103	1	5	2214
7	18	81	2	1	1848
7	19	50	2	2	2077
7	20	47	2	3	2383
8	8	28	2	4	1764
8	9	69			
8	10	84			
8	11	101			
8	12	81			
8	13	92			
8	14	107			
8	15	117			
8	16	90			
8	17	105			
8	18	77			
8	19	52			
8	20	40			
11	8	17			
11	9	51			
11	10	85			
11	11	125			
11	12	128			
11	13	131			
11	14	126			
11	15	145			
11	16	131			
11	17	131			
11	18	140			
11	19	107			
11	20	78			
12	8	20			
12	9	68			

DAY	HR	CALLS	MONTH	WEEK	AVGCALLS
12	10	75			
12	11	101			
12	12	118			
12	13	124			
12	14	126			
12	15	136			
12	16	109			
12	17	130			
12	18	111			
12	19	76			
12	20	55			
13	8	14			
13	9	51			
13	10	65			
13	11	90			
13	12	116			
13	13	104			
13	14	127			
13	15	128			
13	16	104			
13	17	159			
13	18	112			
13	19	113			
13	20	72			
14	8	37			
14	9	49			
14	10	66			
14	11	81			
14	12	96			
14	13	100			
14	14	88			
14	15	109			
14	16	111			
14	17	146			
14	18	132			
14	19	81			
14	20	86			
15	8	28			
15	9	52			
15	10	67			
15	11	89			
15	12	77			
15	13	93			
15	14	67			
15	15	113			
15	16	90			
15	17	109			
15	18	62			

(continues on the following page)

DAY	HR	CALLS	MONTH	WEEK	AVGCALLS
15	19	55			
15	20	33			
18	8	20			
18	9	46			
18	10	84			
18	11	96			
18	12	111			
18	13	112			
18	14	99			
18	15	124			
18	16	120			
18	17	108			
18	18	172			
18	19	144			
18	20	93			
19	8	19			
19	9	48			
19	10	47			
19	11	87			
19	12	109			
19	13	108			
19	14	91			
19	15	105			
19	16	118			
19	17	150			
19	18	142			
19	19	131			
19	20	101			
20	8	22			
20	9	41			
20	10	64			
20	11	84			
20	12	67			
20	13	92			
20	14	108			
20	15	109			
20	16	79			
20	17	112			
20	18	74			
20	19	54			
20	20	36			
21	8	24			
21	9	49			
21	10	71			
21	11	82			
21	12	89			
21	13	81			
21	14	117			
21	15	101			
21	16	65			
21	17	93			
21	18	94			
21	19	65			
21	20	91			
22	8	22			
22	9	36			
22	10	62			
22	11	71			
22	12	60			
22	13	74			
22	14	111			
22	15	115			
22	16	58			
22	17	102			
22	18	69			
22	19	65			
22	20	48			
25	8	22			
25	9	56			
25	10	62			
25	11	143			
25	12	103			
25	13	99			
25	14	79			
25	15	112			
25	16	106			
25	17	123			
25	18	120			
25	19	79			
25	20	56			
26	8	26			
26	9	49			
26	10	92			
26	11	98			
26	12	83			
26	13	93			
26	14	78			
26	15	110			
26	16	82			
26	17	86			
26	18	94			
26	19	94			
27	20	63			
27	8	23			
27	9	48			
27	10	69			

(concludes on the following page)

DAY	HR	CALLS	MONTH	WEEK	AVGCALLS
27	11	86			
27	12	46			
27	13	87			
27	14	103			
27	15	111			
27	16	67			
27	17	105			
27	18	80			
27	19	49			
27	20	32			
28	8	21			
28	9	53			
28	10	64			
28	11	77			
28	12	80			
28	13	86			
28	14	111			
28	15	107			

DAY	HR	CALLS	MONTH	WEEK	AVGCALLS
28	16	54			
28	17	84			
28	18	49			
28	19	25			
28	20	25			
29	8	16			
29	9	48			
29	10	73			
29	11	97			
29	12	69			
29	13	75			
29	14	118			
29	15	94			
29	16	43			
29	17	11			
29	18	76			
29	19	63			
29	20	64			

CASE **66**

HOME EQUITY LOANS

The Bank of East Seattle provides a number of financial services including: checking and savings accounts; mortgage, personal, and home equity loans; safety deposit box rentals; and retirement accounts. Home equity loans are a big part of its business. Home equity loans are those taken out against the equity of a home, and are often used for things such as home improvement or renovation projects.

Seattle has experienced a rapid increase in housing prices in the last few years and customers are applying for home equity loans in record numbers. Some loan applicants are good credit risks and others are not. The Bank of East Seattle currently uses a credit scoring scheme to decide whether to grant a loan. First, the applicant completes a loan application containing information about income, length of time at residence, and so forth. The bank verifies the information on the application from outside sources. A numerical score is assigned to each factor and a total score for that application is derived. A minimum cutoff determines whether or not the application is worth further consideration.

Subjective criteria are also used to make the final decision. A loan officer will often make judgments about things such as character or potential for a future relationship with the bank. Of course, by law, certain characteristics of the applicant, such as gender, cannot be used in the decision making. Even so, unquantifiable factors are sometimes used to sway a borderline score.

It's not as easy as it might sound. Even with lots of data about the applicant and the home, banks make errors. Sometimes loans that are granted end up in default; sometimes rejected loans are eventually granted, and paid off, at another institution. The purpose of the loan application process is to sort out good from bad credit risks.

These data were collected and analyzed by students who wish to remain unnamed. The name of the bank and its location have been changed.

The Bank of East Seattle has asked you to examine data on 59 loan application records. The bank's underwriters have already studied the data and decided whether to accept or reject the loan. Because home equity loans often take several years to pay off, we don't know yet whether any particular loan that was granted was paid off in full and on time. Yet we do have some information. The household income, gender, and age of the applicant, years on the job and at the current residence, equity in the property, and net worth are available from the loan application. These numbers are verified by contacting employers and examining assessments. Whether the loan was accepted or not, as well as whether the applicant was a customer of the bank at the time of the loan, was determined from bank records. Total debt, revolving debt, and number of delinquent credit accounts were calculated from information obtained from TRW, a credit-reporting agency.

Income, equity in the home, and net worth should indicate the applicant's ability to pay off the loan. Number of delinquent credit accounts and amount of revolving debt indicate the financial history and worthiness of the applicant. Years on the job and in the residence may hint at character and stability of the applicant. Whereas each of these factors probably relates to the bank's willingness to grant the loan, nothing is certain.

Your job is to provide a summary of the data. Which factors appear to influence the acceptance or rejection of a loan application? Are there any particularly interesting aspects of the data worth noting? Do the data hint at any problems or inequities that might be arising in the bank's decision to accept or reject loans? The data set described below is available in the file named LOANS.

DEFINITIONS

Age	Age of the applicant in years
Gender	Gender of the applicant: 1 = Male, 0 = Female
Income	Household income in dollars
Del Acct	Number of delinquent credit accounts at the time of the application
Equity	Fair market value of the property minus outstanding debt ($)
Loan Amt	The requested amount of the loan in thousands of dollars
Reside	Years of residence in the home at the time of the application
Year Job	Number of years on the job at the time of the application
Revolve	Amount of the applicant's revolving debt in dollars (including credit card balances, installment debt, and lines of credit; excluding things such as auto loans and student loans)
Tot Debt	Total debt of the applicant in dollars (e.g., mortgage, revolving debt, auto loans, and other personal loans)
NetWorth	Total market value of assets minus total debt at the time of the application in dollars
Customer	Number of years the applicant has been a customer of the bank
Accept?	Whether the loan was accepted (Accept? = 1) or rejected (Accept? = 0)

DATA SET

AGE	GENDER	INCOME	DEL ACCT	EQUITY	LOAN AMT	RESIDE	YEAR JOB	REVOLVE	TOT DEBT	NETWORTH	CUSTOMER	ACCEPT?
34	0	92600	0	97000	14.00	5.000	5.000	6694	153028	110972	4.60	1
28	1	39000	1	64000	15.00	6.000	8.000	16616	107387	88613	5.33	1
38	0	42888	0	28790	20.00	6.750	6.000	9379	89192	14675	1.42	1
43	1	24960	0	54653	14.00	10.000	17.000	434	75781	84219	7.66	1
32	1	40000	1	66000	25.00	3.000	3.000	130	55009	97517	3.08	1
34	1	83914	0	50500	40.00	3.000	17.000	9405	96496	114704	17.83	1
29	1	50000	1	29500	17.70	3.000	3.200	14394	110575	30925	6.92	1
50	1	210000	0	190000	20.00	2.250	27.000	981	507691	889509	1.16	1
28	1	81000	0	50300	13.00	3.167	4.500	59	216619	108131	5.42	1
45	0	107400	0	79837	18.00	8.750	25.167	27739	176380	305620	2.92	1
39	1	72820	7	91866	15.00	8.000	2.600	11755	142119	179881	11.08	1
38	0	98500	1	46569	38.00	7.250	5.000	4363	133777	120223	2.66	1
38	1	60400	2	44000	25.00	8.000	0.250	8822	102201	24099	9.25	1
32	1	50000	0	40875	25.00	0.667	0.000	4506	130569	34437	14.83	1
41	1	89000	1	52177	8.00	1.000	6.500	6333	268340	88660	0.75	1
60	1	64744	0	83287	12.00	2.000	0.000	4826	151539	721461	10.08	1
36	1	134000	3	97285	25.00	1.500	1.500	17167	321785	153215	3.75	1
50	1	73476	0	96592	45.00	10.000	12.000	16811	177998	162702	0.42	1
35	1	81030	0	66421	25.00	7.500	15.500	2240	117553	174447	4.50	1
51	1	70000	0	101500	12.00	5.500	32.000	23620	177271	95364	21.50	1
44	1	46100	2	61398	10.00	6.000	18.250	21463	148065	43635	8.16	1
28	1	105900	0	74068	50.00	2.000	4.500	7703	178236	99864	2.56	1
48	1	127249	0	66250	25.00	0.500	2.500	14143	276792	296208	2.16	1
26	0	63000	1	29474	12.00	2.500	2.800	3677	100602	48998	2.75	1
39	1	52244	0	36151	20.00	4.000	4.000	16146	146989	78642	4.50	1
52	1	179657	2	173259	25.00	2.000	0.250	29696	312652	311498	0.25	1
42	1	108969	0	110724	12.00	0.500	14.000	2427	202427	305573	10.33	1
28	1	56000	1	41358	10.00	0.420	0.420	2838	162967	34033	0.42	1
40	1	66400	0	91796	5.00	3.500	5.000	3539	111017	177483	5.75	1
58	1	65350	0	269545	25.00	2.000	0.000	1790	326443	280457	2.00	1
34	0	63333	1	43061	20.00	1.000	1.000	6390	141147	116353	6.50	1
34	1	200000	0	71087	11.15	3.000	2.000	7674	196674	147576	1.16	1
58	1	42000	0	161095	4.50	3.000	20.000	2021	103516	161384	3.80	1

(concludes on the following page)

AGE	GENDER	INCOME	DEL ACCT	EQUITY	LOAN AMT	RESIDE	YEAR JOB	REVOLVE	TOT DEBT	NETWORTH	CUSTOMER	ACCEPT?
48	1	100040	0	76520	7.00	9.500	26.000	7426	187156	356344	2.83	1
47	1	58000	0	67942	15.00	3.000	13.000	605	62663	137887	4.25	1
29	1	49040	0	23470	12.00	1.250	5.000	438	104929	26971	7.25	1
31	1	68000	2	44173	22.00	3.000	10.000	10504	131528	54472	7.66	1
41	1	91000	0	134216	15.00	2.000	18.000	5256	203696	366224	1.92	1
43	1	71000	0	43631	12.00	15.000	15.000	2786	95369	157931	3.16	1
33	1	120000	2	59126	18.00	3.000	2.000	19950	167097	59903	2.92	1
36	1	160000	1	75450	20.00	0.750	0.125	5529	413563	150437	15.50	1
44	0	40400	0	36919	6.00	3.000	6.250	3502	117083	51517	1.75	1
38	0	55083	1	44362	25.00	4.000	4.500	13212	94531	59371	6.16	1
46	1	50800	3	52642	15.00	5.000	5.000	15746	108104	59696	4.50	1
33	1	35000	3	37307	12.00	3.000	5.000	6322	108208	101892	0.25	1
43	1	59757	2	93290	7.50	8.500	11.000	5233	93339	183311	14.25	1
39	1	202800	0	151641	50.00	5.000	2.000	8013	281272	600528	4.92	1
63	1	61691	0	188074	30.00	1.500	0.000	1595	73412	302621	0.83	1
50	1	92420	0	46298	12.00	0.500	20.000	9485	312687	142463	0.50	0
39	0	33600	0	73500	25.00	4.500	0.500	8159	198908	32552	4.50	0
42	0	43903	1	53660	10.00	0.167	2.500	3577	175317	123366	3.00	0
26	1	27600	4	24474	15.00	3.000	1.000	11293	110153	847	5.50	0
34	1	69000	0	37500	25.00	1.500	3.500	34195	194159	24641	0.50	0
48	1	72000	6	100398	25.00	8.000	6.000	22126	145616	81383	0.00	0
51	0	30000	2	70000	4.50	7.000	0.500	1974	119200	99300	2.40	0
29	1	45000	6	23600	12.00	1.600	4.000	7705	115705	42295	0.00	0
37	0	14700	4	28000	10.00	7.000	0.167	7414	72414	24586	0.00	0
36	1	100000	24	104084	15.00	7.750	10.000	3302	235344	92156	6.75	0
44	1	101400	12	108500	20.00	2.500	0.080	3200	328700	108300	0.00	0

CASE **67**

GAS STATIONS

Petro Corporation is a large petroleum company that owns and operates gasoline stations throughout Canada. The company wants to expand revenue by building new stations and wants to understand what factors influence fuel sales of gasoline stations in Canada.

The data set named FUEL contains information from 192 self-service gas stations for 1997. GALLONS is the average monthly fuel sales (both gasoline and diesel) in number of gallons. A station that operates 24 hours a day is indicated when HOURS = 1; HOURS = 0 for those open 18 hours per day. CARWASH is 1 if the station has a car wash on site and 0 if not. When LOCATION = 1, that station is located along a highway either in a metropolitan or rural area. LOCATION = 2 indicates a station located at an inner-city or metropolitan location but not along a highway. Finally, rural stations not alongside highways are noted by LOCATION = 3.

By how much do location, hours of operation, and on-site car washes influence fuel sales?

DATA SET

HOURS	GALLONS	CARWASH	LOCATION	HOURS	GALLONS	CARWASH	LOCATION
1	114697	0	2	1	151701	0	2
1	163445	0	2	1	112633	0	2
1	69216	0	3	0	100714	0	2
0	100967	0	3	1	203192	0	2
1	120872	0	2	0	96067	0	2

(continues on the following page)

These data were collected and studied by Mike A. Kurowski and others. The company name has been changed.

HOURS	GALLONS	CARWASH	LOCATION
1	118505	0	3
1	66051	0	3
0	60305	0	2
1	144336	0	2
1	148071	0	2
1	102362	0	2
1	125291	0	2
1	151394	0	2
1	162777	0	2
1	93153	0	2
1	189206	0	2
1	84663	0	2
1	169497	0	2
1	128828	0	2
1	141562	0	2
1	113118	0	2
1	134355	0	1
1	119172	0	2
1	173902	0	2
1	106367	0	2
0	90534	0	2
1	123414	0	3
0	106491	0	2
1	110968	0	2
1	136465	0	2
1	128252	0	2
1	152485	0	2
1	109158	0	2
1	107604	0	2
1	229404	0	2
0	146435	0	2
1	160639	0	2
0	95883	0	2
1	102154	0	1
1	165124	0	2
1	97797	0	2
1	251346	0	1
1	186091	0	2
1	124854	0	2
1	131587	0	1
1	194732	0	2
1	175054	0	2
1	140357	0	1
1	139911	1	3
1	157679	0	2
1	277823	0	1
0	155904	0	2
1	151809	0	2

HOURS	GALLONS	CARWASH	LOCATION
1	150464	0	0
1	154559	0	2
1	156151	0	2
1	115560	0	2
1	152842	0	2
0	133815	0	2
1	153472	0	1
1	185860	0	1
0	119287	0	2
1	140529	0	2
1	129997	0	2
1	192850	1	2
1	141361	0	1
1	148093	0	2
1	163568	0	2
1	213409	0	2
1	231592	1	2
1	155271	0	2
1	128626	0	2
1	181823	0	2
1	169513	1	2
1	192564	0	2
1	135808	0	2
1	275438	1	2
1	256209	0	2
1	170729	0	2
1	194222	0	3
0	119046	0	2
0	87284	0	2
0	121615	0	3
0	106435	0	2
0	87453	0	2
1	110244	0	2
1	166050	0	3
1	93565	0	3
0	179159	0	2
1	131137	0	2
0	97689	0	2
0	125342	0	2
1	125947	0	1
1	324049	0	1
1	144502	0	1
1	370774	0	1
1	204132	0	1
1	192684	0	1
1	343020	0	1
0	72907	1	1
1	127770	0	1

(concludes on the following page)

HOURS	GALLONS	CARWASH	LOCATION
1	91359	0	1
1	113865	0	3
1	119219	0	1
1	132316	0	1
1	145490	0	1
1	270821	0	1
1	194330	0	1
1	132368	0	1
1	420904	0	1
1	106618	0	1
1	342251	0	1
1	160447	0	1
1	129956	0	1
1	133935	0	2
1	147899	0	2
1	172422	0	1
1	129794	0	1
1	125602	0	2
1	110978	0	2
1	243723	0	1
1	77580	0	2
1	96678	0	1
1	149398	0	1
1	225170	0	1
1	193951	0	1
1	203392	0	1
1	86758	0	1
1	228659	0	1
1	95964	0	1
1	147958	0	1
1	82515	0	2
1	158866	0	1
1	76103	0	2
1	80374	0	2
1	165755	0	1
1	129052	0	1
1	117422	0	2
1	243341	0	1
1	183087	0	1
1	181567	0	2
1	169198	0	1
1	81390	0	2
1	126188	0	1

HOURS	GALLONS	CARWASH	LOCATION
1	138759	0	2
1	112970	0	2
1	137558	0	2
1	107985	0	2
1	118438	0	2
1	214032	0	1
1	126762	0	2
1	161782	0	1
1	95000	0	2
1	369822	0	1
1	181984	0	1
1	140814	0	2
1	381807	0	1
1	220070	0	1
1	348662	0	1
1	276852	0	1
1	271942	0	1
1	230982	0	1
1	117174	0	1
1	500607	0	1
1	187112	0	1
1	283747	0	1
1	97538	0	1
1	112651	0	1
1	151770	0	2
1	119360	0	1
1	103406	0	1
1	232166	0	1
1	194772	0	1
1	182023	0	1
1	171902	0	1
1	257530	0	1
1	214396	0	1
1	200472	0	1
1	97363	0	1
1	227323	0	1
1	221886	0	1
1	215838	0	1
1	169311	0	1
1	136999	0	1
1	289523	0	1
1	175550	0	1
1	172789	0	1

CASE **68**

ATTITUDES TOWARD NUCLEAR POWER

Public attitudes toward nuclear facilities generally and nuclear power facilities in particular play big roles in the political processes by which various operating permits and licenses are granted for different kinds of power generation facilities. This is particularly true in Japan. As part of a regular survey, the Institute of Nuclear Safety System, Incorporated and the Institute of Social Research surveyed permanent residents in the service area of the Kansai Electric Power Co., Inc., 18 to 79 years of age.[1] One of the questions asked was:

> Many of the questions above concern nuclear power generation. Which of the following statements is closest to your opinion on the whole. Please choose only one.
>
> **1** It is good to have nuclear power generation.
> **2** It is inevitable to have nuclear power generation.
> **3** It is good to rely on safe power generation methods.
> **4** Nuclear power generation should not be conducted.

Based on the survey responses, the tabulation on the following page was prepared. It lists the number of people choosing each response, classified by location in the country, sex, age, education, occupation, and household type. The data are also available in file NUCPOWER.

[1]The Kansai area is the district surrounding Osaka bay. Its major cities are Osaka, Kobe, and Kyoto. The question (number 40) and the results reported here are taken from an English-language report: Chikio Hayashi, "National Character and Communication: Structure of Attitudes toward Nuclear Power Generation and Reponses by Electric Power Companies," Institute of Nuclear Safety System, Incorporated, Institute of Social Research, May 1995. The survey itself was conducted in Japanese. Reprinted by permission of the author, Professor Emeritus Chikio Hayashi, of the Institute of Statistical Mathematics, Tokyo.

Which of the demographic variables (if any) are related to the response to this question, and how?

DATA SET

		Response				
Characteristic	Category	1	2	3	4	No reply
Location in country	Urban	465	2255	700	470	110
	Provincial	80	360	126	70	40
Sex	Male	335	1298	330	204	47
	Female	210	1317	496	336	103
Age	18–24 years old	74	323	166	63	16
	25–29 years old	22	273	93	74	17
	30–34 years old	62	272	87	85	7
	35–39 years old	52	246	92	52	12
	40–44 years old	78	344	105	64	27
	45–49 years old	66	325	78	48	21
	50–54 years old	64	234	71	45	14
	55–59 years old	24	227	53	38	11
	60–64 years old	40	181	47	56	9
	65–79 years old	63	190	34	15	16
Education	Elementary school graduate	83	395	118	54	14
	High school graduate	302	1348	489	249	79
	University graduate	158	858	214	227	48
Occupation	Employee	255	1094	374	219	43
	Self-employed	72	357	49	78	12
	Family business	36	91	21	15	6
	Home maker	116	778	258	185	56
	Student	30	125	70	26	4
	None	36	165	44	10	13
Household type	One-person household	15	114	40	10	2
	One-generation household	81	435	122	71	26
	Two-generation household	333	1605	461	343	81
	Three-generation household	91	396	197	95	30

CASE **69**

ALICIA'S UNION DATA

Today at 4:30 P.M., your boss asked you to prepare a brief report for him to use at a 10:00 A.M. meeting tomorrow. You are to present it to him at 9:30 A.M. It is now 5:15 P.M., and you must decide what kind of response to give him. You have no particular desire to work late on the problem, but it is important, and you've no real excuse not to do a good job.

For the last year, you have worked as staff assistant to the vice president for human resources in a fairly large division (5000 employees) of a major company. It has generally been a good year: you've learned a lot about the business, been noticed by higher management, and discovered that you like personnel work. Right now the future looks pretty good, and you have every reason to think that in another few months when your current assignment as staff assistant ends, you'll get a nice chance for a line management position either in this division or in another.

In the last few months, your boss has been under some heavy pressure from his boss because of an approaching union representation election in your division. The election is next month, but the lobbying for support has been hot and heavy for some six months now, and presumably will intensify in the next couple of weeks. The company is about half union, half nonunion, but the new corporate management (new about six months ago, following some complicated merger and takeover activity) is firmly antiunion, and has made it clear to your boss that vice presidents in whose divisions new unions appear are not the sorts of vice presidents who should look forward to long and lucrative careers in the new, taken-over company. The entire personnel wing of your division has thus been walking on eggshells for several months now, trying to make the company's case for nonunion operation of the division, while not running afoul of any of various state and local labor laws and regulations. More than a few late nights and hearty debates among the management have resulted.

One of some two or three influential employee committees is scheduled to meet tomorrow at 12:00 P.M. to decide whether to recommend support of the union to its constituency. They have scheduled a preliminary meeting at 10:00 A.M. for final comments by both union supporters and management to give any final information they wish before the committee takes its vote at noon. It is this meeting for which your boss wants the report.

At 4:30 P.M. yesterday, your boss called you into the office, and said that he had just talked with Alicia, one of the records analysts in Personnel, and she reported an interesting industry study, which showed that the higher the percentage of union employees in a company, the less life insurance coverage was provided to employees. "A correlation perhaps as strong as −0.9, she said," he reports. "I want you to go over to her office first thing tomorrow, get the details of that study, and write them up in a paragraph that the employee committee will understand, because that will bolster our case to keep the company nonunion. I need it for the meeting tomorrow at 10:00 A.M., so plan on presenting it to me at 9:30 A.M., so that I can ask any questions of you beforehand." With that, he left for the regular Chamber of Commerce meeting.

What do you tell your boss? The data from the study were provided by Alicia (before she left for the day), and they are also available in file ALICIA. The variables included are: UNION, the number of union employees per thousand employees; AVESAL, the average salary (in thousands); and LIFEINS, the average company-paid life insurance per employee (in thousands). The data come from 50 companies in the same business as yours.

DATA SET

UNION	AVESAL	LIFEINS	UNION	AVESAL	LIFEINS	UNION	AVESAL	LIFEINS
850	14	31	180	29	96	350	24	78
400	22	70	750	15	33	870	14	22
600	18	45	140	28	102	670	17	44
650	18	40	700	17	45	400	23	73
750	16	40	580	20	59	150	29	101
550	20	65	330	24	75	50	35	140
600	19	57	200	27	96	600	19	52
800	15	26	700	18	48	480	21	65
800	14	28	90	30	100	440	22	68
650	19	54	530	19	49	250	27	97
500	20	70	400	24	80	670	17	45
750	15	32	260	27	97	800	15	32
700	16	36	700	16	35	300	26	92
500	22	60	240	27	92	20	34	107
700	17	47	410	23	75	300	26	69
650	17	38	330	25	80	400	24	80
835	14	27	220	26	90			

CASE **70**

HEART SURGERY

"So many things affect the outcomes!" Dr. O'Malley explained, "The data capture only part of the picture. Every case is different, and yet I have a responsibility to review things in the aggregate, too. Can you help?"

Dr. John O'Malley held a senior administrative position with a chain of hospitals in the southern United States. Trained in pediatric psychiatry, he had spent most of his career in Boston and Philadelphia until taking an administrative post 10 years ago. Now he was responsible for performance improvement, utilization management, case management medical staff relations, and participating in strategic planning for the hospital chain.

He was convinced that systematic use of the information available in the various databases within the organization could yield important improvements in financial performance and in patient care, but the "devil was in the details." The report in front of him provided an example. It listed in summary form the results of all the heart bypass and related surgeries (378 operations) for one hospital in the chain for the most recent year. More detailed data for each such case was included in the organization's database, using forms and procedures provided by the Society of Thoracic Surgeons.

The primary administrative result was length of stay in the hospital. The hospital's reimbursement was generally a fixed amount based on the diagnosis. Actual expenses were roughly proportional to the length of stay, so understanding it was an important part of financial planning for the hospital.

From a clinical point of view, the primary data in the current report consisted of an indication of whether or not complications (such as operative complications;

Based on data and descriptions provided by John E. O'Malley, M.D., MSHA. Codes related to individual patients and physicians have been disguised.

infections; neurological, renal, or pulmonary problems; atrial fibrillation) arose, and a "risk stratification score," RS, designed to measure the condition of the patient going into the operation. RS was computed by filling out a checklist indicating whether the patient had such conditions as obesity, was dependent on dialysis, etc., and adding up the number of check marks, along with some other scores estimated by the physician.

Dr. O'Malley felt he should be asking such questions as:

- Do results differ by physician (some 15 different physicians referred such cases to the hospital)?
- How do results (rates of complications or length of stay) differ by physician? By time of year? Specific diagnosis? Type of operation? Severity of condition?
- Was the risk stratification score a useful guide to length of stay or other outcomes?

Reviewing each case in detail was impractical (and heart surgeries were of course only one of many procedures he had to review), and the summary information he had in front of him was imperfect. However, his responsibilities to the board of directors, third-party payers, medical staff, and patients required him to learn whatever he could from the data; at least *he* felt so. He wanted whatever message he took away from the data to be fair to the individual physicians, recognizing the different case mixes and severity of patient conditions, but also fair to management and the chain's financial stakeholders. That's why he asked for help. Can you advise him?

The report consisted of the data on the following pages and in file CARDIO. The following variables are included:

REC	Record identification number
MO	Month of surgery (1 = January, etc.)
DA	Day of surgery
ID	Patient ID number
DRG	Diagnosis related group, a classification scheme indicating the nature of the medical condition. The DRG codes included: 104 Heart valve surgeries with heart catheterization 105 Heart valve surgeries without heart catheterization 106 Coronary artery bypass graft with heart catheterization 107 Coronary artery bypass graft without heart catheterization 483 Tracheotomy
CARD	Referring cardiologist: (0 = None, other numbers are codes for individual physicians)
STAT	Operational status (1 = Elective surgery, 2 = Immediate surgery required, 3 = Urgent, 4 = Emergency/salvage, a last-ditch effort to save a person)
CAT	Operative category: 1 = coronary artery bypass (CAB), 2 = CAB and other, 3 = CAB and valve surgery, 4 = CAB and valve surgery and other, 5 = Other, 6 = Valve surgery, 7 = Valve and other)
INC	Incidence: 1 = This is the first operation for the condition on this patient, 2 = The first reoperation, 3 = Second reoperation, etc.
RS	Risk stratification score, as described above
LS	Length of hospital stay, in days
CMP	0 if no complications, 1 if any complications.

DATA SET

REC	MO	DA	ID	DRG	CARD	STAT	CAT	INC	RS	LS	CMP
1	1	3	173	106	1	1	1	1	17	20	0
2	1	3	208	106	15	2	1	1	10	4	0
3	1	4	179	105	1	1	3	1	19	32	0
4	1	4	207	106	15	1	1	1	4	6	0
5	1	5	204	105	10	1	6	1	13	14	1
6	1	6	12	106	4	2	1	1	7	10	0
7	1	6	108	106	4	1	1	1	20	10	0
8	1	6	209	106	4	1	2	1	12	13	0
9	1	9	206	107	10	1	1	1	7	6	0
10	1	10	210	107	1	1	1	1	5	7	0
11	1	11	39	106	1	1	1	1	5	20	0
12	1	11	186	106	4	1	1	1	4	7	0
13	1	11	202	105	10	1	6	1	17	14	1
14	1	12	189	106	6	1	1	1	7	7	0
15	1	12	211	106	15	1	1	1	14	9	0
16	1	12	212	106	10	3	1	1	4	5	0
17	1	13	79	106	1	1	1	1	12	8	0
18	1	16	53	107	15	1	2	1	16	8	1
19	1	16	120	107	11	1	1	2	8	6	0
20	1	18	25	104	1	1	6	1	27	2	1
21	1	19	13	104	6	1	3	1	26	19	1
22	1	19	144	106	6	1	1	1	14	10	0
23	1	19	181	107	0	1	1	1	10	7	0
24	1	20	171	107	0	1	1	1	10	6	0
25	1	23	107	104	1	3	3	1	10	12	0
26	1	23	214	107	1	1	1	1	16	8	1
27	1	24	127	107	4	1	1	1	7	5	0
28	1	24	192	107	9	1	1	1	7	6	0
29	1	24	215	106	1	1	1	1	37	2	1
30	1	25	213	483	14	3	1	1	18	23	1
31	1	26	109	107	1	1	1	1	27	9	1
32	1	26	217	106	1	3	1	1	17	7	0
33	1	27	135	106	15	1	1	1	17	9	0
34	1	27	219	107	12	1	1	1	8	7	1
35	1	30	104	107	10	1	1	1	17	11	1
36	1	31	201	107	10	1	1	1	10	7	0
37	2	1	52	106	1	1	1	1	8	7	1
38	2	1	145	106	15	1	1	1	5	5	0
39	2	2	370	106	10	2	1	1	15	9	1
40	2	3	92	107	0	1	1	1	7	4	0
41	2	3	172	106	15	1	1	1	7	11	0
42	2	3	221	105	14	3	3	2	33	15	1
43	2	3	367	104	10	1	3	1	22	9	0
44	2	6	80	107	14	1	1	2	12	4	1
45	2	6	142	107	15	1	1	1	14	10	1
46	2	6	180	106	11	3	1	1	27	10	1

(continues on the following page)

REC	MO	DA	ID	DRG	CARD	STAT	CAT	INC	RS	LS	CMP
47	2	7	216	107	15	1	1	1	15	6	0
48	2	8	218	106	6	1	1	1	15	8	1
49	2	8	222	106	15	1	1	1	23	7	0
50	2	10	59	106	4	1	1	1	14	9	1
51	2	11	225	107	4	1	*	1	6	6	0
52	2	12	226	106	4	4	1	1	18	2	1
53	2	13	224	104	4	2	4	1	32	12	1
54	2	13	372	106	10	1	1	1	14	10	1
55	2	15	121	483	4	1	2	1	16	58	1
56	2	17	72	104	10	1	4	1	26	10	1
57	2	17	228	106	14	1	1	1	18	12	1
58	2	21	118	106	10	1	1	1	19	9	1
59	2	21	130	106	10	1	1	1	11	9	0
60	2	22	223	107	0	1	1	1	11	8	0
61	2	23	116	107	1	1	1	1	16	7	0
62	2	24	36	107	10	1	1	1	7	7	0
63	2	27	4	107	10	1	1	1	7	8	0
64	2	27	229	107	10	1	1	1	4	8	0
65	2	27	230	107	0	1	1	1	7	8	0
66	3	1	100	107	1	1	1	1	31	10	0
67	3	2	175	106	1	1	1	1	14	6	0
68	3	2	233	106	1	1	1	1	9	6	0
69	3	6	376	107	0	1	1	1	4	5	0
70	3	7	24	106	4	1	1	1	27	19	1
71	3	7	124	106	1	1	1	1	14	6	0
72	3	7	160	107	0	1	1	1	14	7	0
73	3	7	232	107	4	1	1	1	14	18	1
74	3	7	235	107	1	1	1	1	14	7	0
75	3	8	234	106	4	1	1	1	14	15	1
76	3	8	236	105	4	1	3	1	19	14	1
77	3	9	50	106	9	1	1	1	0	8	0
78	3	10	35	106	4	1	1	1	10	9	0
79	3	10	147	104	1	1	6	1	9	7	0
80	3	13	40	108	1	1	2	1	15	18	1
81	3	14	237	106	4	1	1	1	20	7	0
82	3	15	238	107	13	1	1	1	18	21	1
83	3	20	31	107	10	1	1	1	10	5	0
84	3	20	239	107	6	1	1	1	10	10	0
85	3	21	242	107	0	1	1	1	10	5	0
86	3	22	155	107	4	1	1	1	4	5	0
87	3	22	243	106	9	1	1	1	11	0	1
88	3	24	136	106	9	1	1	1	23	9	0
89	3	27	49	106	6	3	1	1	10	12	1
90	3	28	168	107	1	1	1	1	4	5	0
91	3	28	246	106	6	1	1	1	7	5	0
92	3	28	247	106	1	2	1	1	7	3	0
93	3	29	14	107	4	1	1	1	20	7	1

(continues on the following page)

REC	MO	DA	ID	DRG	CARD	STAT	CAT	INC	RS	LS	CMP
94	3	30	66	483	1	1	3	2	55	34	1
95	4	5	248	107	13	1	1	1	23	9	1
96	4	5	249	107	0	1	1	1	31	6	0
97	4	6	116	442	1	3	5	2	16	11	1
98	4	7	47	106	10	1	1	1	8	9	0
99	4	7	251	105	0	3	3	1	23	2	1
100	4	9	184	105	4	1	6	1	5	5	0
101	4	10	250	107	9	1	1	1	25	20	1
102	4	11	48	106	1	1	1	1	19	7	0
103	4	11	74	106	10	1	1	1	7	6	0
104	4	13	205	105	3	1	6	1	4	8	0
105	4	13	252	106	14	1	1	1	27	9	1
106	4	18	253	106	15	1	1	1	0	9	0
107	4	19	95	106	1	1	1	1	19	11	1
108	4	19	244	106	15	1	1	1	7	13	1
109	4	19	255	106	6	1	1	1	10	5	0
110	4	19	256	106	15	1	1	1	4	4	0
111	4	20	117	106	10	2	1	1	14	14	1
112	4	21	257	106	6	1	1	1	7	5	0
113	4	22	188	107	6	2	2	1	4	14	1
114	4	23	134	107	6	3	1	1	20	7	0
115	4	24	220	106	4	3	1	1	30	9	0
116	4	24	258	483	1	1	1	1	13	25	1
117	4	25	254	104	15	1	7	1	5	15	0
118	4	25	375	104	1	1	6	1	4	8	0
119	4	26	259	104	15	1	3	1	30	14	1
120	4	27	260	107	1	1	1	1	10	6	0
121	4	28	245	106	6	1	1	1	5	7	0
122	4	28	262	105	0	1	3	1	12	5	1
123	5	1	166	106	10	2	1	1	7	11	0
124	5	1	261	106	1	1	1	1	8	11	0
125	5	1	263	106	6	3	1	1	14	10	1
126	5	1	371	106	1	3	1	1	17	13	0
127	5	3	151	106	6	2	1	1	15	8	0
128	5	3	264	106	4	1	1	1	3	6	0
129	5	4	266	106	15	1	1	1	10	6	0
130	5	5	265	107	10	1	1	1	20	6	0
131	5	6	123	107	1	1	1	1	16	6	0
132	5	9	200	106	1	2	1	1	20	8	0
133	5	9	268	107	0	1	1	1	19	10	1
134	5	10	26	106	4	1	1	1	7	9	1
135	5	11	152	106	1	1	1	2	31	38	1
136	5	11	269	106	1	1	1	1	8	7	1
137	5	12	6	106	4	1	1	1	11	7	0
138	5	12	272	107	1	1	1	1	4	8	1
139	5	15	89	106	6	1	1	1	20	15	1
140	5	16	11	106	1	1	1	1	7	10	1

(continues on the following page)

REC	MO	DA	ID	DRG	CARD	STAT	CAT	INC	RS	LS	CMP
141	5	16	185	105	0	1	3	1	19	7	0
142	5	17	271	106	4	3	1	1	8	7	0
143	5	18	270	106	6	1	3	1	15	6	0
144	5	18	275	104	10	1	7	1	19	6	1
145	5	19	105	106	10	1	1	1	16	10	0
146	5	19	106	107	4	1	1	1	8	5	0
147	5	19	149	106	15	1	1	1	17	5	0
148	5	22	98	106	4	2	1	1	25	8	0
149	5	22	274	106	9	1	1	1	4	7	0
150	5	24	102	106	10	1	1	1	19	11	0
151	5	24	138	107	4	1	1	1	11	7	1
152	5	25	276	106	9	1	1	1	6	6	0
153	5	26	278	106	14	2	1	1	22	12	1
154	5	30	277	106	15	1	1	1	12	11	0
155	6	1	69	106	10	3	1	1	7	8	1
156	6	6	182	106	15	1	1	1	6	10	0
157	6	9	21	106	4	1	1	1	4	10	0
158	6	9	279	106	0	1	1	1	4	10	0
159	6	12	196	108	14	1	1	1	5	11	1
160	6	12	281	107	0	1	1	1	8	6	1
161	6	13	78	106	4	3	1	1	10	7	0
162	6	13	131	106	6	1	1	1	30	10	0
163	6	13	159	106	1	1	1	1	7	6	0
164	6	13	280	104	6	1	1	1	30	12	0
165	6	14	84	106	15	1	1	1	7	8	0
166	6	16	284	106	6	2	1	1	30	8	0
167	6	17	282	483	0	2	2	1	23	19	1
168	6	20	38	105	1	1	6	1	9	10	1
169	6	20	67	106	15	3	1	1	20	14	0
170	6	20	81	106	1	1	1	2	13	14	1
171	6	21	285	107	0	1	1	1	8	6	1
172	6	23	286	108	1	3	2	1	29	7	1
173	6	26	91	107	1	3	1	1	17	5	1
174	6	27	287	107	9	1	1	1	4	4	1
175	6	28	7	106	11	1	1	1	14	9	0
176	6	28	289	107	0	1	1	1	7	3	0
177	6	29	231	105	1	1	6	1	12	13	1
178	6	30	290	108	0	1	2	1	15	6	1
179	7	3	97	104	9	2	6	3	53	19	1
180	7	3	288	105	1	1	6	1	23	11	1
181	7	6	167	107	12	1	1	1	15	9	1
182	7	7	366	106	4	1	1	1	7	9	0
183	7	8	292	106	9	2	1	1	3	7	0
184	7	9	30	106	15	2	1	1	3	9	0
185	7	10	293	106	15	3	1	1	23	7	0
186	7	11	291	106	15	1	2	1	15	19	1
187	7	11	297	106	1	1	1	1	7	5	0

(continues on the following page)

REC	MO	DA	ID	DRG	CARD	STAT	CAT	INC	RS	LS	CMP
188	7	12	33	107	0	1	1	2	7	4	0
189	7	12	99	106	10	1	1	1	0	9	0
190	7	13	75	106	10	1	1	2	16	9	0
191	7	13	294	106	15	1	1	1	1	9	0
192	7	13	298	108	12	1	5	1	5	7	1
193	7	13	300	106	6	1	1	1	23	11	1
194	7	14	296	104	1	1	3	1	19	13	1
195	7	14	299	105	12	1	6	1	3	8	0
196	7	15	293	106	15	4	1	2	32	17	1
197	7	16	304	106	4	2	1	1	27	4	0
198	7	17	190	106	6	1	1	1	10	16	0
199	7	17	302	107	0	1	1	1	5	5	0
200	7	18	57	107	0	1	1	2	15	7	0
201	7	18	303	106	4	3	1	1	10	7	0
202	7	18	305	106	5	1	1	1	10	7	0
203	7	18	306	106	9	1	1	1	3	6	1
204	7	19	63	107	1	2	1	1	37	6	1
205	7	20	19	104	15	1	6	1	22	9	1
206	7	24	309	107	9	1	1	1	10	6	1
207	7	25	61	106	1	1	1	1	4	6	0
208	7	25	308	106	4	1	1	1	14	10	0
209	7	25	310	107	0	1	1	1	8	6	0
210	7	26	101	108	4	1	2	1	5	25	1
211	7	27	311	106	1	1	1	1	7	5	0
212	7	28	133	104	2	3	6	1	20	7	0
213	7	28	313	106	1	1	1	1	13	7	0
214	7	28	315	106	15	2	1	1	18	6	0
215	8	1	111	104	1	1	3	1	24	18	0
216	8	1	307	107	4	1	1	1	14	6	0
217	8	1	312	106	15	2	1	1	8	21	1
218	8	1	316	106	15	1	1	1	10	6	0
219	8	2	60	106	1	3	1	1	25	17	1
220	8	4	27	105	15	1	3	1	16	6	0
221	8	4	112	106	2	1	1	1	7	8	0
222	8	7	156	106	6	1	1	1	14	11	0
223	8	7	177	468	13	1	1	1	21	16	1
224	8	8	150	106	10	3	1	1	7	6	0
225	8	9	103	106	10	1	1	1	17	5	0
226	8	9	113	106	6	1	1	1	7	6	0
227	8	11	90	106	6	3	1	2	41	8	0
228	8	12	41	106	4	2	1	1	4	7	1
229	8	14	154	107	10	1	1	1	5	4	0
230	8	16	178	106	2	1	1	2	3	9	0
231	8	17	318	107	0	1	1	1	13	6	0
232	8	17	319	106	10	2	1	1	27	7	0
233	8	18	28	107	10	1	1	1	17	6	1
234	8	18	187	106	6	1	1	1	11	10	0

(continues on the following page)

REC	MO	DA	ID	DRG	CARD	STAT	CAT	INC	RS	LS	CMP
235	8	21	317	107	0	1	1	1	19	5	1
236	8	22	119	106	10	1	1	1	0	5	1
237	8	22	132	106	4	2	1	1	27	18	1
238	8	23	122	106	10	2	1	1	23	9	1
239	8	25	240	106	2	1	1	1	19	7	0
240	8	25	301	104	1	1	3	1	12	13	1
241	8	28	56	106	4	1	1	1	34	14	1
242	8	28	165	107	15	1	1	1	14	6	0
243	8	29	15	107	7	1	1	1	13	5	0
244	8	29	321	107	10	1	1	1	7	5	0
245	8	29	365	104	13	2	3	1	40	8	0
246	8	30	153	106	6	1	1	2	29	38	1
247	8	30	283	106	6	1	1	1	7	5	0
248	8	31	20	106	4	3	1	1	17	20	1
249	8	31	326	106	6	2	1	1	17	7	1
250	9	1	96	107	10	1	1	1	18	8	1
251	9	1	327	107	0	1	1	1	27	9	0
252	9	1	328	106	4	1	1	1	10	13	1
253	9	5	325	106	9	1	*	1	19	13	0
254	9	6	93	106	10	1	1	1	4	5	0
255	9	6	163	108	10	1	2	1	10	7	0
256	9	8	148	106	6	1	1	1	12	9	1
257	9	8	329	106	6	1	1	1	20	5	0
258	9	11	64	483	4	1	3	2	26	30	1
259	9	12	157	106	6	1	1	1	25	8	0
260	9	13	54	106	9	1	1	1	14	8	1
261	9	15	191	104	1	1	6	2	4	10	0
262	9	18	9	106	4	1	1	1	3	11	0
263	9	19	115	107	10	1	1	1	12	4	0
264	9	19	330	107	10	1	1	1	15	6	0
265	9	20	323	107	0	1	1	1	18	12	1
266	9	22	143	106	6	1	1	1	14	7	1
267	9	25	37	106	10	1	1	1	7	6	0
268	9	25	195	106	10	1	1	1	8	4	0
269	9	25	332	106	1	1	2	1	4	9	0
270	9	26	51	107	10	1	1	1	22	6	0
271	9	26	193	106	15	1	1	1	8	5	0
272	9	26	203	107	0	1	1	1	28	14	0
273	9	27	139	106	1	1	1	1	32	9	1
274	9	28	10	104	10	1	6	1	27	9	0
275	9	28	333	106	15	1	1	1	7	7	0
276	9	28	334	106	1	3	1	1	8	7	0
277	10	2	70	106	1	1	1	1	7	8	0
278	10	3	320	107	1	3	2	1	13	21	1
279	10	3	331	105	4	1	6	1	17	10	1
280	10	4	295	106	15	1	1	1	10	7	0
281	10	5	198	106	14	1	1	1	11	5	0
282	10	5	241	106	14	2	1	1	26	7	1

(continues on the following page)

REC	MO	DA	ID	DRG	CARD	STAT	CAT	INC	RS	LS	CMP
283	10	6	335	106	2	1	1	1	19	15	1
284	10	6	338	106	10	1	1	1	7	8	0
285	10	9	125	106	4	1	1	1	4	4	0
286	10	9	197	106	2	1	1	1	10	5	1
287	10	9	339	106	6	2	1	1	5	7	0
288	10	9	369	105	5	1	2	1	7	4	1
289	10	11	46	106	4	1	1	1	11	11	1
290	10	11	337	104	6	2	3	1	7	35	1
291	10	12	340	104	8	1	6	1	5	7	0
292	10	13	29	106	6	1	1	2	13	12	1
293	10	13	85	106	6	2	1	1	8	11	0
294	10	13	88	106	15	1	1	1	11	7	0
295	10	17	32	106	2	1	1	1	7	4	0
296	10	17	62	104	10	1	3	1	16	5	0
297	10	17	83	106	10	1	1	2	4	5	0
298	10	18	76	107	10	1	1	2	7	4	1
299	10	19	82	105	10	1	6	1	8	7	0
300	10	19	176	106	1	1	1	1	10	14	0
301	10	20	314	106	4	1	1	1	4	11	0
302	10	20	341	106	4	3	1	1	10	7	0
303	10	24	158	107	9	1	1	1	17	4	0
304	10	24	227	107	1	1	1	1	7	5	0
305	10	24	342	106	10	2	1	1	7	11	1
306	10	25	18	106	15	1	1	1	13	5	0
307	10	25	42	107	15	1	1	1	16	6	0
308	10	25	344	106	4	1	1	1	10	7	1
309	10	26	194	483	9	2	6	1	28	32	1
310	10	27	43	106	1	2	1	1	11	6	1
311	10	27	345	106	9	2	1	1	7	14	1
312	10	30	114	107	4	1	1	1	10	4	0
313	10	30	129	107	14	1	1	1	10	4	0
314	10	30	343	106	4	1	1	1	11	10	1
315	10	31	3	106	1	1	1	1	27	7	0
316	10	31	141	106	2	1	1	1	10	19	0
317	10	31	368	106	10	2	1	1	30	8	1
318	11	1	22	106	10	1	1	1	7	11	0
319	11	1	110	106	1	1	1	1	10	8	0
320	11	1	347	106	9	1	1	1	19	11	0
321	11	1	349	106	15	3	1	1	17	6	0
322	11	2	77	107	2	1	1	2	8	7	0
323	11	2	348	106	4	1	1	2	7	13	1
324	11	3	346	483	2	1	2	4	10	16	1
325	11	3	350	107	0	1	1	1	6	6	0
326	11	6	1	107	10	1	1	1	14	4	0
327	11	6	58	106	4	1	1	1	4	9	0
328	11	7	170	106	14	3	1	1	31	10	1
329	11	7	183	106	10	3	1	1	7	7	1
330	11	7	336	104	6	1	6	1	20	9	0

(concludes on the following page)

REC	MO	DA	ID	DRG	CARD	STAT	CAT	INC	RS	LS	CMP
331	11	9	16	107	13	1	1	2	14	7	0
332	11	9	34	106	9	1	1	1	21	7	0
333	11	10	45	106	4	1	1	2	7	9	0
334	11	11	324	106	9	2	1	1	14	8	0
335	11	13	23	105	10	1	6	1	20	15	0
336	11	14	128	107	14	1	1	1	17	7	1
337	11	15	55	106	6	2	1	11	10	4	0
338	11	15	174	105	1	1	*	1	5	6	0
339	11	16	140	107	4	1	1	2	18	7	0
340	11	16	161	106	6	1	1	1	10	7	0
341	11	16	273	106	6	1	1	1	6	15	0
342	11	20	322	106	4	1	1	1	14	6	1
343	11	21	94	106	4	1	1	1	11	6	0
344	11	21	352	106	2	1	1	1	10	5	0
345	11	22	65	106	4	1	1	1	4	8	1
346	11	22	169	106	13	1	1	1	7	13	0
347	11	24	86	106	15	1	1	1	13	8	0
348	11	24	162	107	10	1	1	1	10	5	0
349	11	27	71	105	10	1	6	1	10	7	0
350	11	28	5	106	10	1	1	1	22	7	1
351	11	29	353	106	8	1	1	1	15	8	0
352	11	29	354	106	15	1	1	1	7	7	0
353	11	30	355	104	8	1	6	1	7	6	0
354	12	1	44	106	0	1	1	1	15	7	0
355	12	1	356	106	9	1	1	1	10	14	0
356	12	4	146	107	10	1	1	1	4	3	0
357	12	4	199	107	0	1	1	1	4	3	0
358	12	5	87	106	4	2	1	1	7	13	0
359	12	5	357	106	4	2	1	1	7	9	0
360	12	6	358	107	14	1	1	2	15	6	0
361	12	7	8	107	10	1	1	1	27	6	1
362	12	7	267	106	2	1	1	1	19	12	1
363	12	8	359	106	4	1	1	1	7	7	0
364	12	11	68	104	1	1	1	1	30	13	1
365	12	12	373	106	6	1	1	1	3	5	0
366	12	13	17	107	15	1	1	1	10	6	0
367	12	13	137	106	15	1	1	1	7	6	1
368	12	13	164	106	4	1	1	1	17	6	1
369	12	14	351	105	8	1	3	1	11	5	0
370	12	14	360	106	6	1	1	1	15	9	1
371	12	14	361	104	15	1	1	1	14	9	1
372	12	15	2	106	6	2	1	3	8	7	0
373	12	19	363	104	6	1	6	1	10	6	0
374	12	20	374	106	10	1	1	1	31	15	0
375	12	25	73	106	10	2	1	1	11	6	1
376	12	27	126	107	4	1	1	1	7	7	0
377	12	27	362	106	10	1	1	1	8	5	0
378	12	29	364	106	6	3	1	1	22	7	0

CASE **71**

THE BAY AREA ATHLETIC CLUB

The Bay Area Athletic Club, located in San Francisco, California, offers the full range of athletic activities and equipment: indoor and outdoor pool and tennis, a weight room with free weights and Cybex machines, basketball, aerobics classes, yoga classes, an outdoor running track, and a senior center. By popular demand, yoga and aerobics classes are only offered during peak hours. Members can purchase two types of memberships. A Gold membership, costing $100 per month per family, allows the user full access to all activities and facilities. Silver memberships allow access to all facilities but not the aerobic and yoga classes; it costs $90 per family per month. Individual memberships are half the monthly family rates.

Even though the club has the latest equipment and fully trained staff, it has a problem. Management would like to increase club revenues, but dares not increase monthly rates. An alternative is to expand its membership base. However, current members are already complaining about insufficient access to parking, some even threatening to cancel their memberships. The club is surrounded on all sides by commercial property and cannot solve its parking problem by extending its existing lots to adjacent property. Management at Bay Area has come up with a solution. If new members can be encouraged to use the facilities during off-peak hours, revenues can be increased without compounding an already serious parking problem.

But what sort of individual can attend the club during off-peak hours? This is important to know since Bay Area has limited funds for advertising and would like to target its marketing efforts to potential nonpeak users.

When a user checks into the club, the bar code on his or her membership card is scanned and the time of entry is recorded. Management has compiled a data set by pulling

These data and scenario are real, although the name of the health club and its location have been changed.

every 20th record from the membership filing cabinets and matching the scanned entry data. The resulting data set, containing 94 observations, is described below and available in the file named PEAK. Use the data to determine a profile of the club's nonpeak members and provide appropriate recommendations to Bay Area Club management.

DEFINITIONS

Age	Age category of the member: 1 = No more than 19 years old, 2 = 20 to 29 years old, 3 = 30 to 39 years old, 4 = 40 to 49 years old, 5 = 50 to 59 years old, 6 = 60 or older
Marital	Marital status: 0 = Single, 1 = Married
Gender	0 = Male, 1 = Female
Kids	0 = No dependent children living in the home, 1 = One or more dependent children living in the home
Location	Location of member's home address, grouped according to five zip codes; Location = 0 is the closest to the club, Location = 4 is farthest from the club
Freq	Frequency of use of the club: number of times the member used the club during the month prior to the collection of the data
Type	Type of membership: 0 = Gold, 1 = Silver
Length	Number of years of membership with Bay Area
NonPeak	Percentage of time the member used the club during off-peak hours for the month prior to the collection of the data. This number was calculated as the number of times the member used the club during off-peak hours divided by the total number of times the member used the club (both peak and off-peak) times 100. Peak usage times are: 6:00 to 9:00 A.M. and 4:30 to 7:30 P.M. during the weekdays and 8:00 A.M. to noon on weekends. All other times are off-peak.

DATA SET

AGE	MARITAL	GENDER	KIDS	LOCATION	FREQ	TYPE	LENGTH	NONPEAK
1	0	0	0	0	1	1	1	100.0
2	1	0	1	0	1	1	3	0.0
2	1	0	0	0	1	1	2	0.0
3	0	0	1	2	1	0	1	0.0
3	0	1	0	0	1	0	2	100.0
3	1	0	1	4	1	1	1	0.0
4	0	0	1	0	1	0	1	0.0
4	0	1	0	0	1	1	3	0.0
2	0	0	1	0	2	0	1	0.0
3	0	0	0	4	2	0	5	0.0
3	0	1	0	4	2	0	4	0.0
3	1	0	1	2	2	0	1	0.0
4	1	0	1	4	2	1	2	0.0
5	0	0	0	0	2	1	2	0.0
2	0	1	0	0	3	1	1	33.3
2	0	0	0	3	3	1	5	33.3

(continues on the following page)

AGE	MARITAL	GENDER	KIDS	LOCATION	FREQ	TYPE	LENGTH	NONPEAK
2	1	0	0	0	3	1	9	33.3
3	1	1	0	4	3	1	3	66.7
3	0	1	1	2	3	1	1	100.0
4	0	1	0	4	3	1	3	66.7
5	1	0	0	2	3	1	3	33.3
2	0	0	1	0	4	1	1	25.0
2	1	0	1	0	4	0	1	0.0
3	0	0	1	0	4	1	2	0.0
3	0	1	1	3	4	0	5	0.0
4	1	1	1	3	4	1	1	0.0
4	1	0	1	4	4	1	4	0.0
4	1	0	1	4	4	0	4	50.0
4	0	0	0	2	4	1	5	50.0
5	1	1	1	2	4	0	6	100.0
6	0	1	1	3	4	0	3	25.0
1	0	1	1	0	5	0	3	20.0
2	0	1	0	4	5	0	1	20.0
2	0	1	1	4	5	0	2	20.0
2	1	0	0	0	5	0	1	0.0
2	0	0	0	0	5	1	3	80.0
2	0	0	1	4	5	1	2	20.0
2	0	1	0	0	5	0	1	20.0
4	0	0	0	4	5	0	6	0.0
4	0	1	0	3	5	1	5	0.0
4	1	0	1	0	5	1	1	0.0
2	0	0	0	1	6	0	2	16.7
2	1	1	1	1	6	0	3	66.7
2	0	0	0	1	6	1	4	16.7
2	0	0	0	1	6	1	1	66.7
2	0	1	0	1	6	0	2	16.7
2	0	1	1	4	6	1	5	83.3
2	0	1	0	4	6	1	3	16.7
2	1	0	0	3	6	1	1	50.0
3	0	1	0	0	6	0	2	16.7
3	1	0	1	4	6	1	2	0.0
3	1	1	1	3	6	1	6	16.7
3	1	0	1	1	6	1	6	16.7
3	0	0	0	4	6	0	2	50.0
5	1	0	0	1	6	1	5	50.0
2	0	1	0	0	7	0	2	71.4
2	0	1	1	2	7	0	1	42.9
3	1	1	0	1	7	1	3	0.0
3	0	1	1	1	7	1	3	0.0
2	1	1	1	2	8	0	2	0.0
2	0	1	0	1	8	1	3	0.0
3	0	0	1	2	8	1	2	0.0
3	0	0	0	4	8	1	2	75.0

(concludes on the following page)

AGE	MARITAL	GENDER	KIDS	LOCATION	FREQ	TYPE	LENGTH	NONPEAK
3	1	1	0	0	8	0	4	0.0
3	1	0	1	1	8	1	5	0.0
2	0	1	0	2	9	1	1	22.2
2	0	0	0	1	9	0	2	44.4
3	0	1	0	3	9	1	4	0.0
3	1	1	0	0	9	1	1	11.1
2	0	0	0	4	10	1	1	40.0
4	0	1	0	2	10	0	1	100.0
4	0	1	1	3	10	1	1	10.0
4	0	0	0	3	10	0	2	40.0
5	0	1	0	2	10	0	1	20.0
5	0	1	0	0	10	0	1	100.0
1	0	0	0	1	11	1	1	54.4
2	1	0	0	0	11	1	1	18.2
3	0	0	1	2	11	1	3	0.0
2	1	1	0	0	12	0	1	0.0
2	0	0	0	2	12	1	1	33.3
4	0	1	0	1	12	1	5	8.3
4	1	1	1	1	12	1	1	16.7
5	1	1	1	1	12	1	5	16.7
5	1	0	0	0	12	0	6	100.0
2	0	0	1	2	14	1	1	21.4
3	1	0	1	1	14	0	1	14.3
5	1	0	1	3	14	1	4	14.3
6	1	0	0	2	14	0	3	100.0
1	0	1	0	3	15	1	1	86.7
6	0	0	0	2	15	1	2	93.3
1	0	0	0	0	16	1	1	50.0
6	0	0	0	1	15	0	4	0.0
4	0	1	0	4	2	1	2	0.0
2	1	0	1	1	7	1	6	42.9

CASE **72**

BLOOD PRESSURE

Blood pressure is a common indicator used to assess human health. It consists of two measurements, the systolic and the diastolic blood pressures, measured in millimeters of mercury (mm Hg). Systolic is the higher of the two. When systolic pressure is above roughly 145 mm Hg, the heart must work harder during each beat. Such a condition may eventually lead to heart problems. When the diastolic pressure (the lower of the two) is above 90 mm Hg, the heart muscle is not receiving enough oxygen. Normally this is not as serious as elevated systolic blood pressure, but it can sometimes lead to complications.

Employers concerned about employee health often provide health fairs, or something similar, at which employees can receive basic health screenings either free or inexpensively and often on company property (perhaps near the cafeteria at lunch time). One company provided blood pressure screenings at such a fair, and recorded the measurements given on the next page (and in file BLOODP). SYST is the systolic blood pressure, DIAST the diastolic blood pressure. SEX is 0 for females and 1 for males. The company was interested in knowing whether their employees had any particular tendencies to high blood pressure and whether there appeared to be any patterns of interest.

This case is based upon a real situation. Jana Morris described it for us and provided the data.

DATA SET

SYST	DIAST	SEX	SYST	DIAST	SEX	SYST	DIAST	SEX
108	78	1	110	78	0	130	98	1
108	76	1	126	84	0	104	82	0
108	68	0	136	86	1	118	80	0
106	60	0	136	102	1	96	70	0
124	90	1	102	72	1	142	100	0
120	76	1	146	112	1	110	70	0
142	94	1	122	86	1	132	86	1
130	94	1	126	82	1	114	80	0
110	82	0	104	68	1	106	70	0
112	86	0	112	76	0	108	74	0
130	72	0	102	70	0	124	76	1
124	80	1	154	84	1	118	78	0
118	72	1	136	96	0	108	70	0
124	80	1	106	68	0	172	96	1
130	90	0	138	88	0	120	82	1
120	76	1	138	100	0	124	78	1
146	90	0	122	80	1	116	68	1
118	84	1	116	82	0	108	80	0
130	80	1	104	88	0	110	70	0
120	72	1	106	74	0	118	78	0
124	84	1	100	80	0	126	80	1
172	100	1	120	84	0	152	100	1
130	80	1	130	86	1	102	58	0
124	76	1	118	70	1	110	80	0
126	82	0	120	90	0	104	70	0
136	80	1	114	76	1	126	86	0
138	94	1	112	70	1	172	114	1
130	102	1	106	80	1	150	100	1
104	70	0	120	82	1	136	78	1
100	84	0	122	74	0	112	70	0
128	80	1	108	80	0			

CASE **73**

MATRIMONIAL UNREST

Ted was worried. He and his wife, Maria, just had another horrible fight. Even though they were newly married, Ted and Maria were often engaged in deep conflict. The fact that they couldn't seem to stop the bickering was bad enough; even worse was that neither one of them seemed to know what was causing the disagreements. Their love for each other was irrefutable. So what was causing the disagreements?

The particularly memorable fights seemed to escalate when either Ted or Maria threw accusations at each other. Arguments often ended in a shouting match about who said what or who did what to the other party and the original reason for the disagreement was blurred or lost. Having recognized this pattern, it occurred to Ted that their disagreements might be more productive if they agreed to refocus their discussions about what, and not who, caused the conflict.

To this end, Ted began collecting data on a daily basis about events in his life that might trigger their fights. For instance, Ted suspected that if he was particularly tired or stressed, or had a difficult day at work, he was apt to be argumentative at home. The number of calories he consumed that day and how much he slept the night before might influence whether he was physically exhausted by the end of the day. Ted works a full-time job while also attending an evening MBA program. A good or bad day at work, or preparation for an upcoming class, might make him more irritable. Sometimes unavoidable and unanticipated events overwhelmed Ted; significant events might be a death in the family, a robbery or car theft, or other major traumas. These were recorded in his daily log as well.

As newlyweds with tuition to pay, Ted and Maria were feeling a financial crunch. Because Ted often wondered if money might be the source of their arguments, he collected three additional pieces of information. One was the daily balance on their checking

The data for this case were collected by Marc and Deborah Schneider of San Antonio, Texas.

account according to the bank. The other was the checking account balance recorded in their personal checkbook register. The two balances can differ when a check is recorded in the register but has not yet cleared the bank, or when automatic deposits are made at the bank but not recorded in the household checkbook. Ted suspected that he felt more stress on those days he paid household bills and recorded this information as well. These and other factors collected by Ted (available in the data set named FIGHTS) are described below.

At the end of 75 days of data collection, Ted sat down at his computer to enter the data on his spreadsheet. Caught in the act, Ted confessed to Maria that he had been surreptitiously collecting information about himself and their fights. Maria looked over the data and pointed out that Ted had collected data only about his life or about their joint checking account, but nothing about factors that might influence *her* proclivity to pick a fight. In fact, Maria had often wondered if her notable mood swings prior to menstruation might influence her interactions with Ted. She contributed the column in the data set labeled PMS.

What do the data have to say about Ted and Maria's problem?

DEFINITIONS

Fight	Whether or not a fight occurred that day (1 = Fight, 0 = No fight)
Calories	Number of calories consumed that day in kilocalories
Sleep	Number of hours of sleep the previous night
Work	Good or bad day at work (0 = Good, 1 = Bad)
School	School the next day (0 = No, 1 = Yes)
Coffee	Number of cups of coffee consumed that day
PMS	Uncomfortable symptoms of PMS that day (0 = No, 1 = Yes)
Bank	Bank statement balance that day ($)
Checkbk	Checkbook balance that day ($)
Events	1 = Significant event occurred that day, 0 = No significant event that day
Weekday	Day of the week (1 = Sunday, 2 = Monday, etc.)
Weekend	1 = Saturday or Sunday, 0 = Monday through Friday
Bills	Were bills paid that day? (0 = No, 1 = Yes)

DATA SET

FIGHT	CALORIES	SLEEP	WORK	SCHOOL	COFFEE	PMS	BANK	CHECKBK	EVENTS	WEEKDAY	WEEKEND	BILLS
0	3350	5.50	1	1	2	0	1282	500	0	3	0	0
1	3487	7.00	1	0	1	0	1216	1560	0	4	0	0
0	4020	4.00	0	0	0	0	640	1560	0	5	0	0
0	2447	9.00	1	0	2	0	340	1504	0	6	0	0
0	1995	7.00	0	0	2	0	340	300	1	7	1	1
0	1684	9.00	0	1	0	0	340	300	0	1	1	0
0	3335	6.00	0	0	3	0	1737	300	0	2	0	0
1	2817	7.00	1	1	1	0	1737	156	0	3	0	1
1	2872	7.00	1	0	1	0	1670	156	0	4	0	0
0	2801	6.00	0	0	2	0	1533	156	0	5	0	0

(continues on the following page)

FIGHT	CALORIES	SLEEP	WORK	SCHOOL	COFFEE	PMS	BANK	CHECKBK	EVENTS	WEEKDAY	WEEKEND	BILLS
0	2589	6.00	1	0	1	0	1533	125	0	6	0	0
0	3200	9.50	0	0	2	0	1533	95	0	7	1	0
0	2217	7.50	0	1	2	0	1533	95	0	1	1	0
1	2617	5.00	1	0	3	0	426	95	0	2	0	0
0	3773	6.50	1	1	2	0	405	95	0	3	0	0
0	2559	6.50	1	0	1	0	325	95	0	4	0	0
0	2739	7.00	0	0	0	0	1295	947	0	5	0	0
1	3867	7.00	0	0	2	0	1295	947	0	6	0	0
1	2460	5.50	0	0	2	1	1295	920	0	7	1	0
1	3157	6.50	0	1	2	1	1295	895	0	1	1	0
0	2639	5.50	1	0	2	1	1194	870	0	2	0	0
0	3061	6.50	0	0	0	1	1147	750	0	3	0	0
0	3544	9.00	0	0	0	1	1105	750	0	4	0	0
0	1661	8.50	0	0	0	0	426	885	0	5	0	0
1	1925	8.00	0	0	0	0	1001	885	0	6	0	0
0	1998	9.00	0	0	0	0	984	-595	0	7	1	1
1	1160	8.50	0	1	0	0	984	-595	0	1	1	0
0	3312	5.25	0	0	1	0	984	-113	0	2	0	0
0	3470	6.50	0	1	1	0	984	386	0	3	0	0
0	2617	6.50	0	0	0	0	1501	320	0	4	0	0
0	3465	6.50	0	0	1	0	1347	320	0	5	0	0
0	2044	8.00	0	0	0	0	1274	320	0	6	0	0
0	3337	8.00	0	0	1	0	828	585	0	7	1	1
1	2417	8.50	0	1	1	0	828	396	0	1	1	0
0	2363	5.00	0	0	2	0	828	396	0	2	0	0
0	3925	6.50	0	1	0	0	814	396	0	3	0	0
0	2840	7.00	0	0	1	0	1030	370	0	4	0	0
0	4272	6.50	0	0	0	0	960	370	0	5	0	0
0	1986	6.50	0	0	0	0	905	370	0	6	0	0
1	2087	6.50	0	0	1	0	455	533	0	7	1	0
0	2872	7.50	0	1	1	0	455	450	0	1	1	0
0	2153	4.50	0	0	2	0	455	1401	0	2	0	0
1	3928	6.50	1	1	1	0	95	1401	0	3	0	0
0	2115	4.00	1	0	1	0	64	1380	0	4	0	0
0	3713	7.00	0	0	0	0	326	1380	0	5	0	0
0	3641	8.50	0	0	0	0	305	1380	0	6	0	0
1	3031	6.50	0	0	3	0	1303	1365	0	7	1	0
0	4037	6.00	0	1	1	0	1303	140	1	1	1	1
1	3986	5.00	0	0	0	0	1303	140	0	2	0	0
1	3912	5.50	1	1	0	1	1237	140	0	3	0	0
0	1937	5.00	0	0	2	1	1155	140	0	4	0	0
1	3254	6.25	1	0	2	1	893	140	0	5	0	0
1	2692	6.25	0	0	1	1	893	120	0	6	0	1
1	3549	7.75	0	0	2	1	796	-200	0	7	1	0
0	2302	6.00	0	1	1	0	796	-459	1	1	1	0
0	3760	6.75	0	0	1	0	796	-419	0	2	0	0

(concludes on the following page)

FIGHT	CALORIES	SLEEP	WORK	SCHOOL	COFFEE	PMS	BANK	CHECKBK	EVENTS	WEEKDAY	WEEKEND	BILLS
0	4123	5.50	0	1	1	0	19	-419	0	3	0	0
0	2198	8.50	0	0	0	0	68	-419	0	4	0	0
1	2954	6.00	0	0	2	0	68	-419	0	5	0	0
0	3524	4.50	0	0	2	0	44	-419	0	6	0	0
0	2900	10.00	0	0	0	0	1015	-419	0	7	1	0
0	2950	8.00	0	1	0	0	1015	575	0	1	1	0
0	3368	6.50	0	0	1	0	990	512	0	2	0	0
0	3200	7.25	0	1	1	0	956	416	0	3	0	0
0	2584	5.50	0	0	1	0	901	387	0	4	0	0
0	2253	6.50	0	0	1	0	901	377	0	5	0	0
0	2642	6.50	0	0	1	0	901	350	0	6	0	0
0	4491	7.50	0	0	2	0	756	325	0	7	1	0
0	2681	10.00	0	1	0	0	666	315	0	1	1	0
0	3970	6.50	0	0	1	0	435	-134	0	2	0	0
0	3890	8.75	0	1	0	0	332	-144	0	3	0	0
0	3210	7.00	0	0	0	0	200	-230	0	4	0	0
0	3800	4.00	0	0	1	0	195	-230	0	5	0	0
0	3486	11.00	0	0	0	0	66	-230	0	6	0	0
1	3250	8.50	0	0	0	0	4	-230	0	7	1	0

CASE **74**

PRODUCTIVITY IN THE WORD PROCESSING CENTER

Partner meetings at the law firm of Fisher, Stonewall, and Curtis are usually quite civil. Today, though, there was a heated discussion about turnaround time of document preparation in the word processing center. Partners and law clerks alike are getting concerned about slow response time, perceived favoritism toward certain lawyers, and inconsistent quality of work.

When the firm was first established, each law partner had his or her own secretary who answered and screened phone calls, typed legal documents, and made appointments. There were several notable problems with this arrangement. Staff members were sometimes underutilized when their bosses were away on business. Staff members rarely pitched in for each other because the organizational structure encouraged a sense of territorialism. The firm was unable to retain an assistant for Mr. Curtis for long because of his infamous temper, and he constantly complained about needing competent help. Two years ago, the administrative staff was reorganized into a word processing/secretarial center. All employees in this group were housed in one large room, and tasks were doled out by Sue, the word processing manager. The reorganization seemed to work well for a few months, but then complaints began to surface. This morning's meeting was called specifically to address concerns about the word processing center.

Ms. Conner was sympathetic to the people in the center, but was feeling her own productivity declining as documents were not returned on time and were filled with serious errors. "Every time I walk by there, people seem to be busy. So why can't I get my work done on time?"

"I spoke with Sue about your concerns," replied Mr. Fisher. "She likes the new organizational structure and thinks that her staff are working as efficiently as possible. She traces the problem to a shortage of staff. In the old days, we had one secretary for each partner. Now the ratio is about four to five. Although we have gained some efficiency and cost savings by creating the word processing center, we may have gone too far. Sue believes that hiring an additional full-time employee for the center would solve the problem."

"That's exactly what I would have *expected* Sue to say," said Mr. Curtis. "She's just trying to get some of her friends a job here. After all, we *do* pay *excellent* benefits. If she would manage her employees more efficiently, we wouldn't have to spend money on a new hire. The mark of any successful law firm is profit margin, which depends on . . ."

Ms. Stonewall, the only partner other than Fisher senior to Curtis, felt inclined to interrupt. "We're all aware of your sensitive attitudes about our support staff. Sue's been with this firm a long time and has proven herself a loyal and hard-working employee. If she says we would benefit from hiring another person, I'm willing to give it a try."

"Loyal and hard-working? You've got to be kidding! Sue is surly and uncommunicative." Mr. Curtis was on a roll.

"Sue isn't one to socialize much with us or those in the center," replied Ms. Stonewall. "She wants to come to work, do a good job, and go home. Even so, she is the fastest and most efficient word processor in the group. There's nothing wrong with being aloof as long as the job gets done. Sue is sensitive to patronizing or unfair assessments of her work, and I can't really blame her for that."

"Let's take this Bob guy as another example," Curtis continued. "His work is so inconsistent as to be useless. Every time I walk into the center, he's on the phone. I'm pretty sure he's not doing firm business. More likely, he's talking to his new girlfriend. If Sue would just monitor his behavior, we'd have the equivalent of one part-time employee right there."

"Curtis, Bob is on the phone a lot because he has the nicest temperament in the group. He's ideal for being the source of first contact for the clients and associates who call our firm," Ms. Stonewall said.

Mr. Curtis was not to be dissuaded. "Now, Ted, he's the type of person we need back there. He is respectful and helpful, and he goes out of his way to ask about my problems. Now that I think about it, we should immediately promote Ted to manager of the center."

Ms. Conner couldn't remain silent any longer. "Ted is *not* the ideal employee! The man has no substance and is divisive. The moment your back is turned, he's gossiping about you or someone else. I'm surprised you fell for his act, Curtis."

Although Fisher tried to bring order to the discussion, the meeting continued to deteriorate and closure was nowhere in sight. Half the voting members were in favor of a new hire, while the other half continued to wonder if better management of the current employees would work just as well. Finally, a compromise was reached. Fisher agreed to investigate an efficiency-rating software package. Once installed on the firm's network, the software could measure words typed per minute on each computer. Curtis believed that if employees knew their working habits were being monitored, they would work harder and eliminate extraneous activities. The partners decided to install the new software. If productivity increased, so much the better. If not, it was agreed that the firm would hire a new employee for the center.

Three weeks later, the software had been purchased and installed. Mr. Fisher called another meeting to discuss implementation issues.

"Although we can run the software without any of the staff knowing about it, I'm concerned about the ethics of doing so," said Fisher. "We should treat our employees with fairness and dignity. I'd like to tell them about the software and present this as a means to recognize outstanding achievement rather than to punish."

"That's ridiculous." Mr. Curtis pounded his fist on the conference room table. "We spend all this money on software designed to detect weak performers, and you want to *tell* them about it? They'll just work hard when they know they're being monitored, then return to their bad habits. Let's begin this experiment unannounced, and I'd be happy to fire those who aren't . . ."

Others were concerned about the surreptitious use of the software, too, but agreed with Curtis's assessment. Another compromise was reached. To establish a baseline reading, each employee would be monitored without their knowledge for one week. Friday afternoon of that week, a staff meeting would be held. The capabilities of the new software would be explained, and employees would be told that the intent of the software was to find ways to improve productivity.

Four employees among the 10 staff members in the center were randomly chosen to be monitored. They were, as it happens, Sue, Bob, Ted, and Sandy. Hired only two months ago, Sandy was turning out to be an asset to the center, but still seemed anxious about doing a good job.

"The Friday afternoon staff meeting went pretty well," reported Mr. Fisher the following week. "I explained the software and was quite explicit about how the information would and would not be used. It was odd, though. Sue, who is usually quite outspoken about her job and her staff, asked no questions. She just walked out of the room at the end of the meeting. We'll just have to see how it goes from here."

Chosen employees were monitored each morning from 9:00 to 10:00 for eight weeks, although they did not know exactly when during the day the data were collected. An observation in the data set is the average words per minute (rounded to the nearest minute) typed during the monitoring period on a particular day. For example, the column labeled Sue contains the 40 daily readings for Sue over the experimental period, the first five of which represent the baseline readings.

You've been asked to analyze the data and prepare a report on your findings. The firm wants to know if there is any evidence that the software package has changed behavior of employees in the word processing center, and if so, how. Include in your report any observations about the experimental design and data collection that might have impacted the outcome or that might be improved for subsequent data collection. The data are in the file named WORDS.

DATA SET

OBS	SUE	BOB	SANDY	TED
1	90	68	64	60
2	89	83	65	62
3	94	60	63	60
4	95	78	62	59
5	99	81	64	62
6	80	65	64	74
7	78	58	65	77
8	75	72	66	75
9	78	61	68	74
10	77	70	65	75
11	63	80	67	69

(concludes on the following page)

OBS	SUE	BOB	SANDY	TED
12	75	67	68	67
13	73	64	67	75
14	75	60	67	70
15	74	81	74	70
16	77	58	73	69
17	75	62	69	68
18	75	80	70	70
19	75	77	77	72
20	69	63	79	75
21	73	84	83	78
22	75	60	83	76
23	72	64	84	74
24	77	76	85	75
25	0	75	84	77
26	73	84	80	74
27	74	65	79	72
28	70	70	77	77
29	69	77	80	70
30	72	73	75	75
31	74	77	78	77
32	77	69	72	79
33	69	82	75	75
34	69	70	70	72
35	0	63	70	75
36	74	86	72	74
37	72	58	69	70
38	66	68	77	68
39	68	84	75	72
40	65	76	72	70